MATHEMATICAL ANALYSIS

MATHEMATICAL ANALYSIS

A STRAIGHTFORWARD APPROACH

K. G. BINMORE

Professor of Mathematics
London School of Economics and Political Science

SECOND EDITION

CAMBRIDGE
UNIVERSITY PRESS

PUBLISHED BY THE PRESS SYNDICATE OF THE UNIVERSITY OF CAMBRIDGE
The Pitt Building, Trumpington Street, Cambridge CB2 1RP, United Kingdom

CAMBRIDGE UNIVERSITY PRESS
The Edinburgh Building, Cambridge CB2 2RU, United Kingdom
40 West 20th Street, New York, NY 10011–4211, USA
10 Stamford Road, Oakleigh, Melbourne 3166, Australia

First published 1977
Reprinted 1978, 1980 (with corrections), 1982
Second edition 1982
Reprinted 1983, 1984, 1985, 1986, 1987, 1988, 1990, 1991, 1993, 1995, 1997

Printed in the United Kingdon by Athenæum Press Ltd, Gateshead, Tyne & Wear

British Library Cataloguing in Publication data
Binmore, K. G.
   Mathematical analysis. – 2nd ed.
   1. Calculus
   I. Title
   515   QA303

ISBN 0 521 28882 7 paperback

CONTENTS

PREFACE TO THE FIRST EDITION

This book is intended as an easy and unfussy introduction to mathematical analysis. Little formal reliance is made on the reader's previous mathematical background, but those with no training at all in the elementary techniques of calculus would do better to turn to some other book.

An effort has been made to lay bare the bones of the theory by eliminating as much unnecessary detail as is feasible. To achieve this end and to ensure that all results can be readily illustrated with concrete examples, the book deals only with 'bread and butter' analysis on the real line, the temptation to discuss generalisations in more abstract spaces having been reluctantly suppressed. However, the need to prepare the way for these generalisations has been kept well in mind.

It is vital to adopt a systematic approach when studying mathematical analysis. In particular, one should always be aware at any stage of what may be assumed and what has to be proved. Otherwise confusion is inevitable. For this reason, the early chapters go rather slowly and contain a considerable amount of material with which many readers may already be familiar. To neglect these chapters would, however, be unwise.

The exercises should be regarded as an integral part of the book. There is a great deal more to be learned from attempting the exercises than can be obtained from a passive reading of the text. This is particularly the case when, as may frequently happen, the attempt to solve a problem is unsuccessful and it is necessary to turn to the solutions provided at the end of the book.

To help those with insufficient time at their disposal to attempt all the exercises, the less vital exercises have been marked with the symbol †. (The same notation has been used to mark one or two passages in the text which can be omitted without great loss at a first reading.) The symbol * has been used to mark exercises which are more demanding than most but which are well worth attempting.

The final few chapters contain very little theory compared with the number of exercises set. These exercises are intended to illustrate the power of the techniques introduced earlier in the book and to provide the opportunity of some revision of these ideas.

This book arises from a course of lectures in analysis which is given at the London School of Economics. The students who attend this course are mostly not specialist mathematicians and there is little uniformity in their previous mathematical training. They are, however, quite well-motivated. The course is a 'one unit' course of approximately forty lectures supplemented by twenty informal problem classes. I have found it possible to cover the material of this book in some thirty lectures. Time is then left for some discussion of point set topology in simple spaces. The content of the book provides an ample source of examples for this purpose while the more general theorems serve as reinforcement for the theorems of the text.

Other teachers may prefer to go through the material of the book at a more leisurely pace or else to move on to a different topic. An obvious candidate for further discussion is the algebraic foundation of the real number system and the proof of the Continuum Property. Other alternatives are partial differentiation, the complex number system or even Lebesgue measure on the line.

I would like to express my gratitude to Elizabeth Boardman and Richard Holmes for reading the text for me so carefully. My thanks are also due to 'Buffy' Fennelly for her patience and accuracy in preparing the typescript. Finally, I would like to mention M.C. Austin and H. Kestelman from whom I learned so much of what I know.

July 1976   K.G.B.

PREFACE TO THE SECOND EDITION

It is a pleasure to write a preface for the second edition of *Mathematical Analysis: A Straightforward Approach*. The first edition was well-received and I have therefore thought it wise to leave its text substantially unaltered except for one or two minor points of clarification and the correction of misprints. The major change is the addition of two long chapters on analysis in vector spaces for which there has been a considerable demand. These get as far as the idea of a derivative as a matrix and the use of the second order derivative of a real-valued function in classifying stationary points. More advanced material than this would seem to me better delayed until after the basic topological notions have been mastered. As far as the material covered is concerned, it does not involve the proof of many theorems and the necessary proofs involve no new analytic ideas. However, the material does require a certain facility with algebraic and geometric ideas and students with only a very limited knowledge of linear algebra may find it heavy going in spite of the fact that some discussion of the necessary concepts from linear algebra is included where appropriate. Another innovation is the inclusion of a collection of further problems for which the solutions are not given. I am grateful to John Erdös for some of these as well as other helpful suggestions. Teachers using this book as part of a taught course may find these problems helpful in setting work but I hope that they will not distract attention from the importance of working carefully through the exercises given in the main body of the text.

Finally, I would like to express my appreciation to those who have commented favourably on the first edition and to Mimi Bell for her patient help in preparing the typescript for the second edition.

October 1981   K.G.B.

1 REAL NUMBERS

1.1 Set notation

A *set* is a collection of objects which are called its *elements*. If x is an element of the set S, we say that x *belongs* to S and write

$$x \in S.$$

If y does not belong to S, we write $y \notin S$.

The simplest way of specifying a set is by listing its elements. We use the notation

$$A = \{\tfrac{1}{2}, 1, \sqrt{2}, e, \pi\}$$

to denote the set whose elements are the real numbers $\tfrac{1}{2}, 1, \sqrt{2}, e$ and π. Similarly

$$B = \{\text{Romeo, Juliet}\}$$

denotes the set whose elements are Romeo and Juliet.

This notation is, of course, no use in specifying a set which has an infinite number of elements. Such sets may be specified by naming the property which distinguishes elements of the set from objects which are not in the set. For example, the notation

$$C = \{x : x > 0\}$$

(which should be read 'the set of all x such that $x > 0$') denotes the set of all positive real numbers. Similarly

$$D = \{y : y \text{ loves Romeo}\}$$

denotes the set of all people who love Romeo.

It is convenient to have a notation for the *empty* set $\emptyset$. This is the set which has *no* elements. For example, if x denotes a variable which ranges over the set of all real numbers, then

$$\{x : x^2 + 1 = 0\} = \emptyset.$$

This is because there are no real numbers x such that $x^2 = -1$.

If S and T are two sets, we say that S is a *subset* of T and write

$$S \subset T$$

if every element of S is also an element of T.

As an example, consider the sets $P = \{1, 2, 3, 4\}$ and $Q = \{2, 4\}$. Then $Q \subset P$. Note that this is *not* the same thing as writing $Q \in P$, which means that Q is an element of P. The elements of P are simply $1, 2, 3$ and 4. But Q is not one of these.

The sets A, B, C and D given above also provide some examples. We have $A \subset C$ and (presumably) $B \subset D$.

1.2   The set of real numbers

It will be adequate for this book to think of the real numbers as being points along a straight line which extends indefinitely in both directions. The line may then be regarded as an ideal ruler with which we may measure the lengths of line segments in Euclidean geometry.

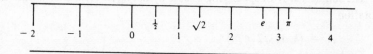

The set of all real numbers will be denoted by $\mathbb{R}$. The table below distinguishes three important subsets of $\mathbb{R}$.

Subset	Notation	Elements
Natural numbers (or whole numbers)	$\mathbb{N}$	$1, 2, 3, 4, 5, \ldots$
Integers	$\mathbb{Z}$	$\ldots -2, -1, 0, 1, 2, 3, \ldots$
Rational numbers (or fractions)	$\mathbb{Q}$	$0, 1, 2, -1, \frac{1}{2}, \frac{3}{4}, \frac{5}{3}, -\frac{1}{2}, -\frac{3}{7}, \ldots$

Not all real numbers are rational. Some examples of irrational numbers are $\sqrt{2}, e$ and π.

While we do not go back to first principles in this book, the treatment will be rigorous in so far as it goes. It is therefore important to be clear, at every stage, about what our assumptions are. We shall then know what has to be proved and what may be taken for granted. Our most vital assumptions are concerned with the properties of the real number system. The rest of this chapter and the following two chapters are consequently devoted to a description of the properties of the real number system which we propose to assume and to some of their immediate consequences. A very much more systematic account of these assumptions is given in the author's book *Logic, Sets and Numbers* (see pp. 44–77).

1.3 Arithmetic

The first assumption is that the real numbers satisfy all the usual laws of addition, subtraction, multiplication and division.

The rules of arithmetic, of course, include the proviso that division by zero is not allowed. Thus, for example, the expression

$$\frac{2}{0}$$

makes no sense at all. In particular, it is *not* true that

$$\frac{2}{0} = \infty.$$

We shall have a great deal of use for the symbol ∞, but it must clearly be understood that ∞ does *not* represent a real number. Nor can it be treated as such except in very special circumstances.

1.4 Inequalities

The next assumptions concern inequalities between real numbers and their manipulation.

We assume that, given any two real numbers a and b, there are three mutually exclusive possibilities:

(i) $a > b$ (a is greater than b)

(ii) $a = b$ (a equals b)

(iii) $a < b$ (a is less than b).

Observe that $a < b$ means the same thing as $b > a$. We have, for example, the following inequalities.

$$1 > 0; \ 3 > 2; \ 2 < 3; \ -1 < 0; \ -3 < -2.$$

There is often some confusion about the statements

(iv) $a \geqslant b$ (a is greater than *or* equal to b)

(v) $a \leqslant b$ (a is less than *or* equal to b).

To clear up this confusion, we note that the following are all true statements.

$$1 \geqslant 0; \ 3 \geqslant 2; \ 1 \geqslant 1; \ 2 \leqslant 3; \ -1 \leqslant 0; \ -3 \leqslant -3.$$

We assume four basic rules for the manipulation of inequalities. From these the other rules may be deduced.

(I) If $a > b$ and $b > c$, then $a > c$.

(II) If $a > b$ and c is any real number, then

$$a + c > b + c.$$

(III) If $a > b$ and $c > 0$, then $ac > bc$ (i.e. inequalities can be multiplied through by a *positive* factor).

(IV) If $a > b$ and $c < 0$, then $ac < bc$ (i.e. multiplication by a *negative* factor reverses the inequality).

1.5   Example If $a > 0$, prove that $a^{-1} > 0$.

Proof We argue by contradiction. Suppose that $a > 0$ but that $a^{-1} \leqslant 0$. It cannot be true that $a^{-1} = 0$ (since then $0 = 0.a = 1$). Hence

$$a^{-1} < 0.$$

By rule III we can multiply this inequality through by a (since $a > 0$). Hence

$$1 = a^{-1}.a < 0.a = 0.$$

But $1 < 0$ is a contradiction. Therefore the assumption $a^{-1} \leqslant 0$ was false. Hence $a^{-1} > 0$.

1.6   Example If x and y are positive, then $x < y$ if and only if $x^2 < y^2$.

Proof We have to show *two* things. First, that $x < y$ implies $x^2 < y^2$, and secondly, that $x^2 < y^2$ implies $x < y$.

(i) We begin by assuming that $x < y$ and try to deduce that $x^2 < y^2$. Multiply the inequality $x < y$ through by $x > 0$ (rule III). We obtain

$$x^2 < xy.$$

Similarly

$$xy < y^2.$$

But now $x^2 < y^2$ follows from rule I.

(ii) We now assume that $x^2 < y^2$ and try and deduce that $x < y$. Adding $-x^2$ to both sides of $x^2 < y^2$ (rule II), we obtain

$$y^2 - x^2 > 0$$

i.e. $$(y - x)(y + x) > 0. \tag{1}$$

Since $x + y > 0$, $(x + y)^{-1} > 0$ (example 1.5). We can therefore multiply through inequality (1) by $(x + y)^{-1}$ to obtain

$$y - x > 0$$

i.e. $x < y$.

(Alternatively, we could prove (ii) as follows. Assume that $x^2 < y^2$ but that $x \geqslant y$. From $x \geqslant y$ it follows (as in (i)) that $x^2 \geqslant y^2$, which is a contradiction.)

1.7   *Example* Suppose that, for *any* $\epsilon > 0$, $a < b + \epsilon$. Then $a \leqslant b$.

Proof Assume that $a > b$. Then $a - b > 0$. But, for *any* $\epsilon > 0$, $a < b + \epsilon$. Hence $a < b + \epsilon$ in the particular case when $\epsilon = a - b$. Thus

$$a < b + (a - b)$$

and so $a < a$.

This is a contradiction. Hence our assumption $a > b$ must be false. Therefore $a \leqslant b$.
(Note: The symbol ϵ in this example is the Greek letter *epsilon*. It should be carefully distinguished from the 'belongs to' symbol $\in$ and also from the symbol ξ which is the Greek letter *xi*.)

1.8   *Exercise*

(1) If x is any real number, prove that $x^2 \geqslant 0$. If $0 < a < 1$ and $b > 1$, prove that

(i) $0 < a^2 < a < 1$ (ii) $b^2 > b > 1$.

(2) If $b > 0$ and $B > 0$ and

$$\frac{a}{b} < \frac{A}{B},$$

prove that $aB < bA$. Deduce that

$$\frac{a}{b} < \frac{a + A}{b + B} < \frac{A}{B}.$$

(3) If $a > b$ and $c > d$, prove that $a + c > b + d$ (i.e. inequalities can be added). If, also, $b > 0$ and $d > 0$, prove that $ac > bd$ (i.e. inequalities between *positive* numbers can be multiplied).

(4) Show that each of the following inequalities may fail to hold even though $a > b$ and $c > d$.

(i) $a - c > b - d$

(ii) $\dfrac{a}{c} > \dfrac{b}{d}$

(iii) $ac > bd$.

What happens if we impose the extra condition that $b > 0$ and $d > 0$?

(5) Suppose that, for *any* $\epsilon > 0$, $a - \epsilon < b < a + \epsilon$. Prove that $a = b$.

(6) Suppose that $a < b$. Show that there exists a real number x satisfying $a < x < b$.

1.9   Roots

Let n be a natural number. The reader will be familiar with the notation $y = x^n$. For example, $x^2 = x.x$ and $x^3 = x.x.x$.

Our next assumption about the real number system is the following. Given any $y \geqslant 0$ there is exactly one value of $x \geqslant 0$ such that

$$y = x^n.$$

(Later on we shall see how this property may be deduced from the theory of continuous functions.)

If $y \geqslant 0$, the value of $x \geqslant 0$ which satisfies the equation $y = x^n$ is called the nth *root* of y and is denoted by

$$x = y^{1/n}.$$

When $n = 2$, we also use the notation $\sqrt{y} = y^{1/2}$. Note that, with this convention, it is always true that $\sqrt{y} \geqslant 0$. If $y > 0$, there are, of course, *two* numbers whose square is y. The positive one is $\sqrt{y}$ and the negative one is $-\sqrt{y}$. The notation $\pm \sqrt{y}$ means '$\sqrt{y}$ *or* $-\sqrt{y}$'.

If $r = m/n$ is a positive rational number and $y \geqslant 0$, we define

$$y^r = (y^m)^{1/n}.$$

If r is a negative rational, then $-r$ is a positive rational and hence y^{-r} is defined. If $y > 0$ we can therefore define y^r by

$$y^r = \frac{1}{y^{-r}}.$$

We also write $y^0 = 1$. With these conventions it follows that, if $y > 0$, then y^r is defined for all rational numbers r. (The definition of y^x when x is an irrational real number must wait until a later chapter.)

1.10   Quadratic equations

If $y > 0$, the equation $x^2 = y$ has two solutions. We denote the *positive* solution by $\sqrt{y}$. The *negative* solution is therefore $-\sqrt{y}$. We note again that there is no ambiguity about these symbols and that $\pm \sqrt{y}$ simply means '$\sqrt{y}$ or $-\sqrt{y}$'.

The general quadratic equation has the form

$$ax^2 + bx + c = 0$$

where $a \neq 0$. Multiply through by $4a$. We obtain

$$4a^2x^2 + 4abx + 4ac = 0$$

$$(2ax + b)^2 - b^2 + 4ac = 0$$

$$(2ax + b)^2 = b^2 - 4ac.$$

It follows that the quadratic equation has no real solutions if $b^2 - 4ac < 0$, one real solution if $b^2 - 4ac = 0$ and two real solutions if $b^2 - 4ac > 0$. If $b^2 - 4ac \geqslant 0$,

$$2ax + b = \pm \sqrt{(b^2 - 4ac)}$$

$$x = \frac{-b \pm \sqrt{(b^2 - 4ac)}}{2a}.$$

The roots of the equation $ax^2 + bx + c = 0$ are therefore

$$\alpha = \frac{-b - \sqrt{(b^2 - 4ac)}}{2a} \quad \text{and} \quad \beta = \frac{-b + \sqrt{(b^2 - 4ac)}}{2a}.$$

It is a simple matter to check that, for all values of x,

$$ax^2 + bx + c = a(x - \alpha)(x - \beta).$$

With the help of this formula, we can sketch the graph of the equation $y = ax^2 + bx + c$.

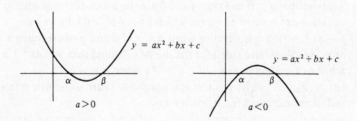

1.11   *Example* A nice application of the work on quadratic equations described above is the proof of the important *Cauchy–Schwarz inequality*. This asserts that, if $a_1, a_2, \ldots, a_n$ and $b_1, b_2, \ldots, b_n$ are any real numbers, then

$$(a_1b_1 + a_2b_2 + \ldots + a_nb_n)^2 \leqslant (a_1^2 + a_2^2 + \ldots + a_n^2)(b_1^2 + b_2^2 + \ldots + b_n^2).$$

Proof For any x,

$$0 \leqslant (a_1 x + b_1)^2 + (a_2 x + b_2)^2 + \ldots + (a_n x + b_n)^2$$

$$= (a_1^2 + \ldots + a_n^2)x^2 + 2(a_1 b_1 + \ldots + a_n b_n)x + (b_1^2 + \ldots + b_n^2)$$

$$= Ax^2 + 2Bx + C.$$

Since $y = Ax^2 + 2Bx + C \geqslant 0$ for *all* values of x, it follows that the equation $Ax^2 + 2Bx + C = 0$ cannot have two (distinct) roots. Hence

$$(2B)^2 - 4AC \leqslant 0$$

i.e. $B^2 \leqslant AC$

which is what we had to prove.

1.12   Exercise

(1) Suppose that n is an *even* natural number. Prove that the equation $x^n = y$ has no solutions if $y < 0$, one solution if $y = 0$ and two solutions if $y > 0$.

Suppose that n is an *odd* natural number. Prove that the equation $x^n = y$ always has one and only one solution.

Draw graphs of $y = x^2$ and $y = x^3$ to illustrate these results.

(2) Simplify the following expressions:

(i) $8^{2/3}$ (ii) $27^{-4/3}$ (iii) $32^{6/5}$.

(3) If $y > 0$, $z > 0$ and r and s are any rational numbers, prove the following:

(i) $y^{r+s} = y^r y^s$ (ii) $y^{rs} = (y^r)^s$ (iii) $(yz)^r = y^r z^r$.

(4) Suppose that $a > 0$ and that α and β are the roots of the quadratic equation $ax^2 + bx + c = 0$ (in which $b^2 - 4ac > 0$). Prove that $y = ax^2 + bx + c$ is negative when $\alpha < x < \beta$ and positive when $x < \alpha$ or $x > \beta$. Show also (without the use of calculus) that $y = ax^2 + bx + c$ achieves a minimum value of $c - b^2/4a$ when $x = -b/2a$.

(5) Let $a_1, a_2, \ldots, a_n$ be positive real numbers. Their arithmetic mean A_n and harmonic mean H_n are defined by

$$A_n = \frac{a_1 + a_2 + \ldots + a_n}{n} \qquad H_n^{-1} = \frac{1}{n}\left(\frac{1}{a_1} + \frac{1}{a_2} + \ldots + \frac{1}{a_n}\right).$$

Deduce from the Cauchy–Schwarz inequality that $H_n \leqslant A_n$.

(6) Let $a_1, a_2, \ldots, a_n$ and $b_1, b_2, \ldots, b_n$ be any real numbers. Prove Minkowski's inequality, i.e.

$$\left\{\sum_{k=1}^{n} (a_k + b_k)^2\right\}^{1/2} \leqslant \left\{\sum_{k=1}^{n} a_k^2\right\}^{1/2} + \left\{\sum_{k=1}^{n} b_k^2\right\}^{1/2}.$$

For the case $n = 2$ (or $n = 3$) this inequality amounts to the assertion that the length of one side of a triangle is less than or equal to the sum of the lengths of the other two sides. Explain this.

1.13   Irrational numbers

In § 1.2 we mentioned the existence of irrational real numbers. That such numbers exist is by no means obvious. For example, one may imagine the process of marking all the rational numbers on a straight line. First one would mark the integers. Then one would move on to the multiples of $\frac{1}{2}$ and then to the multiples of $\frac{1}{3}$ and so on. Assuming that this program could ever be completed, one might very well be forgiven for supposing that there would be no room left for any more points on the line.

But our assumption about the existence of nth roots renders this view untenable. This assumption requires us to accept the existence of a positive real number x (namely $\sqrt{2}$) which satisfies $x^2 = 2$. If x were a rational number it would be expressible in the form

$$x = \frac{m}{n}$$

where m and n are natural numbers with no common divisor (other than 1). It follows that

$$m^2 = 2n^2$$

and so m^2 is even. This implies that m is even. (If m were odd, we should have $m = 2k + 1$. But then $m^2 = 4k^2 + 4k + 1$ which is odd.) We may therefore write $m = 2k$. Hence

$$4k^2 = 2n^2$$

$$n^2 = 2k^2.$$

Thus n is even. We have therefore shown that both m and n are divisible by 2. This is a contradiction and it follows that x cannot be rational, i.e. $\sqrt{2}$ must be an irrational real number.

Of course, $\sqrt{2}$ is not the only irrational number and the ability to extract nth roots allows us to construct many others. But it should not be supposed that all irrational numbers can be obtained in this way. It is not even true that every irrational number is a root of an equation of the form

$$a_0 + a_1x + a_2x^2 + \ldots + a_nx^n = 0$$

in which the coefficients $a_0, a_1, \ldots, a_n$ are rational numbers. Real numbers which are not the roots of such an equation are called transcendental, notable examples being e and π. This fascinating topic, however, lies outside the scope of this book.

1.14   Modulus

Suppose that x is a real number. Its *modulus* (or *absolute value*) $|x|$ is defined by

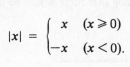

$$|x| = \begin{cases} x & (x \geqslant 0) \\ -x & (x < 0). \end{cases}$$

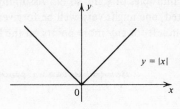

$y = |x|$

Thus $|3| = 3$, $|-6| = 6$ and $|0| = 0$. Obviously $|x| \geqslant 0$ for all values of x.

It is sometimes useful to note that $|x| = \sqrt{x^2}$.

1.15   Theorem   For any real number x,

$$-|x| \leqslant x \leqslant |x|.$$

Proof Either $x \geqslant 0$ or $x < 0$. In the first case, $-|x| \leqslant 0 \leqslant x = |x|$. In the second case, $-|x| = x < 0 < |x|$.

1.16   Theorem   For any real numbers a and b

$$|ab| = |a| \cdot |b|$$

Proof The most elegant proof is the following.

$$|ab| = \sqrt{((ab)^2)} = \sqrt{(a^2b^2)} = \sqrt{(a^2)} \cdot \sqrt{(b^2)} = |a| \cdot |b|.$$

The next theorem is of great importance and is usually called the *triangle inequality*. This nomenclature may be justified by observing that the theorem is the special case of Minkowski's inequality (exercise 1.12(6)) with $n = 1$. We give a separate proof.

1.17   Theorem (triangle inequality)   For any real numbers a and b,

$$|a + b| \leqslant |a| + |b|.$$

Proof We have

$$|a + b|^2 = (a + b)^2 = a^2 + 2ab + b^2$$
$$= |a|^2 + 2ab + |b|^2$$
$$\leqslant |a|^2 + 2|ab| + |b|^2 \quad \text{(theorem 1.15)}$$
$$= |a|^2 + 2|a| \cdot |b| + |b|^2 \quad \text{(theorem 1.16)}$$
$$= (|a| + |b|)^2.$$

It follows (from example 1.6) that

$$|a + b| \leqslant |a| + |b|.$$

1.18 **Theorem** For any real numbers c and d,

$$|c - d| \geqslant |c| - |d|.$$

Proof Take $a = d$ and $b = c - d$ in the triangle inequality. Then $a + b = c$ and so

$$|c| = |a + b| \leqslant |a| + |b| = |d| + |c - d|.$$

1.19 **Example**

(i) $2 = |-1 + 3| \leqslant |-1| + |3| = 1 + 3 = 4.$
(ii) $7 = |6 - (-1)| \geqslant |6| - |-1| = 6 - 1 = 5.$

1.20 **Exercise**

(1) Prove that $|a| < b$ if and only if $-b < a < b$.
[Recall that you have *two* things to prove as in example 1.6.]

(2) Prove that

$$|c - d| \geqslant ||c| - |d||.$$

(3) The *distance* $d(x, y)$ between two real numbers x and y is defined by $d(x, y) = |x - y|$. Show that, for any x, y and z,
   (i) $d(x, y) \geqslant 0$ (iii) $d(x, y) = 0$ if and only if $x = y$
   (ii) $d(x, y) = d(y, x)$ (iv) $d(x, y) \leqslant d(x, z) + d(z, y).$

† (4) If r and s are rational numbers, prove that

$$r + s\sqrt{2}$$

is irrational unless $s = 0$.

† (5) Suppose that the coefficients $a \neq 0, b$ and c of the quadratic equation

$$ax^2 + bx + c = 0$$

are all rational numbers and that $\alpha = r + s\sqrt{2}$ is a root of this equation, where r and s are also rational numbers. Prove that $\beta = r - s\sqrt{2}$ is also a root.

† (6) Prove that $3^{1/2}$ and $2^{1/3}$ are irrational numbers.

2 CONTINUUM PROPERTY

2.1 Achilles and the tortoise

The following is one of the famous paradoxes of Zeno. Achilles is to race a tortoise. Since Achilles runs faster than the tortoise, the tortoise is given a start of x_0 feet. When Achilles reaches the point where the tortoise started, the tortoise will have advanced a bit, say x_1 feet. Achilles soon reaches the tortoise's new position, but, by then, the tortoise will have advanced a little bit more, say x_2 feet. This argument may be continued indefinitely and so Achilles can never catch the tortoise.

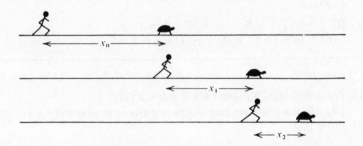

The simplest way to resolve this paradox is to say that Achilles catches the tortoise after he has run a distance of x feet, where x is 'the smallest real number larger than all of the numbers $x_0, x_0 + x_1, x_0 + x_1 + x_2 \ldots$'. Zeno's argument then simply reduces to subdividing a line segment of length x into an infinite number of smaller line segments of respective lengths $x_0, x_1, x_2, \ldots$

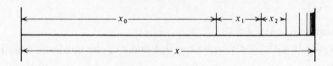

Formulated in this way, the paradox loses its sting.

This solution, of course, depends very strongly on the existence of the real number x, i.e. the smallest real number larger than all of the numbers x_0, $x_0 + x_1, x_0 + x_1 + x_2, \ldots$

A similar lesson is to be learned from Archimedes' method of exhaustion. This was invented by Archimedes as a means of evaluating the area of regions with curved boundaries.

In the case of a circle, the method involves inscribing a sequence of larger and larger regular polygons inside the circle and then calculating their areas $A_1, A_2, A_3, \ldots$

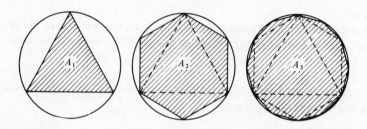

The area A of the circle is then identified with the 'smallest real number larger than each of the numbers $A_1, A_2, A_3, \ldots$'. By taking the radius of the circle to be 1 and calculating the value of A_n for a large enough value of n, Archimedes was able to obtain as close an approximation to π as he chose.

Note that this method again depends strongly on the existence of a smallest real number larger than all of the real numbers $A_1, A_2, A_3, \ldots$

2.2   The Continuum Property

The discussion of the previous section indicates the necessity of making a further assumption about the real number system. It would, after all, be most unsatisfactory if Achilles never caught the tortoise or if a circle had no area. We shall call our new assumption the *Continuum Property* of the real numbers.

The Continuum Property is not usually mentioned explicitly in a school algebra course, but it is very important for what follows in these notes. The rest of this chapter is therefore devoted to its consideration. We begin with some terminology.

A set S of real numbers is *bounded above* if there exists a real number H which is greater than or equal to every element of the set, i.e. if, for some H,

$$x \leqslant H$$

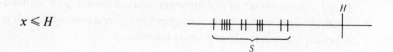

for any $x \in S$.

The number H (if such a number exists) is called an *upper bound* of the set S.

A set S of real numbers is *bounded below* if there exists a real number h

which is less than or equal to every element of the set, i.e. if, for some h,

$$x \geqslant h$$

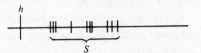

for any $x \in S$.

The number h (if such a number exists) is called a *lower bound* of the set S.

A set which is both bounded above *and* bounded below is just said to be *bounded*.

2.3   Proposition A set S of real numbers is bounded if and only if there exists a real number K such that

$$|x| \leqslant K$$

for any $x \in S$. (This is easily proved with the help of exercise 1.20(1).)

2.4   Examples

(i) The set $\{1, 2, 3\}$ is bounded above. Some upper bounds are 100, 10, 4 and 3. The set is also bounded below. Some lower bounds are -27, 0 and 1.

(ii) The set $\{x : 1 \leqslant x < 2\}$ is bounded above. Some upper bounds are 100, 10, 4 and 2. The set is also bounded below. Some lower bounds are -27, 0 and 1.

(iii) The set $\{x : x > 0\}$ is *unbounded above*. If $H > 0$ is proposed as an upper bound, one has only to point to $H + 1$ to obtain an element of the set larger than the supposed upper bound. However, the set $\{x : x > 0\}$ is bounded below. Some lower bounds are -27 and 0.

We now state the Continuum Property which will be fundamental for the remainder of these notes.

> **Continuum Property**
> Every non-empty set of real numbers which is bounded above has a smallest upper bound. Every non-empty set of real numbers which is bounded below has a largest lower bound.

Thus, if S is a non-empty set which is bounded above, then S has an upper bound B such that, given any other upper bound H of S,

$$B \leqslant H.$$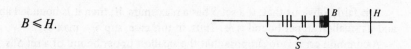

Similarly, if S is a non-empty set which is bounded below, then S has a lower bound b such that, given any other lower bound h,

$$b \geqslant h.$$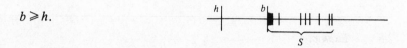

2.5 Examples

(i) The smallest upper bound of the set $\{1, 2, 3\}$ is 3. The largest lower bound of the set $\{1, 2, 3\}$ is 1.

(ii) The smallest upper bound of the set $\{x : 1 \leqslant x < 2\}$ is 2. The largest lower bound is 1.

(iii) The set $\{x : x > 0\}$ has no upper bounds at all. The largest lower bound of the set $\{x : x > 0\}$ is 0.

2.6 Supremum and infimum

If a non-empty set S is bounded above, then, by the Continuum Property, it has a smallest upper bound B. This smallest upper bound B is sometimes called the *supremum* of the set S. We write $B = \sup S$ or

$$B = \sup_{x \in S} x.$$

Similarly, a set which is bounded below has a largest lower bound b. We call b the *infimum* of the set S and write $b = \inf S$ or

$$b = \inf_{x \in S} x.$$

Sometimes you may encounter the notation $\sup S = +\infty$. This simply means that S is unbounded above. Similarly, $\inf S = -\infty$ means that S is unbounded below.

2.7 Maximum and minimum

If a set S has a largest element M, we call M the *maximum* of the set S and write $M = \max S$. If S has a smallest element m, we call m the *minimum* of S and write $m = \min S$.

It is fairly obvious that, if a set S has a maximum M, then it is bounded above and its smallest upper bound is M. Thus, in this case, sup S = max S.

A common error is to suppose that the smallest upper bound of a set S is always the maximum of the set. However, some sets which are bounded above (and hence have a smallest upper bound) do *not* have a maximum. (See example 2.8(ii).)

Similar remarks, of course, apply to largest lower bounds and minima of sets.

2.8   Examples

(i) The set $\{1, 2, 3\}$ has a maximum 3 and this is equal to its smallest upper bound. The set $\{1, 2, 3\}$ has minimum 1 and this is equal to its largest lower bound.

(ii) The set $\{x : 1 \leqslant x < 2\}$ has no maximum. The number 2 cannot be the largest element of the set because it does not belong to the set. On the other hand, any x in the set satisfies $1 \leqslant x < 2$. But then

$$y = \frac{x + 2}{2}$$

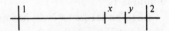

is an element of the set which is larger than x. Hence x cannot be the largest element of the set. However, $\{x : 1 \leqslant x < 2\}$ has smallest upper bound 2.

The set $\{x : 1 \leqslant x < 2\}$ has minimum 1 (why?). This is equal to its largest lower bound.

(iii) The set $\{x : x > 0\}$ has no maximum, nor does it have any upper bounds. The set $\{x : x > 0\}$ has no minimum (why not?). Its largest lower bound is 0.

2.9   Intervals

An *interval I* is a set of real numbers with the property that, if $x \in I$ and $y \in I$ and $x \leqslant z \leqslant y$, then $z \in I$, i.e. if two numbers belong to I, then so does every number between them.

In describing intervals we use the following notation:

$$(a, b) = \{x : a < x < b\}$$
$$[a, b] = \{x : a \leqslant x \leqslant b\}$$
$$[a, b) = \{x : a \leqslant x < b\}$$
$$(a, b] = \{x : a < x \leqslant b\}.$$

These are the bounded intervals (classified by whether or not they have a maximum and whether or not they have a minimum). We also need to consider the unbounded intervals. For these we use the notation below.

$$(a, \infty) = \{x : x > a\}$$

$$[a, \infty) = \{x : x \geqslant a\}$$

$$(-\infty, b) = \{x : x < b\}$$

$$(-\infty, b] = \{x : x \leqslant b\}$$

(Do not be misled by this notation into supposing that ∞ or $-\infty$ are real numbers or that they can be treated as such.)

All intervals (with the exception of the empty set and the set of all real numbers) fall into one of the categories described above. We call the intervals (a, b), (a, ∞) and $(-\infty, b)$ *open* and the intervals $[a, b]$, $[a, \infty)$ and $(-\infty, b]$ *closed*. (The intervals $[a, b)$ and $(b, a]$ are sometimes called half-open.)

Of particular importance are the closed, bounded intervals $[a, b]$. We shall call such an interval *compact*.

2.10   Exercise

(1) Which of the following statements are true?
(i) $3 \in (1, 2)$ (ii) $2 \in (-\infty, 3]$ (iii) $1 \in [1, 2)$
(iv) $2 \in (0, 2)$ (v) $3 \in [1, 3]$.

(2) If ξ is a real number and $\delta > 0$, prove that

$$\{x : |\xi - x| < \delta\} = (\xi - \delta, \xi + \delta).$$

(3) Decide in each of the following cases whether or not the given set is bounded above. For those sets which are bounded above, write down three different upper bounds including the smallest upper bound. Decide which of the sets have a maximum and, where this exists, write down its value.
(i) $(0, 1)$ (ii) $(-\infty, 2]$ (iii) $\{-1, 0, 2, 5\}$
(iv) $(3, \infty)$ (v) $[0, 1]$.

(4) Repeat the above question but with the words 'above, upper and maximum' replaced by 'below, lower and minimum'.

(5) Give an example of a set which has smallest upper bound 4 but contains no element x satisfying $3 < x < 4$.

(6) Show that, given any element x of the set $(0, \infty)$, there is another element $y \in (0, \infty)$ with the property that $y < x$. Deduce that $(0, \infty)$ has no minimum.

2.11   Manipulations with sup and inf

We prove one result and list some others as exercises.

2.12   Theorem Suppose that S is a non-empty set of real numbers which is bounded above and $\xi > 0$. Then

$$\sup_{x \in S} \xi x = \xi \sup_{x \in S} x.$$

Proof Let $B = \sup S$. Then B is the smallest number such that, for any $x \in S$,

$$x \leqslant B.$$

Let $T = \{\xi x : x \in S\}$. Since $\xi > 0$,

$$\xi x \leqslant \xi B$$

for any $x \in S$. Hence T is bounded above by ξB. By the Continuum Property, T has a smallest upper bound (or supremum) C. We have to prove that $C = \xi B$. Since ξB is *an* upper bound for T and C is the *smallest* upper bound for T,

$$C \leqslant \xi B.$$

Now repeat the argument with the roles of S and T reversed. We know that C is the smallest number such that, for any $y \in T$,

$$y \leqslant C.$$

Since $\xi > 0$, it follows that

$$\xi^{-1} y \leqslant \xi^{-1} C$$

for any $y \in T$. But $S = \{\xi^{-1} y : y \in T\}$. Hence $\xi^{-1} C$ is *an* upper bound for S. But B is the *smallest* upper bound for S. Thus

$$B \leqslant \xi^{-1} C$$

$$\xi B \leqslant C.$$

We have shown that $C \leqslant \xi B$ and also that $\xi B \leqslant C$. Hence $\xi B = C$.

2.13   Exercise

(1) Suppose that S is bounded above and that $S_0 \subset S$. Prove that $\sup S_0 \leqslant \sup S$.

(2) Suppose that S is bounded above and that ξ is any real number. Prove that

$$\sup_{x \in S} (x + \xi) = \sup_{x \in S} x + \xi.$$

(3) Suppose that S is bounded above. Prove that

$$\inf_{x \in S} (-x) = - \sup_{x \in S} x.$$

Hence obtain analogies of theorem 2.12 and exercises 2.13(1 and 3) in which inf replaces sup.

(4) The distance $d(\xi, S)$ between a real number ξ and a non-empty set S of real numbers is defined by

$$d(\xi, S) = \inf_{x \in S} |\xi - x|$$

(see exercise 1.20(3)). Find the distance between the real number $\xi = 3$ and the sets

    (i) $S = \{0, 1, 2\}$ (ii) $S = (0, 1)$

    (iii) $S = [1, 2]$ (iv) $S = (2, 3)$.

(5) (i)   If $\xi \in S$, prove that $d(\xi, S) = 0$. Give an example of a real number ξ and a non-empty set S for which $d(\xi, S) = 0$ but $\xi \notin S$.

    (ii)   If S is bounded above and $\xi = \sup S$, prove that $d(\xi, S) = 0$. Deduce that the same is true if S is bounded below and $\xi = \inf S$.

    †(iii)   If I is a closed interval, prove that $d(\xi, I) = 0$ implies that $\xi \in I$. If I is an open interval (other than $\mathbb{R}$ or $\emptyset$), show that a $\xi \notin I$ can always be found for which $d(\xi, I) = 0$.

* (6) Suppose that every element of an interval I belongs to one or the other of two non-empty subsets S and T. Show that one of the two sets S or T contains an element at zero distance from the other. (Do *not* assume that S and T are necessarily intervals. The set S, for example, could consist of all rational numbers in I and the set T of all irrational numbers in I.)

[Hint: Suppose that $s < t$, where $s \in S$ and $t \in T$. Let $T_0 = \{x : x \in T$ and $x > s\}$. Show that $T_0 \neq \emptyset$ and is bounded below. Write $b = \inf T_0$ and consider the two cases $b \in T$ and $b \notin T$.]

3 NATURAL NUMBERS

3.1 Introduction

In this chapter we explore some implications of the Continuum Property for sets of natural numbers. As always, arithmetical properties are taken for granted. But where a proof hinges on such a property, the property will be explicitly stated.

3.2 Archimedean property

The Archimedean property is the assertion of the following theorem.

3.3 Theorem The set $\mathbb{N}$ of natural numbers is unbounded above.

Proof For this proof we need to assume that, if n is a natural number, then so is $n + 1$.

Suppose that the theorem is false. Then $\mathbb{N}$ is bounded above. By the Continuum Property it therefore has a smallest upper bound B. Since B is the *smallest* upper bound for $\mathbb{N}$, $B - 1$ is *not* an upper bound for $\mathbb{N}$. Thus, for some $n \in \mathbb{N}$,

$$n > B - 1$$
$$\therefore \quad n + 1 > B.$$

But then $n + 1$ is an element of the set $\mathbb{N}$ which is greater than B. But B is an upper bound of the set $\mathbb{N}$. This is a contradiction.

Hence $\mathbb{N}$ is unbounded above.

3.4 Example Prove that the set

$$S = \{n^{-1} : n \in \mathbb{N}\}$$

is bounded below with largest lower bound 0.

Proof Since the natural numbers are all positive,

$$n^{-1} > 0$$

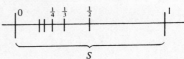

for all $n \in \mathbb{N}$. Hence the set S is bounded below by 0. We have to prove that 0 is the *largest* lower bound.

Suppose that $h > 0$ is a lower bound for S. Then, for each $n \in \mathbb{N}$,

$$n^{-1} \geqslant h.$$

Hence, for each $n \in \mathbb{N}$,

$$n \leqslant h^{-1}.$$

But then h^{-1} is an upper bound for the set $\mathbb{N}$ which we know is unbounded above. It follows from this contradiction that no $h > 0$ can be a lower bound for S. Thus 0 is the largest lower bound for S.

Note that $0 \notin S$ and hence S has *no* minimum.

3.5   *Theorem* Every set of natural numbers which is not empty has a minimum.

Proof For this proof we need to assume that the distance between distinct natural numbers is at least 1.

Let S be a non-empty set of natural numbers. Given that all natural numbers are positive, 0 is a lower bound for S. By the Continuum Property, it follows that S has a largest lower bound b. Since b is the *largest* lower bound, $b + 1$ is *not* a lower bound. Therefore, for some $n \in S$,

$$n < b + 1.$$

If n is the minimum of S, there is nothing to prove. If not, then, for some $m \in S$, $m < n$. We obtain the inequality

$$b \leqslant m < n < b + 1$$

from which follows the contradiction $0 < n - m < 1$.

3.6   *Exercise*

(1) Prove that the set

$$S = \left\{ \frac{n-1}{n} : n \in \mathbb{N} \right\}$$

is bounded above with smallest upper bound 1. Does S have a maximum?

(2) Let $X > 1$. Prove that the set

$$S = \{ X^n : n \in \mathbb{N} \}$$

is unbounded above. [Hint: if B were the smallest upper bound, then BX^{-1} could not be an upper bound.] Show that, if $0 < x < 1$, then the set

$$T = \{x^n : n \in \mathbb{N}\}$$

is bounded below with largest lower bound 0.

(3) Prove that any non-empty set of integers which is bounded above has a maximum and that any non-empty set of integers which is bounded below has a minimum.

(4) Let $a < b$. Prove that there exists a *rational* number r satisfying $a < r < b$. [Hint: justify the existence of a natural number n satisfying $n > (b - a)^{-1}$ and consider the rational number $r = m/n$ where m is the smallest integer satisfying $m > an$.]

(5) Let S be the set of all *rational* numbers r which satisfy $0 < r < 1$. Show that S has no maximum and no minimum. Prove that S is bounded and has largest lower bound 0 and smallest upper bound 1. [Hint: for the last part, use the previous question.]

(6) Let $a < b$. Prove that there exists an *irrational* number ξ satisfying $a < \xi < b$. [Hint: exercise 1.20(4).]

3.7   Principle of induction

Suppose that a line of dominoes is arranged so that, if the nth one falls, it will knock over the $(n + 1)$th. If the first domino is pushed over, most people would agree that *all* the dominoes would then fall down.

The principle of induction is an idealisation of this simple notion. We show below how it may be deduced from theorem 3.5.

3.8   *Theorem (principle of induction)*

Suppose that, for each $n \in \mathbb{N}$, $P(n)$ is a statement about the natural number n. Suppose also that

(i) $P(1)$ is true

(ii) if $P(n)$ is true, then $P(n + 1)$ is true.

Then $P(n)$ is true for *every* $n \in \mathbb{N}$.

Proof For this proof we need to assume that, if n is a natural number other than 1, then $n - 1$ is a natural number as well.

Let $S = \{n : P(n) \text{ is false}\}$. We want to prove that S is empty. Suppose that S is not empty. Then from theorem 3.5 it follows that S has a minimum m. Since $P(1)$ is true, $1 \notin S$. Hence $m \neq 1$. Therefore $m - 1$ is a natural number. But m is the *minimum* of S. Thus $m - 1 \notin S$ and so $P(m - 1)$ is true. But then $P(m)$ is true (because of hypothesis (ii) of the theorem). Hence $m \notin S$ which is a contradiction.

3.9   *Example* For each $n \in \mathbb{N}$

$$1 + 2 + 3 + \ldots + n = \tfrac{1}{2}n(n + 1).$$

Proof Let $S_n = 1 + 2 + 3 + \ldots + n = \Sigma_{k=1}^{n} k$. Let $P(n)$ be the statement $S_n = \tfrac{1}{2}n(n + 1)$.

We have $S_1 = 1$, and $\tfrac{1}{2}. 1(1 + 1) = 1$. Hence $P(1)$ is true. We now wish to show that 'If $P(n)$ is true, then $P(n + 1)$ is true'. To do this, we assume that $P(n)$ is true and try and deduce that $P(n + 1)$ is true. Thus we assume that $S_n = \tfrac{1}{2}n(n + 1)$ and try and deduce that $S_{n+1} = \tfrac{1}{2}(n + 1)(n + 2)$. But

$$\begin{aligned}
S_{n+1} &= 1 + 2 + 3 + \ldots + n + (n + 1) \\
&= S_n + (n + 1) \\
&= \tfrac{1}{2}n(n + 1) + (n + 1) \\
&= \tfrac{1}{2}(n + 1)(n + 2)
\end{aligned}$$

as required. The result now follows by induction.

3.10   *Example* The real number

$$A_n = \frac{1}{n}(a_1 + a_2 + \ldots + a_n)$$

is the *arithmetic mean* of the numbers $a_1, a_2, \ldots, a_n$. If $a_1, a_2, \ldots, a_n$ are positive, their *geometric mean* G_n is defined by

$$G_n = (a_1 a_2 \ldots a_n)^{1/n}.$$

The 'inequality of the arithmetic and geometric means' asserts that

$$G_n \leqslant A_n.$$

Proof This inequality has an entertaining proof using 'backwards induction' (see exercise 3.11(5)).

Let $P(n)$ be the statement 'For any positive numbers $a_1, a_2, \ldots, a_n$, $G_n \leqslant A_n$'. We shall show that

(i) $P(2^n)$ is true for each $n \in \mathbb{N}$.

(ii) If $P(n)$ is true, then $P(n - 1)$ is true.

The result then follows by backwards induction.

(i) We begin by proving $P(2)$. Now,

$$0 \leqslant (\sqrt{a_1} - \sqrt{a_2})^2 = a_1 - 2\sqrt{(a_1 a_2)} + a_2$$

$$(a_1 a_2)^{1/2} \leqslant \tfrac{1}{2}(a_1 + a_2).$$

This is simply $P(2)$. We now show that $P(2^n)$ implies $P(2^{n+1})$. Let $m = 2^n$. Then $2^{n+1} = 2m$. We have to assume $P(m)$ and try and deduce $P(2m)$. Since $P(m)$ is true,

$$(a_1 a_2 \ldots a_m)^{1/m} \leqslant \frac{1}{m}(a_1 + a_2 + \ldots + a_m)$$

and $(a_{m+1} a_{m+2} \ldots a_{2m})^{1/m} \leqslant \dfrac{1}{m}(a_{m+1} + a_{m+2} + \ldots + a_{2m}).$

But we know that $P(2)$ is true. Hence

$$\{(a_1 a_2 \ldots a_m)^{1/m}(a_{m+1} a_{m+2} \ldots a_{2m})^{1/m}\}^{1/2}$$

$$\leqslant \frac{1}{2}\left\{\frac{a_1 + \ldots + a_m}{m} + \frac{a_{m+1} + \ldots + a_{2m}}{m}\right\}.$$

Thus

$$(a_1 a_2 \ldots a_{2m})^{1/2m} \leqslant \frac{1}{2m}(a_1 + a_2 + \ldots + a_{2m})$$

which is $P(2m)$.

We have shown that $P(2^1)$ is true and that, if $P(2^n)$ is true, then $P(2^{n+1})$ is true. Hence $P(2^n)$ is true for all $n \in \mathbb{N}$ by straightforward induction.

(ii) We now show that, if $P(n)$ is true, then $P(n-1)$ is true. Suppose that $P(n)$ is true. Then

$$\{a_1 a_2 \ldots a_{n-1} G_{n-1}\}^{1/n} \leqslant \frac{a_1 + a_2 + \ldots + a_{n-1} + G_{n-1}}{n}$$

$$\{G_{n-1}^{n-1} G_{n-1}\}^{1/n} \leqslant \frac{(n-1)A_{n-1} + G_{n-1}}{n}$$

$$nG_{n-1} \leqslant (n-1)A_{n-1} + G_{n-1}$$

$$G_{n-1} \leqslant A_{n-1}.$$

Thus $P(n-1)$ is true.

This completes the proof.

3.11   Exercise

(1) Prove by induction that

(i) $\displaystyle\sum_{k=1}^{n} k^2 = 1^2 + 2^2 + \ldots + n^2 = \tfrac{1}{6}n(n+1)(2n+1)$

(ii) $\displaystyle\sum_{k=1}^{n} k^3 = 1^3 + 2^3 + \ldots + n^3 = \tfrac{1}{4}n^2(n+1)^2.$

(2) Prove by induction that, if $x \neq 1$,

$$\sum_{k=0}^{n} x^k = 1 + x + x^2 + \ldots + x^n = \frac{1-x^{n+1}}{1-x}.$$

Deduce that, for any real numbers a and b,

$$a^n - b^n = (a - b)(a^{n-1} + a^{n-2}b + a^{n-3}b^2 + \ldots + ab^{n-2} + b^{n-1}).$$

(3) A *polynomial* $P(x)$ is an expression of the form

$$P(x) = a_n x^n + a_{n-1} x^{n-1} + \ldots + a_1 x + a_0.$$

If $a_n \neq 0$, the polynomial is said to be of *degree n*.

Suppose that $P(x)$ is a polynomial of degree n and that, for some ξ, $P(\xi) = 0$. Prove that $P(x) = (x - \xi)Q(x)$ where $Q(x)$ is a polynomial of degree $n - 1$. [Hint: use the previous question.] If $\xi_1, \xi_2, \ldots, \xi_n$ are distinct real numbers and $P(\xi_1) = P(\xi_2) = \ldots = P(\xi_n) = 0$, prove that

$$P(x) = k(x - \xi_1)(x - \xi_2) \ldots (x - \xi_n)$$

where k is a constant.

(4) Assuming that $n! = 1 \, . \, 2 \, . \, 3 \ldots n$ and $0! = 1$, we define

$$\binom{n}{r} = \frac{n!}{r!(n-r)!} \qquad (r = 0, 1, 2, \ldots, n).$$

Prove that

$$\binom{n}{r} + \binom{n}{r-1} = \binom{n+1}{r} \qquad (r = 1, 2, \ldots, n).$$

Using this result and the principle of induction, obtain the binomial theorem in the form

$$(a + b)^n = a^n + na^{n-1}b + \frac{n(n-1)}{2} a^{n-2}b^2 + \ldots + b^n$$

$$= \sum_{r=0}^{n} \binom{n}{r} a^{n-r}b^r.$$

(5) Let $P(n)$ be a statement about the natural number n and suppose that
  (i) $P(2^n)$ is true for each $n \in \mathbb{N}$
  (ii) if $P(n)$ is true, then $P(n - 1)$ is true.
  Prove that $P(n)$ is true for *every* $n \in \mathbb{N}$. [Hint: the set $\{2^n : n \in \mathbb{N}\}$ is unbounded above (exercise 3.6(2)).]

(6) Suppose that $0 < X \leq Y$ and that x and y belong to the interval $[X, Y]$. Prove that, for each $n \in \mathbb{N}$,

$$XY^{1/n} | x - y | \leq nXY | x^{1/n} - y^{1/n} | \leq YX^{1/n} | x - y |.$$

[Hint: exercise 3.11(2).]

4 CONVERGENT SEQUENCES

4.1 The bulldozers and the bee

Two bulldozers are moving towards each other at a speed of one mile per hour on a collision course. When they started they were one mile apart and a bee was perched on the front of one of them. The bee began to fly back and forward between the bulldozers at a constant speed of two miles per hour, vainly seeking to avoid his fate. How far from his starting point will the unfortunate insect be crushed?

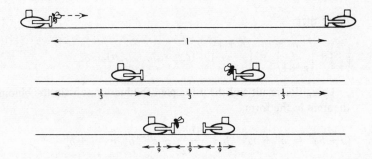

This riddle is usually posed in the hope that the victim will embark on some complicated calculations involving the flight of the bee. But it is quite obvious that the bee will be crushed when the bulldozers collide. Since the bulldozers travel at the same speed, this will happen halfway between their starting points. The answer is therefore one half mile.

Let us, however, take up the role of the riddler's victim and examine the flight of the bee. Let x_n denote the distance the bee is from his starting point when he makes his nth landing. Then

$$x_1 = \tfrac{2}{3}$$
$$x_2 = \tfrac{2}{3} - \tfrac{2}{9}$$

and, in general,

$$x_n = \tfrac{2}{3} - \tfrac{2}{9} + \tfrac{2}{27} - \ldots + (-1)^{n-1} 2(\tfrac{1}{3})^n$$

$$= \tfrac{2}{3}\{1 + (-\tfrac{1}{3}) + (-\tfrac{1}{3})^2 + \ldots + (-\tfrac{1}{3})^{n-1}\}$$

$$= \frac{2}{3}\left\{\frac{1 - (-\tfrac{1}{3})^n}{1 - (-\tfrac{1}{3})}\right\} = \tfrac{1}{2}(1 - (-\tfrac{1}{3})^n).$$

How is the answer $\tfrac{1}{2}$ to be extracted from this formula?

It seems clear that, as n gets larger and larger, $(-\tfrac{1}{3})^n$ gets closer and closer to zero and so x_n gets closer and closer to $\tfrac{1}{2}$. We say that 'x_n tends to $\tfrac{1}{2}$ as n tends to infinity' – or, in symbols, '$x_n \to \tfrac{1}{2}$ as $n \to \infty$'.

This idea is of the greatest importance, but, before we can make proper use of it, it is necessary to give a *precise* formulation of what it means to say that $x_n \to l$ as $n \to \infty$.

4.2   Sequences

A *sequence* may be regarded as a list of numbers $x_1, x_2, x_3, x_4, \ldots$
More precisely, we can say that a sequence is determined by a rule which assigns to each natural number n a unique real number x_n. We call x_n the nth *term* of the sequence.

The notation

$$\langle x_n \rangle$$

means the sequence whose nth term is x_n.

The set $\{x_n : n \in \mathbb{N}\}$ is called the *range* of the sequence. We say that a sequence is bounded above (or below) if its range is bounded above (or below). Thus $\langle x_n \rangle$ is bounded above by H if and only if $x_n \leqslant H$ $(n = 1, 2, 3, \ldots)$.

4.3   Examples

(i) $\langle (-1)^n \rangle = -1, +1, -1, +1, \ldots$ (ii) $\langle n^{-1} \rangle = 1, \tfrac{1}{2}, \tfrac{1}{3}, \tfrac{1}{4}, \ldots$

(iii) $\langle 1 \rangle = 1, 1, 1, 1, \ldots$ (iv) $\langle 2^n \rangle = 2, 4, 8, 16, \ldots$

(v) The sequence $\langle x_n \rangle$ defined inductively by $x_1 = 2$ and

$$x_n = \frac{1}{2}\left(x_{n-1} + \frac{2}{x_{n-1}}\right) \quad (n = 2, 3, 4, \ldots).$$

The first few terms are $2, \tfrac{3}{2}, \tfrac{17}{12}, \tfrac{577}{408}, \ldots$

4.4   Definition of convergence

A sequence $\langle x_n \rangle$ is said to *converge* to the *limit* l if and only if the following criterion is satisfied.

> Given any $\epsilon > 0$, we can find an N such that, for any $n > N$,
> $|x_n - l| < \epsilon$.

We write $x_n \to l$ as $n \to \infty$ or

$$\lim_{n \to \infty} x_n = l.$$

It may be helpful to think of $x_1, x_2, x_3, \ldots$ as successive approximations to the number l. The *distance* $|x_n - l|$ between x_n and l is then the error involved in approximating l by x_n. The definition of convergence then simply asserts that we can make this error as small as we choose by taking n large enough.

The diagram is supposed to represent a sequence $\langle x_n \rangle$ with the property that $x_n \to l$ as $n \to \infty$.

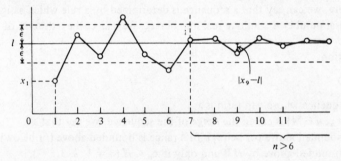

For the value of ϵ indicated in the diagram, a suitable value for N in the definition is $N = 6$. For each value of $n > 6$,

$$|x_n - l| < \epsilon.$$

In particular, the value of $|x_9 - l|$ has been noted in the diagram.

Notice that the definition of convergence begins 'Given *any* $\epsilon > 0$, we can find an $N \ldots$' Here the emphasis is on the very small values of $\epsilon > 0$. It is clear from the diagram above that, if we had chosen to look at a very much smaller value of $\epsilon > 0$, then we should have had to have picked out a very much larger value of N. In general, the *smaller* the value of $\epsilon > 0$, the *bigger* must be the corresponding value of N.

4.5   *Example*

$$1 + \frac{1}{n} \to 1 \text{ as } n \to \infty.$$

Proof Let $\epsilon > 0$ be given. We must find a value of N such that, for any $n > N$,

$$\left| \left(1 + \frac{1}{n} \right) - 1 \right| < \epsilon.$$

But $\qquad \left| \left(1 + \frac{1}{n} \right) - 1 \right| = \left| \frac{1}{n} \right| = \frac{1}{n}$

and so we simply need to find an N such that, for any $n > N$,

$$\frac{1}{n} < \epsilon$$

i.e. $\quad n > \dfrac{1}{\epsilon}.$

But this solves the problem. We simply chose $N = 1/\epsilon$. Then, for any $n > N$,

$$n > N = \frac{1}{\epsilon}$$

i.e. $\quad \dfrac{1}{n} < \epsilon$

i.e. $\quad \left| \left(1 + \dfrac{1}{n} \right) - 1 \right| < \epsilon.$

We have shown that, given *any* $\epsilon > 0$, we can find an N (namely $N = 1/\epsilon$) such that, for any $n > N$,

$$\left| \left(1 + \frac{1}{n} \right) - 1 \right| < \epsilon.$$

It is important to observe that N *depends on* ϵ, i.e. for each $\epsilon > 0$ we use a *different* N. Some values of ϵ and the corresponding values of N are entered in the table below.

ϵ	N
0·4	2·5
0·1	10
0·000 001	1 000 000

(Note: Some authors insist that N be a natural number. This makes the definition of convergence a little more elegant but renders examples like that above marginally more complicated. If we wanted N to be a natural number in the example above, we could not simply write $N = 1/\epsilon$. Instead we should have to choose N to be some natural number larger than $1/\epsilon$.)

4.6   Exercise

(1) Prove that

$$\lim_{n \to \infty} \left(\frac{n^2 - 1}{n^2 + 1} \right) = 1.$$

(2) Let r be any *positive* rational number. Prove that

$$\frac{1}{n^r} \to 0 \text{ as } n \to \infty.$$

(3) Let λ be any real number. If $x_n \to l$ as $n \to \infty$, prove that $\lambda x_n \to \lambda l$ as $n \to \infty$.

4.7   Criteria for convergence

In the example and exercises above it is fairly easy to decide what value of N is appropriate to any given value of $\epsilon > 0$. But this is by no means always the case. It is therefore natural to look around for some shortcuts which will enable us to determine whether a sequence converges (and what its limit is) without our having to indulge in the painful process of appealing to the definition.

In this and the next section we give some useful results of this sort.

4.8   Proposition (combination theorem) Let $x_n \to l$ as $n \to \infty$ and $y_n \to m$ as $n \to \infty$ and let λ and μ be any real numbers. Then

(i) $\lambda x_n + \mu y_n \to \lambda l + \mu m$ as $n \to \infty$

(ii) $x_n y_n \to lm$ as $n \to \infty$

(iii) $\dfrac{x_n}{y_n} \to \dfrac{l}{m}$ as $n \to \infty$ (provided that $m \neq 0$).

The proofs of (ii) and (iii) can be found in the appendix. These proofs are somewhat technical and we prefer not to hold up the discussion by presenting them at this stage. Readers who prefer to omit the proofs of (ii) and (iii) altogether will not suffer greatly as these results may be deduced from (i) once the theory of the exponential and logarithm functions has been developed.

Proof of 4.8(i). After exercise 4.6(3) we need only show that, if $x_n \to l$ as $n \to \infty$ and $y_n \to m$ as $n \to \infty$, then $x_n + y_n \to l + m$ as $n \to \infty$.

Let $\epsilon > 0$ be given. Then $\frac{1}{2}\epsilon > 0$. Since $x_n \to l$ as $n \to \infty$, it follows that we can find an N_1 such that, for any $n > N_1$,

$$|x_n - l| < \tfrac{1}{2}\epsilon. \tag{1}$$

Similarly, we can find an N_2 such that, for any $n > N_2$,

$$|y_n - m| < \tfrac{1}{2}\epsilon. \tag{2}$$

Let N be whichever of N_1 and N_2 is the larger, i.e. $N = \max \{N_1, N_2\}$. Then, if $n > N$, both inequalities (1) and (2) are true simultaneously. Thus, for any $n > N$,

$$
\begin{aligned}
|(x_n + y_n) - (l + m)| &= |(x_n - l) + (y_n - m)| \\
&\leqslant |x_n - l| + |y_n - m| \quad \text{(triangle inequality)} \\
&< \tfrac{1}{2}\epsilon + \tfrac{1}{2}\epsilon = \epsilon.
\end{aligned}
$$

Given any $\epsilon > 0$ we have found a value of N (namely $N = \max \{N_1, N_2\}$) such that, for any $n > N$, $|(x_n + y_n) - (l + m)| < \epsilon$. Hence $x_n + y_n \to l + m$ as $n \to \infty$.

4.9   *Example* Prove that

$$
\lim_{n \to \infty} \left(\frac{2n^3 - 3n}{5n^3 + 4n^2 - 2} \right) = \frac{2}{5}.
$$

Proof We write

$$
\frac{2n^3 - 3n}{5n^3 + 4n^2 - 2} = \frac{2 - 3n^{-2}}{5 + 4n^{-1} - 2n^{-3}} \to \frac{2 - 0}{5 + 0 - 0} = \frac{2}{5} \text{ as } n \to \infty.
$$

The supporting argument is as follows. It is obvious from the definition of convergence that $2 \to 2$ as $n \to \infty$. From exercise 4.6(2), $n^{-2} \to 0$ as $n \to \infty$. Hence, by proposition 4.8(i),

$$
2 - 3n^{-2} \to 2 - 3.0 = 2 \text{ as } n \to \infty.
$$

Similarly,

$$
5 + 4n^{-1} - 2n^{-3} \to 5 + 4.0 - 2.0 = 5 \text{ as } n \to \infty.
$$

The result then follows from proposition 4.8(iii).

4.10   *Theorem (the sandwich theorem)* Suppose that $y_n \to l$ as $n \to \infty$ and that $z_n \to l$ as $n \to \infty$. If $y_n \leqslant x_n \leqslant z_n$ $(n = 1, 2, \ldots)$, then

$$
x_n \to l \text{ as } n \to \infty.
$$

(Here the sequence $\langle x_n \rangle$ is 'sandwiched' between the two sequences $\langle y_n \rangle$ and $\langle z_n \rangle$ – just as the bee of §4.1 was 'sandwiched' between the bulldozers.)

Proof For this proof it is necessary to note that the inequality $|x - l| < \epsilon$ is true if and only if $l - \epsilon < x < l + \epsilon$. (See exercise 1.20(1).)

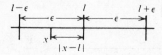

Let $\epsilon > 0$ be given. We have to find an N such that, for any $n > N$, $|x_n - l| < \epsilon$.

Since $y_n \to l$ as $n \to \infty$, we know that there exists an N_1 such that, for any $n > N_1$,

$$|y_n - l| < \epsilon. \tag{3}$$

Similarly, since $z_n \to l$ as $n \to \infty$, there exists an N_2 such that, for any $n > N_2$,

$$|z_n - l| < \epsilon. \tag{4}$$

Let N be whichever of N_1 and N_2 is the larger, i.e. $N = \max\{N_1, N_2\}$. Then, if $n > N$, both inequalities (3) and (4) are true simultaneously. Thus, for any $n > N$,

$$l - \epsilon < y_n < l + \epsilon$$

and $l - \epsilon < z_n < l + \epsilon$.

But $y_n \leqslant x_n \leqslant z_n$ $(n = 1, 2, \ldots)$. Hence, for any $n > N$,

$$l - \epsilon < y_n \leqslant x_n \leqslant z_n < l + \epsilon.$$

Hence $l - \epsilon < x_n < l + \epsilon$

i.e. $|x_n - l| < \epsilon$.

Given any $\epsilon > 0$, we have found a value of N (namely $N = \max\{N_1, N_2\}$) such that, for any $n > N$,

$$|x_n - l| < \epsilon.$$

Hence $x_n \to l$ as $n \to \infty$.

4.11   Corollary Suppose that $y_n \to 0$ as $n \to \infty$ and that

$$|x_n - l| \leqslant y_n \quad (n = 1, 2, \ldots).$$

Then $x_n \to l$ as $n \to \infty$.

Proof The inequality $|x_n - l| \leqslant y_n$ is equivalent to $l - y_n \leqslant x_n \leqslant l + y_n$. But, using proposition 4.8, $l - y_n \to l$ as $n \to \infty$ and $l + y_n \to l$ as $n \to \infty$. Hence $x_n \to l$ as $n \to \infty$, by theorem 4.10.

4.12   Example Suppose that $|x| < 1$. Prove that

$$x^n \to 0 \text{ as } n \to \infty.$$

Proof Write

$$|x|^n = \frac{1}{(1+h)^n} = \frac{1}{1 + nh + \frac{1}{2}n(n-1)h^2 + \ldots + h^n} < \frac{1}{nh}.$$

Thus

$$|x^n - 0| = |x|^n < \frac{1}{nh} \quad (n = 1, 2, \ldots).$$

But $1/n \to 0$ as $n \to \infty$ (by exercise 4.6(2)). Thus, by corollary 4.11,

$$x^n \to 0 \text{ as } n \to \infty.$$

4.13 *Example* In the problem of the bulldozers and the bee, we obtained the sequence

$$x_n = \tfrac{1}{2}\{1 - (-\tfrac{1}{3})^n\}.$$

By example 4.12, $(-\tfrac{1}{3})^n \to 0$ as $n \to \infty$ and hence $x_n \to \tfrac{1}{2}$ as $n \to \infty$.

4.14 *Example* Let $x > 0$. Prove that

$$x^{1/n} \to 1 \text{ as } n \to \infty.$$

Proof In the inequality of the arithmetic and geometric means, take $1 = a_1 = a_2 = \ldots = a_{n-1}$ and $a_n = x$. Then $G_n = x^{1/n}$ and $A_n = (n - 1 + x)/n = (x - 1)/n + 1$.

Hence $\quad x^{1/n} \leqslant \dfrac{x - 1}{n} + 1$

i.e. $\quad x^{1/n} - 1 \leqslant \dfrac{1}{n}(x - 1)$.

(Alternatively, take $X = y = 1$ in exercise 3.11(6).)

Assume to begin with that $x \geqslant 1$. Then

$$0 \leqslant x^{1/n} - 1 \leqslant \frac{1}{n}(x - 1)$$

and hence $x^{1/n} \to 1$ as $n \to \infty$ by the sandwich theorem. If $0 < x < 1$, then $x = y^{-1}$ where $y > 1$. But $y^{1/n} \to 1$ as $n \to \infty$ and hence

$$x^{1/n} = \frac{1}{y^{1/n}} \to \frac{1}{1} \text{ as } n \to \infty$$

by proposition 4.8(iii).

4.15 Monotone sequences

A sequence $\langle x_n \rangle$ is *increasing* if

$$x_{n+1} \geqslant x_n \quad (n = 1, 2, \ldots).$$

Similarly, $\langle x_n \rangle$ is *decreasing* if

$$x_{n+1} \leqslant x_n \quad (n = 1, 2, \ldots).$$

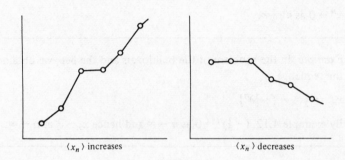

$\langle x_n \rangle$ increases $\langle x_n \rangle$ decreases

A sequence $\langle x_n \rangle$ is *strictly increasing* if $x_{n+1} > x_n$ $(n = 1, 2, \ldots)$. It is *strictly decreasing* if $x_{n+1} < x_n$ $(n = 1, 2, \ldots)$.

A sequence which is either increasing or else decreasing is called *monotone*.

4.16   Example

(i)   The sequence $\langle 2^n \rangle$ is strictly increasing.

(ii) The sequences $\langle n^{-1} \rangle$ and $\langle -n \rangle$ are both strictly decreasing.

(iii) The sequence $\langle 1 \rangle$ is both increasing *and* decreasing.

(iv) The sequence $\langle (-1)^n \rangle$ is *not* monotone. It is neither increasing nor decreasing.

4.17   Theorem

(i)   If $\langle x_n \rangle$ is increasing and bounded above, then it converges to its smallest upper bound.

(ii) If $\langle x_n \rangle$ is decreasing and bounded below, then it converges to its largest lower bound.

Proof If $\langle x_n \rangle$ is decreasing and bounded below, then $\langle -x_n \rangle$ is increasing and bounded above. Hence it is only necessary to prove (i).

Suppose that $\langle x_n \rangle$ increases and is bounded above with smallest upper bound B. We must show that $x_n \to B$ as $n \to \infty$.

Let $\epsilon > 0$ be given. Since $B - \epsilon$ is *not* an upper bound, there exists at least one term x_N of the sequence such that

$$x_N > B - \epsilon.$$

But $\langle x_n \rangle$ increases. Hence, for any $n > N$, $x_n \geqslant x_N$. Thus, for any $n > N$,

$$x_n \geqslant x_N > B - \epsilon.$$

Also B is an upper bound for the sequence. Hence $x_n \leqslant B$ $(n = 1, 2, \ldots)$. Thus, for any $n > N$,

$$B - \epsilon < x_n \leqslant B.$$
$$B - \epsilon < x_n < B + \epsilon$$

i.e. $\qquad |x_n - B| < \epsilon.$

Given any $\epsilon > 0$, we have found an N such that, for any $n > N$, $|x_n - B| < \epsilon$. Hence $x_n \to B$ as $n \to \infty$.

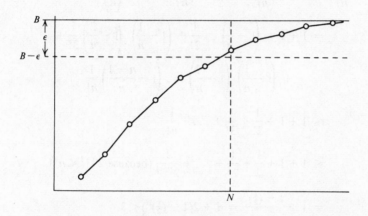

4.18 *Examples* (i) The sequence $\langle (n - 1)/n \rangle$ increases and we know from exercise 3.6(1) that it is bounded above with smallest upper bound 1. By theorem 4.17 it follows that

$$\frac{n - 1}{n} \to 1 \text{ as } n \to \infty.$$

(ii) If $0 < x < 1$, the sequence $\langle x^n \rangle$ decreases and is bounded below with largest lower bound 0 (exercise 3.6(2)). Thus

$$x^n \to 0 \text{ as } n \to \infty.$$

4.19 *Example* The examples above are nothing new. But consider the sequence

$$\langle (1 + n^{-1})^n \rangle.$$

We first show that this sequence *increases*. In the inequality for the geometric and arithmetic means, take $a_1 = a_2 = \ldots = a_{n-1} = 1 + (n - 1)^{-1}$ and $a_n = 1$. Then

$$\left(1 + \frac{1}{n-1}\right)^{(n-1)/n} \leqslant \frac{(n-1)(1 + (n-1)^{-1}) + 1}{n} = \left(1 + \frac{1}{n}\right).$$

Hence

$$\left(1 + \frac{1}{n-1}\right)^{n-1} \leqslant \left(1 + \frac{1}{n}\right)^{n} \quad (n = 2, 3, \ldots)$$

and so the sequence increases. To be able to deduce the convergence of the sequence from theorem 4.17, we also need to prove that the sequence is *bounded above*. By the binomial theorem,

$$\left(1 + \frac{1}{n}\right)^{n} = 1 + n\left(\frac{1}{n}\right) + \frac{n(n-1)}{2}\left(\frac{1}{n}\right)^{2} + \ldots + \left(\frac{1}{n}\right)^{n}$$

$$= 1 + 1 + \left(1 - \frac{1}{n}\right) \cdot \frac{1}{2!} + \left(1 - \frac{1}{n}\right)\left(1 - \frac{2}{n}\right)\frac{1}{3!} + \ldots$$

$$+ \left(1 - \frac{1}{n}\right)\left(1 - \frac{2}{n}\right) \ldots \left(1 - \frac{n-1}{n}\right)\frac{1}{n!}$$

$$\leqslant 1 + 1 + \frac{1}{2!} + \frac{1}{3!} + \ldots + \frac{1}{n!}$$

$$\leqslant 1 + 1 + \frac{1}{2} + \frac{1}{2^2} + \ldots + \frac{1}{2^{n-1}} \text{ (because } 2^{n-1} \leqslant n!)$$

$$= 1 + \frac{1 - (\frac{1}{2})^n}{1 - \frac{1}{2}} = 1 + 2(1 - (\tfrac{1}{2})^n) < 3.$$

Hence the sequence is bounded above by 3. From theorem 4.17 it follows that $\langle(1 + n^{-1})^n\rangle$ converges to a limit B and $B \leqslant 3$. (In fact, $B = e = 2 \cdot 718 \ldots$)

4.20   Exercise

(1) Prove that

$$\frac{n^3 + 5n^2 + 2}{2n^3 + 9} \to \tfrac{1}{2} \text{ as } n \to \infty.$$

(2) Decide for what values of x the limit

$$\lim_{n \to \infty} \left\{\frac{x + x^n}{1 + x^n}\right\}$$

exists. Draw a graph which plots the value of the limit against the value of x.

(3) Use the sandwich theorem to prove that

$$\sqrt{(n+1)} - \sqrt{n} \to 0 \text{ as } n \to \infty.$$

*(4) Let x be a positive number, and let N be the smallest natural number satisfying $N > x$. Prove that

$$\frac{x^n}{n!} \leqslant \frac{x^{N-1}}{(N-1)!} \left(\frac{x}{N}\right)^{n-N+1} \qquad (n \geqslant N).$$

Deduce that $x^n/n! \to 0$ as $n \to \infty$.

*(5) Let α be any positive rational number and let $|x| < 1$. Show that there exists a natural number N such that $(1 + 1/N)^{\alpha+1} |x| \leqslant 1$. Deduce that

$$|n^{\alpha+1}x^n| \leqslant |N^{\alpha+1}x^N| \qquad (n \geqslant N).$$

Hence show that $n^\alpha x^n \to 0$ as $n \to \infty$.

*(6) Prove that the sequence $\langle n^{1/n} \rangle$ decreases for $n \geqslant 3$. [Hint: example 4.19.] Hence prove that the sequence converges.

(The convergence of $\langle n^{1/n} \rangle$ can also be established by a method like that of example 4.12. If $n = (1 + h_n)^n$, then it follows from the binomial theorem that

$$\frac{n(n-1)}{2} h_n^2 < n.$$

Therefore $h_n \to 0$ as $n \to \infty$.)

4.21 Some simple properties of convergent sequences

4.22   *Theorem* A sequence can have at most one limit.

Proof Suppose that $x_n \to l$ as $n \to \infty$ and $x_n \to m$ as $n \to \infty$. Let $\epsilon > 0$ be given. Then

$$|l - m| = |l - x_n + x_n - m| \leqslant |l - x_n| + |x_n - m| < \epsilon + \epsilon = 2\epsilon$$

$$\text{(triangle inequality)}$$

provided that n is sufficiently large. But, from example 1.7 it follows that, if $0 \leqslant \frac{1}{2}|l - m| < \epsilon$ for *every* $\epsilon > 0$, then $|l - m| = 0$, i.e. $l = m$.

4.23   *Theorem* Suppose that $x_n \to l$ as $n \to \infty$.
(i) If $x_n \geqslant a$ $(n = 1, 2, \ldots)$, then $l \geqslant a$.
(ii) If $x_n \leqslant b$ $(n = 1, 2, \ldots)$, then $l \leqslant b$.

Proof If $x_n \leqslant b$ $(n = 1, 2, \ldots)$, then $-x_n \geqslant -b$ $(n = 1, 2, \ldots)$. Hence we need only prove (i).

Let $\epsilon > 0$. Then there exists an N such that, for any $n > N$,

$$|x_n - l| < \epsilon$$

i.e. $l - \epsilon < x_n < l + \epsilon.$

But $x_n \geqslant a \ (n = 1, 2, \ldots)$ and so, for any $n > N, a \leqslant x_n < l + \epsilon$.

Hence, given *any* $\epsilon > 0, a < l + \epsilon$. From example 1.7 it follows that $a \leqslant l$.

4.24   *Example* Suppose that $x_n \to l$ as $n \to \infty$. If $x_n > a \ (n = 1, 2, \ldots)$, it is tempting to conclude that $l > a$. But this may not be true. For example, $1/n \to 0$ as $n \to \infty$ and $1/n > 0 \ (n = 1, 2, \ldots)$. But we certainly cannot conclude that $0 > 0$.

4.25   *Theorem* Any convergent sequence is bounded.

   Proof Let $x_n \to l$ as $n \to \infty$. We have to find a K such that $|x_n| \leqslant K$ $(n = 1, 2, \ldots)$. (See proposition 2.3.)

It is true that, for any $\epsilon > 0$, there exists an N such that, for any $n > N$, $|x_n - l| < \epsilon$. In particular, this is true when $\epsilon = 1$, i.e. there exists an N_1 such that, for any $n > N_1$,

$$|x_n - l| < 1.$$

From theorem 1.18 it follows that, for any $n > N_1$,

$$|x_n| - |l| \leqslant |x_n - l| < 1$$

i.e. $|x_n| < |l| + 1 \quad (n > N_1)$.

The result now follows if we take

$$K = \max \{|x_1|, |x_2|, \ldots, |x_{N_1}|, |l| + 1\}.$$

4.26   Divergent sequences

A *divergent* sequence is a sequence which does not converge. Any unbounded sequence is therefore divergent (theorem 4.25). However *bounded* divergent sequences exist as well.

4.27   *Example* The sequence $\langle (-1)^n \rangle$ diverges.

   Proof The sequence is obviously bounded. Suppose that $(-1)^n \to l$ as $n \to \infty$. Then, given any $\epsilon > 0$, we can find an N such that, for any $n > N$, $|(-1)^n - l| < \epsilon$. But there are both even and odd values of n greater than N and so $|1 - l| < \epsilon$ and $|-1 - l| < \epsilon$. These inequalities must be true for *any* $\epsilon > 0$ and so we obtain the contradiction $l = 1$ and $l = -1$.

We say that a sequence $\langle x_n \rangle$ *diverges to* $+ \infty$ and write $x_n \to + \infty$ as $n \to \infty$ if, for any $H > 0$, we can find an N such that, for any $n > N$,

$$x_n > H.$$

Similarly, a sequence $\langle x_n \rangle$ *diverges to* $-\infty$ and $x_n \to -\infty$ as $n \to \infty$ if, for any $H > 0$, we can find an N such that, for any $n > N$,

$$x_n < -H$$

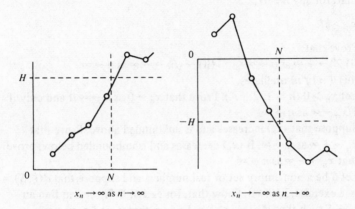

$x_n \to \infty$ as $n \to \infty$ $x_n \to -\infty$ as $n \to \infty$

A divergent sequence $\langle x_n \rangle$ which does *not* satisfy $x_n \to +\infty$ as $n \to \infty$ or $x_n \to -\infty$ as $n \to \infty$ is said to *oscillate*. An example is the sequence $\langle (-1)^n \rangle$.

4.28   *Example* Let α be a positive rational number. Prove that $n^\alpha \to +\infty$ as $n \to \infty$.

     Proof Let $H > 0$ be given. We have to find an N such that, for any $n > N$,

$$n^\alpha > H$$

i.e. $n > H^{1/\alpha}$.

We simply choose $N = H^{1/\alpha}$. Then, for any $n > N$, $n > H^{1/\alpha}$ and hence $n^\alpha > H$.

A common error is to suppose that proposition 4.8 (about combining *convergent* sequences) applies to sequences which *diverge* to $+\infty$ or $-\infty$. For example, it is tempting to suppose that, if $x_n \to \infty$ as $n \to \infty$ and $y_n \to \infty$ as $n \to \infty$, then $x_n - y_n \to \infty - \infty = 0$ as $n \to \infty$. But ∞ is *not* a real number and cannot be treated as such. Indeed, one should immediately be warned that something is wrong by the appearance of the totally *meaningless* expression $\infty - \infty$. To press the point home, we consider the example $x_n = n^2$ $(n = 1, 2, \ldots)$ and $y_n = n$ $(n = 1, 2, \ldots)$. Then $x_n \to \infty$ as $n \to \infty$ and $y_n \to \infty$ as $n \to \infty$, but $x_n - y_n \to \infty$ as $n \to \infty$. Or, again, take $x_n = (n + 1)$ $(n = 1, 2, \ldots)$ and $y_n = n$ $(n = 1, 2, \ldots)$. This time $x_n - y_n \to 1$ as $n \to \infty$.

4.29   Exercise

(1) Suppose that $x_n \to l$ as $n \to \infty$. Prove that
   (i) $|x_n - l| \to 0$ as $n \to \infty$
   (ii) $|x_n| \to |l|$ as $n \to \infty$.

(2) Suppose that $x_n \to l$ as $n \to \infty$. If $l > 0$, prove that there exists an N such that, for any $n > N$,

$$x_n > \tfrac{1}{2}l.$$

(3) Prove that
   (i) $2^n \to +\infty$ as $n \to \infty$ (ii) $-\sqrt{n} \to -\infty$ as $n \to \infty$
   (iii) $\langle (-1)^n n \rangle$ oscillates.

(4) Let $x_n > 0$ $(n = 1, 2, \ldots)$. Prove that $x_n \to 0$ as $n \to \infty$ if and only if $1/x_n \to \infty$ as $n \to \infty$.

(5) Suppose that $\langle x_n \rangle$ increases and is unbounded above. Prove that $x_n \to +\infty$ as $n \to \infty$. If $\langle x_n \rangle$ decreases and is unbounded below, prove that $x_n \to -\infty$ as $n \to \infty$.

*(6) Let S be a non-empty set of real numbers and suppose that $d(\xi, S) = 0$ (see exercise 2.13(4)). Show that, for each $n \in \mathbb{N}$, we can find an $x_n \in S$ such that $|\xi - x_n| < 1/n$. Deduce that $x_n \to \xi$ as $n \to \infty$.

   If S is bounded above, show that a sequence of points of S can be found which converges to its supremum. If S is unbounded above, show that a sequence of points of S can be found which diverges to $+\infty$.

5 SUBSEQUENCES

5.1 Subsequences

Suppose that $\langle x_n \rangle$ is a sequence and that $\langle n_r \rangle$ is a strictly increasing sequence of natural numbers. Then the sequence

$$\langle x_{n_r} \rangle$$

is called a *subsequence* of $\langle x_n \rangle$.

The first few terms of some subsequences of $\langle x_n \rangle$ are listed below.

$$\langle x_{r+1} \rangle = x_2, x_3, x_4, \ldots$$

$$\langle x_{2r} \rangle = x_2, x_4, x_6, \ldots$$

$$\langle x_{3r+5} \rangle = x_8, x_{11}, x_{14}, x_{17}, \ldots$$

$$\langle x_{2^r} \rangle = x_2, x_4, x_8, x_{16}, \ldots$$

Roughly speaking, if we think of a sequence as a list of its terms, then we obtain a subsequence by crossing out some of the terms. For example, $\langle x_{2n} \rangle$ may be obtained from $\langle x_n \rangle$ by crossing out terms as below.

$$\not{x}_1, x_2, \not{x}_3, x_4, \not{x}_5, x_6, \ldots$$

The insistence that $\langle n_r \rangle$ is strictly increasing is of some importance. It means, for example, that the sequence whose first few terms are

$$x_3, x_1, x_5, x_1, x_9, \ldots$$

is *not* a subsequence of $\langle x_n \rangle$.

The fact that $\langle n_r \rangle$ is strictly increasing has the consequence that

$$n_r \geqslant r \quad (r = 1, 2, \ldots).$$

This is easily proved by induction.

5.2 Theorem

Suppose that $x_n \to l$ as $n \to \infty$ and that $\langle x_{n_r} \rangle$ is a subsequence of $\langle x_n \rangle$. Then

$$x_{n_r} \to l \text{ as } r \to \infty.$$

Proof Let $\epsilon > 0$ be given. Since $x_n \to l$ as $n \to \infty$, there exists an N such that, for any $n > N$,

$$|x_n - l| < \epsilon.$$

Let $R = N$. Then, for any $r > R$, $n_r \geqslant r > R = N$ and so $n_r > N$. Thus $|x_{n_r} - l| < \epsilon$.

We have shown that, given any $\epsilon > 0$, we can find an R (namely $R = N$) such that, for any $r > R$,

$$|x_{n_r} - l| < \epsilon.$$

Hence $x_{n_r} \to l$ as $r \to \infty$.

5.3   *Example* We have seen that the bounded sequence $\langle (-1)^n \rangle$ diverges. A simpler proof can be based on theorem 5.2. Obviously the subsequence of odd terms tends to -1 and the subsequence of even terms to $+1$. But, if $\langle (-1)^n \rangle$ converged, all its subsequences would tend to the same limit.

5.4   *Example* In example 4.14 we showed that, if $x > 0$, then $x^{1/n} \to 1$ as $n \to \infty$. We prove this again by a different method. As in example 4.4 we need only consider the case $x \geqslant 1$.

If $x \geqslant 1$, then $x^{1/n} \geqslant 1$ $(n = 1, 2, \ldots)$ and hence the sequence $\langle x^{1/n} \rangle$ is bounded below by 1. Moreover

$$x^{1/n}/x^{1/(n+1)} = x^{(n+1-n)/n(n+1)} = x^{1/n(n+1)} \geqslant 1 \quad (n = 1, 2, \ldots)$$

and therefore $\langle x^{1/n} \rangle$ decreases. From theorem 4.17 it follows that $\langle x^{1/n} \rangle$ converges to a limit l and $l \geqslant 1$. What is the value of l?

By theorem 5.2, we know that all subsequences of $\langle x^{1/n} \rangle$ must also tend to l. Hence

$$x^{1/2n} \to l \text{ as } n \to \infty.$$

Using proposition 4.8, it follows that

$$x^{1/n} = x^{1/2n} . x^{1/2n} \to l.l = l^2 \text{ as } n \to \infty.$$

But a sequence can only have one limit. Thus

$$l = l^2$$

and so $l = 0$ or $l = 1$. But $l \geqslant 1$ and hence $l = 1$.

We have shown that

$$x^{1/n} \to 1 \text{ as } n \to \infty.$$

Note: A warning is appropriate here. Do not use this method of working out the value of a limit until you have shown that the sequence converges. For example, if you thoughtlessly tried to evaluate the limit of the sequence $\langle (-1)^n \rangle$ by looking at its subsequences, you would obtain $l = +1 = -1$.

5.5   *Example* Let $a > 0$ and let $x_1 > 0$. Define the rest of the sequence $\langle x_n \rangle$ inductively by

$$x_{n+1} = \tfrac{1}{2}(x_n + ax_n^{-1}) \quad (n = 1, 2, \ldots).$$

Obviously $x_n > 0 \ (n = 1, 2, \ldots)$. But further

$$2x_{n+1}x_n = x_n^2 + a$$

$$x_n^2 - 2x_{n+1}x_n + a = 0.$$

We know in advance that this quadratic equation in x_n has a real solution. Thus '$b^2 - 4ac \geqslant 0$', i.e.

$$x_{n+1}^2 \geqslant a.$$

Since $x_{n+1} > 0$, it follows that

$$x_{n+1} \geqslant \sqrt{a} \quad (n = 1, 2, \ldots).$$

We wish to prove that $\langle x_n \rangle$ decreases and so we consider

$$x_n - x_{n+1} = x_n - \tfrac{1}{2}(x_n + ax_n^{-1})$$

$$= \frac{1}{2x_n}(x_n^2 - a)$$

$$\geqslant 0 \quad (n = 2, 3, \ldots).$$

Hence the sequence $\langle x_n \rangle$ decreases and is bounded below by $\sqrt{a}$ (provided we omit its first term). From theorem 4.17 it follows that

$$x_n \to l \text{ as } n \to \infty$$

where $l \geqslant \sqrt{a}$. What is the value of l?

By theorem 5.2 we also have $x_{n+1} \to l$ as $n \to \infty$. But

$$x_{n+1} = \tfrac{1}{2}(x_n + ax_n^{-1}).$$

Now $l \neq 0$ (because $l \geqslant \sqrt{a}$). Hence, by proposition 4.8,

$$x_{n+1} = \tfrac{1}{2}(x_n + ax_n^{-1}) \to \tfrac{1}{2}(l + al^{-1}) \text{ as } n \to \infty.$$

Since a sequence can have at most one limit,

$$l = \tfrac{1}{2}(l + al^{-1})$$

i.e. $l^2 = a$.

Thus $l = \sqrt{a}$ or $l = -\sqrt{a}$. But we know that $l \geqslant \sqrt{a}$. Hence $l = \sqrt{a}$. We have therefore shown that

$$x_n \to \sqrt{a} \text{ as } n \to \infty.$$

When studying the convergence properties of a sequence defined inductively, it is not always obvious how to proceed. Sometimes it is possible to prove that the sequence decreases and is bounded below as in the present example. Sometimes it is possible to prove that the sequence increases and is bounded above. Often neither of these possibilities is correct and some other method has to be employed as in example 5.20. The appropriate method to use, of course, depends on the behaviour of the sequence. Before embarking on a proof it is therefore necessary to make an educated guess about the behaviour of the sequence. In making such a guess, it is usually helpful to draw a diagram. In the case of the sequence studied in the current example, a useful diagram is one which incorporates the graphs of the equations $y = \frac{1}{2}(x + ax^{-1})$ and $y = x$.

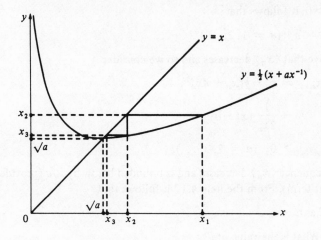

The diagram illustrates very clearly why $\sqrt{a}$ is a lower bound for $\langle x_n \rangle$, why $\langle x_n \rangle$ decreases and why $x_n \to \sqrt{a}$ as $n \to \infty$. But it must be remembered that such a diagram is just an aid in deciding what to try to prove. It is *not* a substitute for a proof.

5.6   *Example* We attack the previous example by a different method. Again let $a > 0$ but this time let x_1 be unspecified for the moment. Define the rest of the sequence inductively by

$$x_{n+1} = \frac{1}{2}(x_n + ax_n^{-1}) \quad (n = 1, 2, \ldots).$$

Now

$$x_{n+1} - \sqrt{a} = \tfrac{1}{2}(x_n + ax_n^{-1}) - \sqrt{a}$$

$$= \frac{1}{2x_n}(x_n^2 - 2x_n\sqrt{a} + a)$$

$$= \frac{1}{2x_n}(x_n - \sqrt{a})^2$$

$$= \frac{1}{2x_n}\left\{\frac{(x_{n-1} - \sqrt{a})^2}{2x_{n-1}}\right\}^2$$

$$= \frac{1}{2x_n}\frac{1}{(2x_{n-1})^2}(x_{n-1} - \sqrt{a})^4$$

$$= \frac{1}{2x_n}\frac{1}{(2x_{n-1})^2}\frac{1}{(2x_{n-2})^4}(x_{n-2} - \sqrt{a})^8$$

$$= \frac{1}{2x_n}\frac{1}{(2x_{n-1})^2}\cdots\frac{1}{(2x_1)^{2^{n-1}}}(x_1 - \sqrt{a})^{2^n}.$$

If we assume that $x_1 \geqslant \sqrt{a}$, then it follows, as in example 5.5, that $x_n \geqslant \sqrt{a}$ $(n = 1, 2, \ldots)$. Hence

$$|x_{n+1} - \sqrt{a}| \leqslant \left(\frac{1}{2\sqrt{a}}\right)^{1+2+2^2+\ldots+2^{n-1}}(x_1 - \sqrt{a})^{2^n}$$

$$= \left(\frac{1}{2\sqrt{a}}\right)^{(2^n-1)/(2-1)}(x_1 - \sqrt{a})^{2^n}$$

$$= 2\sqrt{a}\left\{\frac{x_1 - \sqrt{a}}{2\sqrt{a}}\right\}^{2^n}.$$

We know that, if $|y| < 1$, then $y^n \to 0$ as $n \to \infty$. By theorem 5.2 it follows that $y^{2^n} \to 0$ as $n \to \infty$.

Hence our argument shows that $x_n \to \sqrt{a}$ as $n \to \infty$ provided that

$$\frac{x_1 - \sqrt{a}}{2\sqrt{a}} < 1.$$

We have already assumed that $x_1 \geqslant \sqrt{a}$ and so we have shown that $x_n \to \sqrt{a}$ as $n \to \infty$ provided that $\sqrt{a} \leqslant x_1 < 3\sqrt{a}$. Of course we already know from example 5.5 that this was true under the weaker hypothesis that $x_1 > 0$. However, the argument given here is interesting because it allows us to estimate how 'good' an approximation x_n is to $\sqrt{a}$.

For example, suppose we want to estimate the value of $\sqrt{2}$. It is obvious that $1 < \sqrt{2} < 2$ (because $1^2 = 1 < 2$ and $2^2 = 4 > 2$). For a better estimate take $a = 2$ and $x_1 = 2$ in the argument above. Then

$$|x_4 - \sqrt{2}| \leqslant 2\sqrt{2}\left\{\frac{2 - \sqrt{2}}{2\sqrt{2}}\right\}^8 = 2\sqrt{2}\left\{\frac{\sqrt{2}(\sqrt{2}-1)}{2\sqrt{2}}\right\}^8$$

$$= \frac{2\sqrt{2}}{2^8}(\sqrt{2}-1)^8.$$

Using the crude estimate $\sqrt{2} < 2$ we obtain $|x_4 - \sqrt{2}| \leqslant 2^{-6} = \frac{1}{64} < 0\cdot016$. Hence $x_4 = \frac{577}{408} = 1\cdot414\ldots$ differs from $\sqrt{2}$ by at most $0\cdot016$.

(A better estimate for $\sqrt{2}$ may be obtained by evaluating x_5 or by starting with $x_1 = \frac{3}{2}$ and using the estimate $\sqrt{2} < \frac{3}{2}$.)

5.7   Exercise

(1) Given that $\langle n^{1/n} \rangle$ converges (see exercise 4.20(6)), show that

$n^{1/n} \to 1$ as $n \to \infty$

by considering the subsequence $\langle (2n)^{1/2n} \rangle$.

(2) A sequence $\langle x_n \rangle$ is defined by $x_1 = h$ and

$x_{n+1} = x_n^2 + k$

where $0 < k < \frac{1}{4}$ and h lies between the roots a and b of the equation $x^2 - x + k = 0$.

   Prove that $a < x_{n+1} \leqslant x_n < b$ $(n = 1, 2, \ldots)$. Show that $\langle x_n \rangle$ converges and determine its limit.

*(3) If $k > 0$ and $x_1 > 0$ and $\langle x_n \rangle$ is defined inductively by

$$x_{n+1} = \frac{k}{1 + x_n},$$

show that one of the sequences $\langle x_{2n} \rangle$ and $\langle x_{2n-1} \rangle$ is increasing and the other decreasing.

   Prove that both sequences converge to the limit l which is the positive root of the equation $x^2 + x = k$. What conclusion may be drawn concerning $\langle x_n \rangle$?

*(4) Two sequences $\langle x_n \rangle$ and $\langle y_n \rangle$ are defined inductively by $x_1 = \frac{1}{2}$ and $y_1 = 1$ and

$x_n = \sqrt{(x_{n-1}y_{n-1})}$ $(n = 2, 3, \ldots)$

$$\frac{1}{y_n} = \frac{1}{2}\left(\frac{1}{x_n} + \frac{1}{y_{n-1}}\right) \quad (n = 2, 3, \ldots).$$

Prove that $x_{n-1} < x_n < y_n < y_{n-1}$ $(n = 2, 3, \ldots)$ and deduce that both sequences converge to the same limit l, where $\frac{1}{2} < l < 1$. (Actually $l = \pi/4$.)

† (5) Let y be any real number. Prove that the sequence

$$\langle(1 + yn^{-1})^n\rangle$$

is increasing for those values of n which satisfy $n \geq 1 - y$. [Hint: example 4.19.] Show also that the sequence is bounded above and hence converges. [Hint: From exercise 4.20(4), $(2y)^n/n! \to 0$ as $n \to \infty$ and therefore we can find an N such that, for any $n \geq N$, $y^n/n! \leq (\frac{1}{2})^n$.]

† (6) Show that the product of the limits of the two sequences $\langle(1 + xn^{-1})^n\rangle$ and $\langle(1 - xn^{-1})^n\rangle$ is equal to 1.

$$\left[\text{Hint:} \left(1 + \frac{x}{n}\right)^n \left(1 - \frac{x}{n}\right)^n = \left\{\left(1 - \frac{x^2}{n^2}\right)^{n^2}\right\}^{1/n}\right].$$

5.8 Bolzano–Weierstrass theorem

5.9 *Theorem* Every sequence has a monotone subsequence.

Proof Let $\langle x_n \rangle$ be any sequence of real numbers. We must construct a subsequence $\langle x_{n_r} \rangle$ which is either increasing or decreasing. We distinguish two cases.

(i) Every set $\{x_n : n > N\}$ has a maximum. In this case we can find a sequence $\langle n_r \rangle$ of natural numbers such that

$$x_{n_1} = \max_{n > 1} x_n$$

$$x_{n_2} = \max_{n > n_1} x_n$$

$$x_{n_3} = \max_{n > n_2} x_n$$

and so on. Obviously $n_1 < n_2 < n_3 < \ldots$ and so, at each stage, we are taking the maximum of a *smaller* set than at the previous stage. Hence $\langle x_{n_r} \rangle$ is a *decreasing* subsequence of $\langle x_n \rangle$.

(ii) Suppose that it is *not* true that *all* of the sets $\{x_n : n > N\}$ have a maximum. Then, for some N_1, the set $\{x_n : n > N_1\}$ has no maximum. It follows that, given any x_m with $m > N_1$, we can find an x_n following x_m such that $x_n > x_m$. (Otherwise the biggest of $x_{N_1+1}, \ldots, x_m$ would be a maximum for $\{x_n : n > N_1\}$).

We define $x_{n_1} = x_{N_1+1}$ and then let x_{n_2} be the first term following x_{n_1} for which $x_{n_2} > x_{n_1}$. Now let x_{n_3} be the first term following x_{n_2} for which $x_{n_3} > x_{n_2}$. And so on. We obtain an *increasing* subsequence $\langle x_{n_r} \rangle$ of $\langle x_n \rangle$.

5.10   Theorem (*Bolzano–Weierstrass theorem*). Every bounded sequence of real numbers has a convergent subsequence.

Proof Let $\langle x_n \rangle$ be a bounded sequence. By theorem 5.9, $\langle x_n \rangle$ has a monotone subsequence $\langle x_{n_r} \rangle$. Since $\langle x_n \rangle$ is bounded, so is $\langle x_{n_r} \rangle$. Hence, by theorem 4.17, $\langle x_{n_r} \rangle$ converges.

5.11   Examples

(i) The sequence $\langle (-1)^n \rangle$ is bounded. Two convergent subsequences are the sequence $-1, -1, -1, \ldots$ of odd terms and the sequence $1, 1, 1, 1, \ldots$ of even terms.

(ii) A more sophisticated example is the sequence $\langle r_n \rangle$ whose first few terms are

$$\tfrac{1}{2}, \tfrac{1}{3}, \tfrac{2}{3}, \tfrac{1}{4}, \tfrac{2}{4}, \tfrac{3}{4}, \tfrac{1}{5}, \tfrac{2}{5}, \tfrac{3}{5}, \tfrac{4}{5}, \tfrac{1}{6}, \ldots$$

Every term of the sequence is a rational number between 0 and 1. Hence the sequence is bounded. It has many convergent subsequences. For example

$$\tfrac{1}{2}, \tfrac{2}{4}, \tfrac{3}{6}, \tfrac{4}{8}, \ldots$$

$$\tfrac{1}{2}, \tfrac{1}{3}, \tfrac{1}{4}, \tfrac{1}{5}, \ldots$$

$$\tfrac{1}{2}, \tfrac{2}{3}, \tfrac{3}{4}, \tfrac{4}{5}, \ldots$$

5.12† Lim sup and lim inf

We shall not have a great deal of use for the subject matter of this section. However, it is material which has to be learned eventually and it is natural to deal with it here.

Suppose that $\langle x_n \rangle$ is a bounded sequence and let L denote the set of all real numbers which are the limit of some subsequence of $\langle x_n \rangle$. We know from the Bolzano–Weierstrass theorem that L is not empty (i.e. $L \neq \emptyset$). Of course, if $\langle x_n \rangle$ converges, L will just consist of a single point (theorem 5.2).

5.13† *Proposition* Let $\langle x_n \rangle$ be a bounded sequence and let L be the set of all real numbers which are the limit of some subsequence of $\langle x_n \rangle$. Then L has a *maximum* and a *minimum*.

That L is a bounded set follows from theorem 4.23. Hence L certainly has a supremum and infimum. Proposition 5.13 asserts that these actually belong to the set L. The proof is relegated to the appendix. Readers who prefer to omit the proof altogether will encounter it again at a later stage since it is a special case of a general theorem concerning 'closed sets'.

If $\langle x_n \rangle$ is a bounded sequence, let us denote the maximum of L by $\bar{l}$ and the minimum by $\underline{l}$. Then $\bar{l}$ is the *largest* number to which a subsequence of $\langle x_n \rangle$ converges and $\underline{l}$ is the *smallest* number to which a subsequence of $\langle x_n \rangle$ converges.

We call $\bar{l}$ the *limit superior* of $\langle x_n \rangle$ and write

$$\bar{l} = \limsup_{n \to \infty} x_n.$$

Similarly, $\underline{l}$ is the *limit inferior* of $\langle x_n \rangle$ and we write

$$\underline{l} = \liminf_{n \to \infty} x_n.$$

(Occasionally you will see such expressions as $\limsup_{n \to \infty} x_n = + \infty$. This simply means that $\langle x_n \rangle$ is unbounded above. Similarly $\liminf_{n \to \infty} x_n = - \infty$ means that $\langle x_n \rangle$ is unbounded below.)

5.14[†]   *Examples*

(i) The sequence $\langle 1/n \rangle$ converges with limit 0. By theorem 5.2, all its subsequences converge to 0. Hence L contains the single point 0 and

$$\limsup_{n \to \infty} \frac{1}{n} = \liminf_{n \to \infty} \frac{1}{n} = \lim_{n \to \infty} \frac{1}{n} = 0.$$

(ii) The sequence $\langle (-1)^n \rangle$ is bounded. The set L consists of *two* points $+1$ and -1 (i.e. $L = \{+1, -1\}$). We have

$$\limsup_{n \to \infty} (-1)^n = 1; \liminf_{n \to \infty} (-1)^n = -1.$$

(iii) Let $\langle r_n \rangle$ be the sequence whose first few terms are $\frac{1}{2}, \frac{1}{3}, \frac{2}{3}, \frac{1}{4}, \frac{2}{4}, \frac{3}{4}, \frac{1}{5}, \frac{2}{5}, \frac{3}{5}, \frac{4}{5}, \frac{1}{6}, \ldots$ Obviously $0 < r_n < 1$ $(n = 1, 2, \ldots)$. From theorem 4.23 it follows that L is a subset of $[0, 1]$ (i.e. $L \subset [0, 1]$). But the subsequence $\frac{1}{2}, \frac{1}{3}, \frac{1}{4}, \frac{1}{5}, \ldots$ converges to 0. Thus $\underline{l} = 0$. Also the subsequence $\frac{1}{2}, \frac{2}{3}, \frac{3}{4}, \frac{4}{5}, \ldots$ converges to 1. Thus $\bar{l} = 1$. We have shown that

$$\limsup_{n \to \infty} r_n = 1; \qquad \liminf_{n \to \infty} r_n = 0$$

5.15[†]   *Exercise*

[†] (1)   Calculate

(i) $\limsup_{n \to \infty} \{(-1)^n (1 + n^{-1})\}$ (ii) $\liminf_{n \to \infty} \{(-1)^n (1 + n^{-1})\}$

and show that these are *not* the same as the numbers

$$\sup_{n \geqslant 1} (-1)^n (1 + n^{-1}) \quad \text{and} \quad \inf_{n \geqslant 1} (-1)^n (1 + n^{-1}).$$

† (2) Let $\langle r_n \rangle$ be the sequence whose first few terms are $\frac{1}{2}, \frac{1}{3}, \frac{2}{3}, \frac{1}{4}, \frac{2}{4}, \frac{3}{4}, \frac{1}{5}, \frac{2}{5}, \frac{3}{5},$ $\frac{4}{5}, \frac{1}{6}, \ldots$ Show that *every* point in the interval $[0, 1]$ is the limit of a subsequence of $\langle r_n \rangle$. [Hint: see exercise 3.6(4).]

† (3) Suppose that $\langle x_n \rangle$ is a bounded sequence and that, for any N, we can find an $n > N$ such that $x_n \geqslant b$. Show that $\langle x_n \rangle$ has a subsequence which converges to a limit $l \geqslant b$.

† (4) Let $\langle x_n \rangle$ be a bounded sequence with limit superior $\bar{l}$ and limit inferior $\underline{l}$. Show that, given any $\epsilon > 0$, we can find an N such that, for any $n > N, x_n < \bar{l} + \epsilon$. [Hint: use question 3 with $b = \bar{l} + \epsilon$]. What is the corresponding result for $\underline{l}$?

† (5) Deduce from question 4 that a bounded sequence $\langle x_n \rangle$ converges with limit l if and only if

$$\limsup_{n \to \infty} x_n = \liminf_{n \to \infty} x_n = l.$$

Hence show that a bounded sequence $\langle x_n \rangle$ converges if and only if all its *convergent* subsequences have the same limit.

† (6) Let $\langle x_n \rangle$ be a bounded sequence and let

$$M_n = \sup_{k \geqslant n} x_k.$$

Show that $\langle M_n \rangle$ decreases and is bounded below. Deduce that $\langle M_n \rangle$ converges and denote its limit by M. If l is the limit of some subsequence of $\langle x_n \rangle$, show that $M_n \geqslant l$ $(n = 1, 2, \ldots)$. Deduce that $M \geqslant \bar{l}$. Obtain also the reverse inequality $M \leqslant \bar{l}$ [Hint: use question 4] and hence show that

$$\limsup_{n \to \infty} x_n = \lim_{n \to \infty} \left\{ \sup_{k \geqslant n} x_k \right\}.$$

(This explains the choice of notation for the limit superior. We also have, of course,

$$\liminf_{n \to \infty} x_n = \lim_{n \to \infty} \left\{ \inf_{k \geqslant n} x_k \right\}.)$$

5.16   Cauchy sequences

Suppose that we want to prove that a sequence $\langle x_n \rangle$ converges but we have no idea in advance what its limit l might be. Then there is no point in appealing directly to the definition because we shall certainly not be able to prove that $|x_n - l|$ can be made as small as we choose by taking n sufficiently large if we do not know the value of l.

Theorem 4.17 gives us one way of resolving this difficulty. It tells us that an increasing sequence which is bounded above converges and that a decreasing sequence which is bounded below converges. Thus we are able to deduce the convergence of the sequence without necessarily knowing the value of its limit. But, of course, theorem 4.17 deals only with monotone sequences. In order to deal with sequences which may not be monotone, we introduce the idea of a Cauchy sequence.

We say that $\langle x_n \rangle$ is a *Cauchy sequence* if, given any $\epsilon > 0$, we can find an N such that, for any $m > N$ and any $n > N$,

$$|x_m - x_n| < \epsilon.$$

Very roughly, the terms of a Cauchy sequence get 'closer and closer' together.

5.17   Proposition Any convergent sequence is a Cauchy sequence.

5.18   Proposition Any Cauchy sequence is bounded.

The proofs are easy. The next theorem is what makes Cauchy sequences important.

5.19   Theorem Every Cauchy sequence converges.

Proof Let $\langle x_n \rangle$ be a Cauchy sequence. By proposition 5.18, $\langle x_n \rangle$ is bounded. Hence, by the Bolzano–Weierstrass theorem, it has a convergent sub-sequence $\langle x_{n_r} \rangle$. Suppose that $x_{n_r} \to l$ as $r \to \infty$. We shall show that $x_n \to l$ as $n \to \infty$.

Let $\epsilon > 0$. Then $\frac{1}{2}\epsilon > 0$. Hence there exists an R such that, for any $r > R$,

$$|x_{n_r} - l| < \tfrac{1}{2}\epsilon. \tag{1}$$

Since $\langle x_n \rangle$ is a Cauchy sequence, there also exists an N such that, for any $m > N$ and any $n > N$,

$$|x_m - x_n| < \tfrac{1}{2}\epsilon. \tag{2}$$

Now suppose that $n > N$ and choose r so large that $n_r > N$ and $r > R$. Then (1) is satisfied and also (2) is satisfied with $m = n_r$. Thus, for any $n > N$,

$$\begin{aligned}
|x_n - l| &= |x_n - x_{n_r} + x_{n_r} - l| \\
&\leqslant |x_n - x_{n_r}| + |x_{n_r} - l| \quad \text{(triangle inequality)} \\
&< \tfrac{1}{2}\epsilon + \tfrac{1}{2}\epsilon = \epsilon.
\end{aligned}$$

Given any $\epsilon > 0$, we have found an N such that, for any $n > N$, $|x_n - l| < \epsilon$. Thus $x_n \to l$ as $n \to \infty$.

5.20   *Example* A sequence $\langle x_n \rangle$ is defined by $x_1 = a$, $x_2 = b$ and

$$x_{n+2} = \tfrac{1}{2}(x_{n+1} + x_n) \quad (n = 1, 2, 3, \ldots).$$

Prove that $\langle x_n \rangle$ converges.

Proof We have

$$x_{n+2} - x_{n+1} = \tfrac{1}{2}(x_{n+1} + x_n) - x_{n+1} = \tfrac{1}{2}(x_n - x_{n+1}).$$

Thus

$$|x_{n+2} - x_{n+1}| = \frac{1}{2}|x_n - x_{n+1}| = \frac{1}{2^2}|x_n - x_{n-1}|$$

$$= \ldots = \frac{1}{2^n}|x_2 - x_1| = \frac{1}{2^n}|b - a|.$$

Hence, if $n > m$,

$$|x_n - x_m| = |x_n - x_{n-1} + x_{n-1} - x_{n-2} - \ldots + x_{m+1} - x_m|$$

$$\leqslant |x_n - x_{n-1}| + |x_{n-1} - x_{n-2}| + \ldots + |x_{m+1} - x_m|$$

$$\leqslant \left\{ \frac{1}{2^{n-2}} + \frac{1}{2^{n-3}} + \ldots + \frac{1}{2^{m-1}} \right\} |b - a|$$

$$= \frac{1}{2^{m-1}} \left\{ 1 + \frac{1}{2} + \frac{1}{2^2} + \ldots + \frac{1}{2^{n-m-1}} \right\} |b - a|$$

$$= \frac{1}{2^{m-1}} \frac{1 - (\tfrac{1}{2})^{n-m}}{1 - \tfrac{1}{2}} |b - a| \leqslant \frac{1}{2^{m-2}}|b - a|.$$

Let $\epsilon > 0$ be given. Choose N so large that

$$\frac{1}{2^{N-2}}|b - a| < \epsilon.$$

Then, for any $n > N$ and any $m > N$.

$$|x_n - x_m| \leqslant \frac{1}{2^{N-2}}|b - a| < \epsilon.$$

Thus $\langle x_n \rangle$ is a Cauchy sequence. Therefore, by theorem 5.19, it converges.

(An alternative method would be to show that one of the two sequences $\langle x_{2n-1} \rangle$ and $\langle x_{2n} \rangle$ is increasing and the other decreasing and then to show that they have the same limit.)

5.21   *Exercise*

(1) Suppose that $0 < \alpha < 1$ and that $\langle x_n \rangle$ is a sequence which satisfies $|x_{n+1} - x_n| \leqslant \alpha^n$ $(n = 1, 2, \ldots)$. Prove that $\langle x_n \rangle$ is a Cauchy sequence and hence converges.

Give an example of a sequence $\langle y_n \rangle$ such that $y_n \to +\infty$ as $n \to \infty$ but $|y_{n+1} - y_n| \to 0$ as $n \to \infty$. [Hint: see exercise 4.20].

(2) A sequence $\langle x_n \rangle$ satisfies $0 < a \leqslant x_1 \leqslant x_2 \leqslant b$ and

$$x_{n+2} = \{x_{n+1} x_n\}^{1/2} \quad (n = 1, 2, \ldots).$$

Prove that $a \leqslant x_n \leqslant b$ $(n = 1, 2, \ldots)$. Hence or otherwise show that

$$|x_{n+1} - x_n| \leqslant \frac{b}{a+b} |x_n - x_{n-1}| \quad (n = 2, 3, \ldots).$$

Deduce that $\langle x_n \rangle$ is a Cauchy sequence and hence converges.

(3) For the sequence of example 5.20, prove that

$$x_{n+1} + \tfrac{1}{2}x_n = x_n + \tfrac{1}{2}x_{n-1} = \ldots = x_2 + \tfrac{1}{2}x_1.$$

Deduce that $|x_{n+1} - l| = \tfrac{1}{2}|x_n - l|$, where $l = \tfrac{2}{3}(x_2 + \tfrac{1}{2}x_1)$. What conclusion may be drawn about the convergence of the sequence $\langle x_n \rangle$?

Tackle exercise 5.21(2) by a similar method.

(4) Let $[a, b]$ be a compact interval (see §2.9). Prove that every sequence of points of $[a, b]$ contains a subsequence which converges to a point of $[a, b]$.

† (5) Let I be an interval which has the property that every sequence of points of I contains a subsequence which converges to a point of I. Prove that I is compact. [Hint: First prove that I is bounded by assuming otherwise and appealing to exercise 4.29(6). Then prove that sup I and inf I are both elements of I with another appeal to exercise 4.29(6).]

† (6) Given a set S of real numbers, let

$$S_\xi = \{x : x \in S \text{ and } x \neq \xi\}.$$

We say that ξ is a cluster point (or point of accumulation or 'limit point') of S if ξ is at zero distance from S_ξ. (Note that ξ need not be an element of S). A form of the Bolzano–Weierstrass theorem asserts that every bounded set with an infinite number of elements has at least one cluster point. Prove this.

6 SERIES

6.1   Definitions

Given a sequence $\langle a_n \rangle$ of real numbers the sequence $\langle s_N \rangle$ defined by

$$s_N = \sum_{n=1}^{N} a_n = a_1 + a_2 + \ldots + a_N$$

is called the sequence of *partial sums* of the *series*

$$\sum_{n=1}^{\infty} a_n.$$

If $s_N \to s$ as $N \to \infty$, the series is said to *converge* to the *sum s*. We write

$$s = \sum_{n=1}^{\infty} a_n.$$

It is important to remember that this formula can only make sense when the series converges.

6.2   *Example* Consider the series

$$\sum_{n=0}^{\infty} x^n.$$

If $x \neq 1$, the partial sums satisfy

$$s_N = \sum_{n=0}^{N} x^n = 1 + x + x^2 + \ldots + x^N$$

$$= \frac{1 - x^{N+1}}{1 - x}.$$

If $|x| < 1$, $x^{N+1} \to 0$ as $N \to \infty$ and hence

$$s_N \to \frac{1}{1-x} \text{ as } N \to \infty.$$

It follows that the series $\sum_{n=0}^{\infty} x^n$ converges if $|x| < 1$ and we may write

$$\sum_{n=0}^{\infty} x^n = \frac{1}{1-x} \quad (|x| < 1).$$

If $|x| \geqslant 1$, the series *diverges*.

6.3   *Example* The partial sums of the series

$$\sum_{n=1}^{\infty} \frac{1}{n(n+1)}$$

are given by

$$s_N = \sum_{n=1}^{N} \frac{1}{n(n+1)} = \sum_{n=1}^{N} \left(\frac{1}{n} - \frac{1}{n+1} \right)$$

$$= \left(1 - \frac{1}{2} \right) + \left(\frac{1}{2} - \frac{1}{3} \right) + \left(\frac{1}{3} - \frac{1}{4} \right) + \ldots + \left(\frac{1}{N} - \frac{1}{N+1} \right)$$

$$= 1 - \frac{1}{N+1} \to 1 \text{ as } N \to \infty.$$

Hence the series converges and we may write

$$\sum_{n=1}^{\infty} \frac{1}{n(n+1)} = 1.$$

6.4   Series of positive terms

Series whose terms are all *positive* (or non-negative) are particularly easy to deal with. This is because the sequence of partial sums of such a series is *increasing*. Thus, if we wish to show that a series of positive terms converges, we only need to show that its sequence of partial sums is *bounded above*. If the sequence of partial sums is unbounded above, then the series diverges to $+\infty$ (see exercise 4.29(5)).

6.5   *Theorem* The series

$$\sum_{n=1}^{\infty} \frac{1}{n}$$

diverges to $+\infty$.

Proof Since the partial sums of this series increase, we only need to show that they are unbounded above. But

$$s_{2^N} = 1 + \frac{1}{2} + \frac{1}{3} + \ldots + \frac{1}{2^N}$$

$$= 1 + \frac{1}{2} + \left(\frac{1}{3} + \frac{1}{4}\right) + \left(\frac{1}{5} + \frac{1}{6} + \frac{1}{7} + \frac{1}{8}\right) + \ldots + \left(\frac{1}{2^{N-1}+1} + \ldots + \frac{1}{2^N}\right)$$

$$\geqslant 1 + \frac{1}{2} + \left(\frac{1}{4} + \frac{1}{4}\right) + \left(\frac{1}{8} + \frac{1}{8} + \frac{1}{8} + \frac{1}{8}\right) + \ldots + \left(\frac{1}{2^N} + \ldots + \frac{1}{2^N}\right)$$

$$= 1 + \frac{1}{2} + \frac{2}{4} + \frac{4}{8} + \ldots + \frac{2^{N-1}}{2^N} = 1 + \tfrac{1}{2}N.$$

Thus the partial sums are unbounded above and the theorem follows.

6.6   *Theorem* Let α be a rational number such that $\alpha > 1$. Then the series

$$\sum_{n=1}^{\infty} \frac{1}{n^\alpha}$$

converges.

Proof Since the partial sums are increasing, we need only show that they are bounded above. For $N > 1$,

$$s_N \leqslant s_{2^N-1} = 1 + \frac{1}{2^\alpha} + \frac{1}{3^\alpha} + \ldots + \frac{1}{(2^N-1)^\alpha}$$

$$= 1 + \left(\frac{1}{2^\alpha} + \frac{1}{3^\alpha}\right) + \left(\frac{1}{4^\alpha} + \frac{1}{5^\alpha} + \frac{1}{6^\alpha} + \frac{1}{7^\alpha}\right) + \ldots$$

$$+ \left(\frac{1}{2^{(N-1)\alpha}} + \ldots + \frac{1}{(2^N-1)^\alpha}\right)$$

$$\leqslant 1 + \left(\frac{1}{2^\alpha} + \frac{1}{2^\alpha}\right) + \left(\frac{1}{4^\alpha} + \frac{1}{4^\alpha} + \frac{1}{4^\alpha} + \frac{1}{4^\alpha}\right) + \ldots$$

$$+ \left(\frac{1}{2^{(N-1)\alpha}} + \ldots + \frac{1}{2^{(N-1)\alpha}}\right)$$

$$= 1 + \frac{2}{2^\alpha} + \frac{4}{4^\alpha} + \ldots + \frac{2^{N-1}}{2^{(N-1)\alpha}}$$

$$= 1 + \frac{1}{2^{\alpha-1}} + \left(\frac{1}{2^{\alpha-1}}\right)^2 + \ldots + \left(\frac{1}{2^{\alpha-1}}\right)^{N-1}$$

$$= \frac{1 - (1/2^{\alpha-1})^N}{1 - (1/2^{\alpha-1})} \leqslant \frac{1}{1 - (1/2^{\alpha-1})}$$

provided that $1/2^{\alpha-1} < 1$. But this follows from the assumption that $\alpha > 1$. Hence $\langle s_N \rangle$ is bounded above and the theorem follows.

6.7 Elementary properties of series

6.8 *Theorem* Suppose that the series $\Sigma_{n=1}^{\infty} a_n$ and $\Sigma_{n=1}^{\infty} b_n$ converge to α and β respectively. Then, if λ and μ are any real numbers, the series $\Sigma_{n=1}^{\infty} (\lambda a_n + \mu b_n)$ converges to $\lambda\alpha + \mu\beta$.

Proof We have

$$\sum_{n=1}^{N} (\lambda a_n + \mu \beta_n) = \lambda \sum_{n=1}^{N} a_n + \mu \sum_{n=1}^{N} b_n$$

$$\to \lambda\alpha + \mu\beta \text{ as } N \to \infty$$

by proposition 4.8.

6.9 *Theorem* Suppose that the series $\Sigma_{n=1}^{\infty} a_n$ converges. Then

$a_n \to 0$ as $n \to \infty$.

Proof Let the sum of the series be s. Then

$$s_N = \sum_{n=1}^{N} a_n \to s \text{ as } n \to \infty.$$

Also $s_{N-1} \to s$ as $n \to \infty$.

But then

$$a_N = (a_1 + a_2 + \ldots + a_N) - (a_1 + a_2 + \ldots + a_{N-1})$$

$$= s_N - s_{N-1} \to s - s = 0 \text{ as } N \to \infty.$$

6.10 *Examples* The series

$$\sum_{n=1}^{\infty} (-1)^n$$

diverges. This can be deduced from theorem 6.9 by observing that its terms do not tend to zero.

Note that the converse of theorem 6.9 is *false*. Just because the terms of a series tend to zero it does *not* follow that the series converges. Theorem 6.5 provides an example. The terms of the series

$$\sum_{n=1}^{\infty} \frac{1}{n}$$

tend to zero, but the series *diverges*. One might say that the terms of the series do not tend to zero 'fast enough' to make the series converge.

6.11   Proposition Suppose that the series $\sum_{n=1}^{\infty} a_n$ converges. Then, for each natural number N, the series $\sum_{n=N}^{\infty} a_n$ converges and

$$\sum_{n=N}^{\infty} a_n \to 0 \text{ as } N \to \infty.$$

This result is often referred to by saying that the 'tail' of a convergent series tends to zero. The proof is easy.

6.12   Series and Cauchy sequences

In the series we have studied so far, we have either been blessed with a nice formula for the partial sums (examples 6.2 and 6.3) or else the partial sums increased and so we only had to consider whether or not the partial sums were bounded above (theorems 6.5 and 6.6). What should we do in the absence of such favourable conditions? We ask the question: is the sequence of partial sums a *Cauchy sequence*? This question has some fruitful answers as we shall see below.

6.13   Theorem Suppose that $\langle a_n \rangle$ is a decreasing sequence of positive numbers such that $a_n \to 0$ as $n \to \infty$. Then the series

$$\sum_{n=1}^{\infty} (-1)^{n-1} a_n = a_1 - a_2 + a_3 - a_4 + \ldots$$

converges.

Proof We show that the sequence $\langle s_n \rangle$ of partial sums of the series is a Cauchy sequence. From theorem 5.19, it then follows that the series converges.

The proof depends on the fact that, for each $n > m$, we have the inequality

$$0 \leqslant a_{m+1} - a_{m+2} + a_{m+3} - \ldots a_n \leqslant a_{m+1}.$$

This follows easily from the fact that $a_k - a_{k+1}$ is always non-negative because $\langle a_k \rangle$ decreases.

Let $\epsilon > 0$ be given. Since $a_n \to 0$ as $n \to \infty$ we can find an N such that, for any $n > N, a_n < \epsilon$. But for any $n > m > N$,

$$|s_n - s_m| = |(a_1 - a_2 + a_3 - \ldots a_n) - (a_1 - a_2 + \ldots a_m)|$$

$$= |a_{m+1} - a_{m+2} + a_{m+3} - \ldots a_n|$$

$$\leqslant a_{m+1} < \epsilon \quad \text{(because } m > N).$$

Thus $\langle s_n \rangle$ is a Cauchy sequence and the theorem follows.

6.14 *Example* We saw in theorem 6.5 that the series $\Sigma_{n=1}^{\infty} 1/n$ diverges. But it follows from theorem 6.13 that the series

$$\sum_{n=1}^{\infty} \frac{(-1)^{n-1}}{n} = 1 - \frac{1}{2} + \frac{1}{3} - \frac{1}{4} + \frac{1}{5} - \dots$$

converges. Of course, theorem 6.13 yields no clue as to what the sum is. (It is, in fact, $\log_e 2$.)

Theorem 6.13 is sometimes useful, but its hypotheses are rather restrictive. A more useful theorem is the following.

6.15 *Theorem (comparison test)* Let $\Sigma_{n=1}^{\infty} b_n$ be a *convergent* series of positive real numbers. If

$$|a_n| \leqslant b_n \quad (n = 1, 2, \dots)$$

then the series $\Sigma_{n=1}^{\infty} a_n$ converges.

Proof Let $\epsilon > 0$. Since $\Sigma_{n=1}^{\infty} b_n$ converges, its tail tends to zero (proposition 6.11). Hence we can find an N such that, for any $n > N$,

$$\sum_{k=n+1}^{\infty} b_k < \epsilon.$$

Let the sequence of partial sums of the series $\Sigma_{n=1}^{\infty} a_n$ be $\langle s_n \rangle$. Then, if $n > m > N$,

$$
\begin{aligned}
|s_n - s_m| &= |(a_1 + a_2 + \dots + a_n) - (a_1 + a_2 + \dots + a_m)| \\
&= |a_{m+1} + a_{m+2} + \dots + a_n| \\
&\leqslant |a_{m+1}| + |a_{m+2}| + \dots + |a_n| \quad \text{(triangle inequality)} \\
&\leqslant b_{m+1} + b_{m+2} + \dots + b_n \\
&\leqslant \sum_{k=m+1}^{\infty} b_k < \epsilon.
\end{aligned}
$$

Thus $\langle s_n \rangle$ is a Cauchy sequence and the theorem follows.

Note that the hypothesis $|a_n| \leqslant b_n$ $(n = 1, 2, \dots)$ can be replaced by $|a_n| \leqslant H b_n$ $(n = N, N+1, \dots)$. (Why?) Note further that the hypothesis of the comparison test is *not*

$$\left| \sum_{k=1}^{n} a_k \right| \leqslant \sum_{k=1}^{n} b_k.$$

This condition is satisfied by $a_k = (-1)^k$ and $b_k = 1/k^2$, but we know that $\Sigma_{n=1}^{\infty} (-1)^n$ diverges.

6.16   *Example* Prove that, if α is any positive rational number and $|x| < 1$, then the series

$$\sum_{n=1}^{\infty} n^{\alpha} x^n$$

converges.

Proof From exercise 4.20(5) we know that $n^{\alpha+2} x^n \to 0$ as $n \to \infty$. Hence the sequence $\langle n^{\alpha+2} x^n \rangle$ is bounded (theorem 4.25). It follows that, for some H,

$$|n^{\alpha} x^n| \leqslant H/n^2 \quad (n = 1, 2, \ldots).$$

Since the series $\Sigma_{n=1}^{\infty} 1/n^2$ converges (theorem 6.6), the result follows from the comparison test.

The next two propositions are sometimes helpful. We indicate only very briefly how they are proved.

6.17   *Proposition (ratio test)* Let $\Sigma_{n=1}^{\infty} a_n$ be a series which satisfies

$$\lim_{n \to \infty} \left| \frac{a_{n+1}}{a_n} \right| = l.$$

If $l > 1$, the series diverges and, if $l < 1$, the series converges.

If $l < 1$, we may take $\epsilon > 0$ so small that $l + \epsilon < 1$. Then, for a sufficiently large value of N,

$$|a_n| = \left| \frac{a_n}{a_{n-1}} \right| \left| \frac{a_{n-1}}{a_{n-2}} \right| \cdots \left| \frac{a_{N+2}}{a_{N+1}} \right| |a_{N+1}| < (1 + \epsilon)^{n-N-1} |a_{N+1}|.$$

The series $\Sigma_{n=1}^{\infty} a_n$ then converges by comparison with the geometric series $\Sigma_{n=1}^{\infty} (1 + \epsilon)^n$. If $l > 1$, a similar argument shows that the terms of $\Sigma_{n=1}^{\infty} a_n$ do not tend to zero and so the series diverges.

6.18   *Proposition (nth root test)* Let $\Sigma_{n=1}^{\infty} a_n$ be a series which satisfies

$$\limsup_{n \to \infty} |a_n|^{1/n} = l.$$

If $l > 1$, the series diverges and, if $l < 1$, the series converges.

If $l < 1$, we may take $\epsilon > 0$ so small that $l + \epsilon < 1$. Then, for a sufficiently large value of N,

$$|a_n| < (1 + \epsilon)^n \quad (n > N) \quad \text{(exercise 5.15(4))}$$

and the convergence of $\Sigma_{n=1}^{\infty} a_n$ follows from the comparison test. If $l > 1$, ϵ is chosen so that $l - \epsilon > 1$. Then, for some subsequence $\langle a_{n_k} \rangle$,

$$|a_{n_k}| > (l - \epsilon)^{n_k} \to \infty \text{ as } k \to \infty$$

and so the terms of $\sum_{n=1}^{\infty} a_n$ do not tend to zero.

Note that if the expressions of propositions 6.17 and 6.18 diverge to $+\infty$ (instead of converging to l), then the series diverge. If $l = 1$ however the results yield *no* information about the convergence or divergence of the series at all.

6.19   Example Prove that the series

$$\sum_{n=0}^{\infty} \frac{x^n}{n!}$$

converges for all values of x.

Proof We could use the comparison test. Alternatively, if $x \neq 0$,

$$\left| \frac{x^{n+1}}{(n+1)!} \middle/ \frac{x^n}{n!} \right| = \frac{|x|}{n+1} \to 0 \text{ as } n \to \infty$$

and hence the series converges by the ratio test.

6.20   Absolute and conditional convergence

A series $\sum_{n=1}^{\infty} a_n$ is said to converge *absolutely* if the series $\sum_{n=1}^{\infty} |a_n|$ converges. A series which converges but does *not* converge absolutely is said to be *conditionally* convergent.

6.21   Theorem Every absolutely convergent series is convergent.

Proof Simply take $b_n = |a_n|$ in the comparison test.

6.22   Examples

(i) Let $a_n = \dfrac{(-1)^{n-1}}{n}$ $(n = 1, 2, \ldots)$. Then

$$\sum_{n=1}^{\infty} a_n = \sum_{n=1}^{\infty} \frac{(-1)^{n-1}}{n} = 1 - \frac{1}{2} + \frac{1}{3} - \frac{1}{4} + \ldots$$

$$\sum_{n=1}^{\infty} |a_n| = \sum_{n=1}^{\infty} \frac{1}{n} = 1 + \frac{1}{2} + \frac{1}{3} + \frac{1}{4} + \ldots$$

The first of these series *converges* (example 6.14). The second *diverges* (theorem 6.5). We conclude that the series

$$\sum_{n=1}^{\infty} \frac{(-1)^{n-1}}{n}$$

is *conditionally* convergent.

(ii) On the other hand the series

$$\sum_{n=1}^{\infty} a_n = \sum_{n=1}^{\infty} \frac{(-1)^{n-1}}{n^2}$$

is *absolutely* convergent because

$$\sum_{n=1}^{\infty} |a_n| = \sum_{n=1}^{\infty} \frac{1}{n^2}$$

which *converges*.

It should be noted that the comparison test, the ratio test and the nth root test all demonstrate *absolute* convergence. The only criterion we have given which can establish the convergence of a series which is only *conditionally* convergent is theorem 6.13

6.23   Manipulations with series

Series are 'infinite sums'. It would therefore be optimistic to expect to be able to manipulate them just like 'finite sums'. Indeed, only *absolutely* convergent series may be freely manipulated. If one tries to obtain results by manipulating divergent or conditionally convergent series, only nonsense can be expected in general.

6.24   *Example* Consider the following argument.

$$\begin{aligned}
0 &= 0 + 0 + 0 + \ldots \\
&= (1-1) + (1-1) + (1-1) + \ldots \\
&= 1 - 1 + 1 - 1 + 1 - 1 + \ldots \\
&= 1 + (-1+1) + (-1+1) + (-1+1) + \ldots \\
&= 1 + 0 + 0 + 0 + \ldots \\
&= 1.
\end{aligned}$$

The error is not hard to find. The series $1 - 1 + 1 - 1 + 1 - 1 + \ldots$ is divergent. (Its terms do not tend to zero.)

6.25 *Example* We know that the series

$$1 - \tfrac{1}{2} + \tfrac{1}{3} - \tfrac{1}{4} + \tfrac{1}{5} - \tfrac{1}{6} + \tfrac{1}{7} - \ldots$$

converges conditionally. Denote its sum by s. We rearrange the order in which the terms of the series appear and obtain

$$1 - \tfrac{1}{2} - \tfrac{1}{4} + \tfrac{1}{3} - \tfrac{1}{6} - \tfrac{1}{8} + \tfrac{1}{5} - \tfrac{1}{10} - \tfrac{1}{12} + \tfrac{1}{7} - \ldots$$

Let the nth partial sum of this series be t_n. Then

$$t_{3n} = 1 - \frac{1}{2} - \frac{1}{4} + \frac{1}{3} - \ldots + \frac{1}{2n-1} - \frac{1}{4n-2} - \frac{1}{4n}$$

$$= \left(1 + \frac{1}{3} + \ldots + \frac{1}{2n-1}\right) - \left(\frac{1}{2} + \frac{1}{6} + \ldots + \frac{1}{4n-2}\right)$$

$$\qquad - \left(\frac{1}{4} + \frac{1}{8} + \ldots + \frac{1}{4n}\right)$$

$$= \left(1 + \frac{1}{3} + \ldots + \frac{1}{2n-1}\right) - \frac{1}{2}\left(1 + \frac{1}{3} + \ldots + \frac{1}{2n-1}\right)$$

$$\qquad - \frac{1}{2}\left(\frac{1}{2} + \frac{1}{4} + \ldots + \frac{1}{2n}\right)$$

$$= \frac{1}{2}\left(1 - \frac{1}{2} + \frac{1}{3} - \frac{1}{4} + \ldots + \frac{1}{2n-1} - \frac{1}{2n}\right)$$

$$\to \tfrac{1}{2}s \text{ as } n \to \infty.$$

Notice that the *finite* sum for t_{3n} can validly be rearranged in any way we like.

Since $t_{3n+1} - t_{3n} \to 0$ as $n \to \infty$ and $t_{3n+2} - t_{3n} \to 0$ as $n \to \infty$, it follows that the rearranged series converges to $\tfrac{1}{2}s$ which is *not* the same as the sum s of our original series (because $s \neq 0$).

6.26 *Exercise*

(1) Using partial fractions, prove that

$$\sum_{n=1}^{\infty} \frac{3n-2}{n(n+1)(n+2)} = 1.$$

(2) If $\langle a_n \rangle$ and $\langle b_n \rangle$ are two sequences of positive terms and

$$\frac{a_n}{b_n} \to l \text{ as } n \to \infty$$

where $l \neq 0$, prove that the series

$$\sum_{n=1}^{\infty} a_n \text{ and } \sum_{n=1}^{\infty} b_n$$

either both converge or both diverge.
Discuss the convergence or divergence of the series

(i) $\displaystyle\sum_{n=1}^{\infty} \frac{1}{2n}$ (ii) $\displaystyle\sum_{n=1}^{\infty} \frac{1}{2n-1}$ (iii) $\displaystyle\sum_{n=1}^{\infty} \frac{2}{n^2+3}$.

(3) Suppose that $\langle a_n \rangle$ is a decreasing sequence of positive terms such that $\Sigma_{n=1}^{\infty} a_n$ converges. Prove that $na_n \to 0$ as $n \to \infty$. [Hint: consider $a_{n+1} + a_{n+2} + \ldots + a_{2n}$.]

(4) Let $\langle r_n \rangle$ denote the rational numbers from $(0, 1)$ arranged in the sequence whose first few terms are $\frac{1}{2}, \frac{1}{3}, \frac{2}{3}, \frac{1}{4}, \frac{2}{4}, \frac{3}{4}, \ldots$ Prove that the series

$$\sum_{n=1}^{\infty} r_n$$

diverges.

(5) Determine whether or not the following series converge or diverge.

(i) $\displaystyle\sum_{n=1}^{\infty} \frac{(n!)^2}{(2n)!}$ (ii) $\displaystyle\sum_{n=1}^{\infty} \frac{(n!)^2}{(2n)!} 5^n$ (iii) $\displaystyle\sum_{n=1}^{\infty} \left(\frac{n}{n+1}\right)^{n^2}$

(iv) $\displaystyle\sum_{n=1}^{\infty} \left(\frac{n}{n+1}\right)^{n^2} 4^n$ (v) $\displaystyle\sum_{n=1}^{\infty} \frac{1}{n} \{\sqrt{(n+1)} - \sqrt{n}\}$

(vi) $\displaystyle\sum_{n=1}^{\infty} \frac{(-1)^{n-1}}{\sqrt{n}}$.

(6) If the sum of the conditionally convergent series

$$1 - \tfrac{1}{2} + \tfrac{1}{3} - \tfrac{1}{4} + \tfrac{1}{5} - \ldots$$

is s, prove that the sum of the rearranged series

$$1 + \tfrac{1}{3} - \tfrac{1}{2} + \tfrac{1}{5} + \tfrac{1}{7} - \tfrac{1}{4} + \tfrac{1}{9} + \tfrac{1}{11} - \tfrac{1}{6} + \ldots$$

is $\frac{3}{2}s$. $\left[\text{Hint: } t_{3n} = 1 + \dfrac{1}{3} - \dfrac{1}{2} + \ldots + \dfrac{1}{4n-3} + \dfrac{1}{4n-1} - \dfrac{1}{2n}.\right]$

7 FUNCTIONS

7.1   Notation

A *function f* from a set A to a set B (write $f: A \to B$) defines a rule
which assigns to *each* $x \in A$ a *unique* element $y \in B$. The element y is called the
image of the element x and we write $y = f(x)$.

When A and B are sets of real numbers we can draw the *graph* of the function
as in the diagram below. The defining property of a function ensures that each
vertical line drawn through a point of A cuts the graph in one and only one
place.

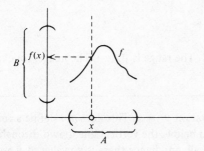

If f is a function from A to B and $S \subset A$, we say that f is *defined* on the set
S. The largest set on which f is defined is, of course, the set A. We call A the
domain of f. For example, a sequence is a function whose domain is the set $\mathbb{N}$
of natural numbers.

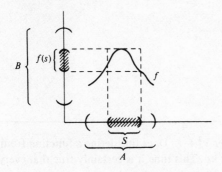

If f is defined on a set S, we use the notation

$$f(S) = \{f(x) : x \in S\}$$

and say that $f(S)$ is the *image* of the set S under the function f.

The set $f(A)$ is called the *range* of f. Note that $f(A)$ need not be the whole of B.

7.2   *Example* Consider the equation $y = x^2$. This defines a function from $\mathbb{R}$ to itself. For *each* $x \in \mathbb{R}$ there exists a *unique* $y \in \mathbb{R}$ which satisfies the rule $y = x^2$. Observe that, in the diagram below, each vertical line cuts the graph in one and only one place.

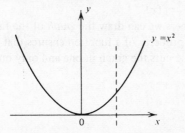

The domain of this function is $\mathbb{R}$. The range is $[0, \infty)$. The image, for example, of the set $[-2, 1]$ is $[0, 4]$. (Why?)

7.3   *Example* Consider the equation $y^2 = x$. This does *not* define a function from $\mathbb{R}$ to itself. In the diagram below, the vertical line drawn through the point x_0 does not meet the graph at all, and hence there is *no* value of y associated with x_0.

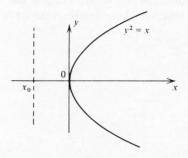

7.4   *Example* Again consider $y^2 = x$. Does this define a function from $[0, \infty)$ to $\mathbb{R}$? Again the answer is *no*. This time it is certainly true that every vertical line drawn through a point of $[0, \infty)$ meets the graph (see §1.9). But all but one of these vertical lines meets the graph in *two* points. Thus, in the diagram, there is not a *unique* value of y associated with x_1.

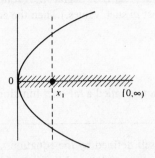

7.5 *Example* Again consider $y^2 = x$. This *does* define a function from $[0, \infty)$ to $[0, \infty)$.

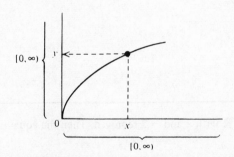

Since y must be in $[0, \infty)$ we omit from the diagram the part of our previous graphs which lies below the x-axis. Then every vertical line drawn through a point of $[0, \infty)$ meets the graph in one and only one point (see §1.9). Thus, given *any* $x \in [0, \infty)$, there is a *unique* $y \in [0, \infty)$ which satisfies $y^2 = x$. Thus a function f is defined from $[0, \infty)$ to $[0, \infty)$. Recalling the content of §1.9, we observe that, for each $x \geqslant 0$,

$$f(x) = \sqrt{x}.$$

7.6 **Polynomial and rational functions**

If $a_0, a_1, a_2, \ldots a_n$ are all real numbers, then the equation

$$y = a_0 + a_1 x + a_2 x^2 + \ldots + a_n x^n$$

defines a function from $\mathbb{R}$ to itself. *Any* value of x which is substituted on the right hand side generates a *unique* corresponding value of y. If $a_n \neq 0$, we call this function a *polynomial* of *degree n*. A polynomial of degree 0 is called a *constant*.

Suppose that P and Q are polynomial functions. Let S denote the set $\mathbb{R}$ with all the values of x for which $Q(x) = 0$ removed. (If Q is of degree m, it follows from exercise 3.11(3) that there can be at most m such values.) Then the equation

$$y = \frac{P(x)}{Q(x)}$$

defines a function from S to $\mathbb{R}$. Such a function is called a *rational* function.

7.7   *Example* The function from $\mathbb{R}$ to itself defined by the equation $y = x^3 - 3x^2 + 2x$ is called a polynomial function of degree 3 (or, more loosely, a 'cubic polynomial'). Its graph is sketched below.

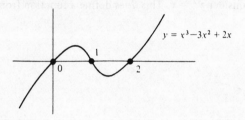

7.8   *Example* Let S be the set $\mathbb{R}$ with 2 and -2 removed. Then the equation

$$y = \frac{x^2 + 4}{x^2 - 4} \quad (x \neq \pm 2)$$

defines a function from S to $\mathbb{R}$. Its graph is sketched below.

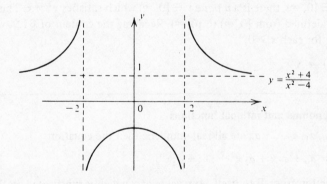

7.9 Combining functions

We begin with some almost obvious notation. If $S \subset \mathbb{R}$ and f and g are two functions from S to $\mathbb{R}$, then we define the function $f + g$ to be that function from S to $\mathbb{R}$ which satisfies

$$(f + g)(x) = f(x) + g(x) \quad (x \in S).$$

Similarly, if λ is any real number, we define λf to be the function from S to $\mathbb{R}$ which satisfies

$$(\lambda f)(x) = \lambda f(x) \quad (x \in S).$$

Again, we define the functions fg and f/g by

$$(fg)(x) = f(x) . g(x) \quad (x \in S)$$

$$(f/g)(x) = f(x)/g(x) \quad (x \in S).$$

For the latter definition to make sense, of course, it is essential that $g(x) \neq 0$ for all $x \in S$.

A somewhat less trivial way of combining functions is to employ the operation of *composition*. Let S and T be subsets of $\mathbb{R}$ and suppose that $g: S \to T$ and $f: T \to \mathbb{R}$. Then we define the *composite function* $f \circ g: S \to \mathbb{R}$ by

$$f \circ g(x) = f(g(x)) \quad (x \in S).$$

Sometimes $f \circ g$ is called a 'function of a function'.

7.10   *Example* Let $f: \mathbb{R} \to \mathbb{R}$ be defined by

$$f(x) = \frac{x^2 - 1}{x^2 + 1} \quad (x \in \mathbb{R})$$

and let $g: \mathbb{R} \to \mathbb{R}$ be defined by

$$g(x) = x^3.$$

Then $f \circ g: \mathbb{R} \to \mathbb{R}$ is given by the formula

$$f \circ g(x) = f(g(x)) = \frac{\{g(x)\}^2 - 1}{\{g(x)\}^2 + 1} = \frac{x^6 - 1}{x^6 + 1}.$$

7.11 Inverse functions

Suppose that A and B are sets and that f is a function from A to B. This means that *each* element $a \in A$ has a *unique* image $b = f(a) \in B$.

We say that f^{-1} is the *inverse* function to f if f^{-1} is a function from B to A which has the property that $x = f^{-1}(y)$ if and only if $y = f(x)$.

Not all functions have inverse functions. In fact, it is clear that a function

$f: A \rightarrow B$ has an inverse function $f^{-1}: B \rightarrow A$ if and only if *each* $b \in B$ is the image of a *unique* $a \in A$. (Otherwise f^{-1} could not be a function). A function which has this property is said to be a *1 : 1 correspondence* between A and B.

In geometric terms, a function $f: A \rightarrow B$ is a 1 : 1 correspondence between A and B (and hence has an inverse function $f^{-1}: B \rightarrow A$) if and only if each vertical line through A meets the graph of f in one and only one point (which makes f a function) *and* each horizontal line through B meets the graph of f in one and only one point (which makes f^{-1} a function).

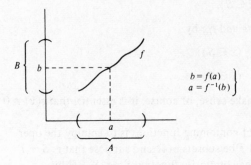

$$\left. \begin{array}{l} b = f(a) \\ a = f^{-1}(b) \end{array} \right\}$$

7.12   *Example* Let $f: (1, \infty) \rightarrow (0, 1)$ be defined by

$$f(x) = \frac{x - 1}{x + 1} \quad (x > 1).$$

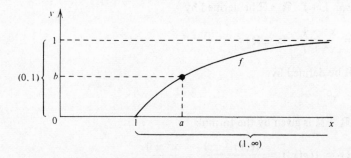

It seems clear from the diagram that f is a 1 : 1 correspondence between $(1, \infty)$ and $(0, 1)$. Thus it has an inverse function. To *prove* this we must show that, given any y satisfying $0 < y < 1$, there is a unique $x > 1$ which satisfies

$$y = \frac{x - 1}{x + 1}. \tag{1}$$

This is easily accomplished by solving (1) for x. We obtain

$$y(x + 1) = x - 1$$

$$x(y-1) = -1-y$$

$$x = \frac{1+y}{1-y}. \qquad (2)$$

Thus, if y satisfies $0 < y < 1$, there is a unique $x > 1$ which satisfies (1) – namely, that given by (2). We have established the existence of an inverse function $f^{-1} : (0, 1) \to (1, \infty)$. It is given by the formula

$$f^{-1}(y) = \frac{1+y}{1-y} \quad (y \in (0, 1)).$$

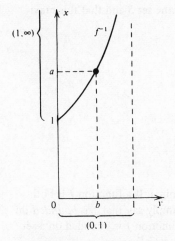

It is instructive to observe how the graph of $x = f^{-1}(y)$ is related to that of $y = f(x)$.

7.13 Bounded functions

Let f be defined on S. We say that f is *bounded above* on S by the upper bound H if and only if, for any $x \in S$,

$$f(x) \leq H.$$

This is the same as saying that the set

$$f(S) = \{ f(x) : x \in S \}$$

is bounded above by H.

If f is bounded above on S, then it follows from the continuum property that it has a smallest upper bound (or supremum) on S. Suppose that

$$B = \sup_{x \in S} f(x) = \sup f(S).$$

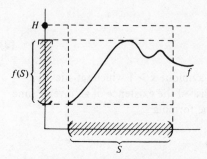

It may or may not be true that, for some $\xi \in S$, $f(\xi) = B$. If such a value of ξ does exist, we say that B is the *maximum* of f on the set S and that this maximum is *attained* at the point ξ.

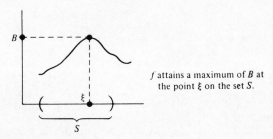

f attains a maximum of B at the point ξ on the set S.

Similar remarks apply to lower bounds and minima. If a function f is both bounded above *and* below on the set S, then we simply say that f is *bounded* on the set S. From proposition 2.3 it follows that a function f is bounded on a set S if and only if, for some K, it is true that, for any $x \in S$,

$$|f(x)| \leqslant K.$$

7.14   *Example* Let $f: (0, \infty) \to \mathbb{R}$ be defined by

$$f(x) = \frac{1}{x} \quad (x > 0).$$

This function is *unbounded* above on $(0, 1]$. It is, however, bounded below on $(0, 1]$ and attains a minimum of 1 at the point $x = 1$.

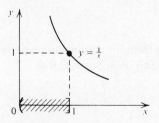

7.15   *Example* Let $f: \mathbb{R} \to \mathbb{R}$ be defined by $f(x) = x$. This function is
bounded above on both the sets $(0, 1)$ and $[0, 1]$ and in both cases its supremum
is 1. But f has *no* maximum on the set $(0, 1)$. There is no ξ satisfying $0 < \xi < 1$
for which $f(\xi) = 1$. On the other hand, f attains a maximum of 1 at the point
$x = 1$ on the set $[0, 1]$.

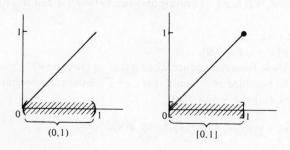

7.16   *Exercise*

(1) Draw a diagram illustrating the set of all (x, y) such that

$$y = \begin{cases} 5 & \text{if } x \geqslant 1 \\ 2 & \text{if } x < 1. \end{cases}$$

Explain why this is a graph of a function from $\mathbb{R}$ to itself. What is the
range of this function? What is the image of the set $[1, 2]$ under this
function?

(2) Draw a diagram illustrating the equation

$$|x| + |y| = 1.$$

[Hint: consider each quadrant separately.] Explain why
   (i) the equation does *not* define a function from $\mathbb{R}$ to itself;
   (ii) the equation does *not* define a function from $[-1, 1]$ to itself;
   (iii) the equation *does* define a function from $[-1, 1]$ to $[0, 1]$.

(3) Let $f: [0, 1] \to [0, 1]$ be defined by

$$f(x) = \frac{1 - x}{1 + x} \quad (0 \leqslant x \leqslant 1)$$

and let $g: [0, 1] \to [0, 1]$ be defined by

$$g(x) = 4x(1 - x) \quad (0 \leqslant x \leqslant 1).$$

Find formulae for $f \circ g$ and $g \circ f$ and hence show that these functions are not the same.

Show that f^{-1} exists but that g^{-1} does not exist. Find a formula for
f^{-1}.

(4) The formula $y = x^2$ may be employed to define a function $f: \mathbb{R} \to \mathbb{R}$.

Explain why f has *no* inverse function. If, instead, we use the formula $y = x^2$ to define a function $g: [0, \infty) \to [0, \infty)$, show that $g^{-1}: [0, \infty] \to [0, \infty)$ exists and that

$$g^{-1}(y) = \sqrt{y} \quad (y \geqslant 0).$$

(5) A function $g: A \to B$ is a $1:1$ correspondence between A and B. Prove that

(i) $g^{-1} \circ g(x) = x \quad (x \in A)$

(ii) $g \circ g^{-1}(y) = y \quad (y \in B)$.

To what do these formulae reduce when g is as in question 4?

(6) Let f and g be bounded above on S. Let c be a constant. Prove that

(i) $\sup\limits_{x \in S} \{f(x) + c\} = c + \sup\limits_{x \in S} f(x)$.

(ii) $\sup\limits_{x \in S} \{f(x) + g(x)\} \leqslant \sup\limits_{x \in S} f(x) + \sup\limits_{x \in S} g(x)$.

Give an example to show that equality need not hold in (ii).

8 LIMITS OF FUNCTIONS

8.1 Limits from the left

Suppose that f is defined on an interval (a, b). We say that $f(x)$ *tends* (or *converges*) to a *limit* l as x tends to b from the left and write

$$f(x) \to l \text{ as } x \to b -$$

or, alternatively,

$$\lim_{x \to b-} f(x) = l$$

if the following criterion is satisfied.

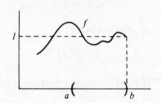

Given any $\epsilon > 0$, we can find a $\delta > 0$ such that

$|f(x) - l| < \epsilon$

provided that $b - \delta < x < b$.

The number $|f(x) - l|$ is the distance between $f(x)$ and l. We can think of it as the error in approximating to l by $f(x)$. The definition of the statement $f(x) \to l$ as $x \to b -$ then amounts to the assertion that we can make the error in approximating to l by $f(x)$ as small as we like by taking x sufficiently close to b on the left.

8.2 Limits from the right

Suppose again that f is defined on an interval (a, b). We say that $f(x)$ tends (or converges) to a limit l as x tends to a from the right and write

$$f(x) \to l \text{ as } x \to a +,$$

or, alternatively,

$$\lim_{x \to a+} f(x) = l$$

if the following criterion is satisfied.

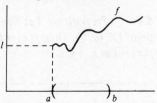

Given any $\epsilon > 0$, we can find a $\delta > 0$ such that

$|f(x) - l| < \epsilon$

provided that $a < x < a + \delta$.

8.3 $f(x) \to l$ as $x \to \xi$

Suppose that f is defined on an interval (a, b) *except possibly* for some point $\xi \in (a, b)$. We say that $f(x)$ tends (or converges) to a limit l as x tends to ξ and write

$$f(x) \to l \text{ as } x \to \xi$$

or, alternatively,

$$\lim_{x \to \xi} f(x) = l$$

if the following criterion is satisfied.

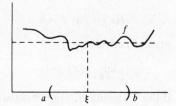

Given any $\epsilon > 0$, we can find a $\delta > 0$ such that

$|f(x) - l| < \epsilon$

provided that $0 < |x - \xi| < \delta$.

If α and β are real numbers, it is often useful to note that the inequality $|\alpha| < \beta$ is equivalent to $-\beta < \alpha < \beta$, or, what is the same thing, $-\beta < -\alpha < \beta$ (see exercise 1.20(1)).

Thus, in the definitions above, the condition $|f(x) - l| < \epsilon$ can be replaced throughout by $-\epsilon < f(x) - l < \epsilon$ or, alternatively, by $-\epsilon < l - f(x) < \epsilon$.

Similarly, the condition $0 < |x - \xi| < \delta$ in the last definition is equivalent to the assertion $-\delta < x - \xi < \delta$ and $x \neq \xi$. Thus to say that $0 < |x - \xi| < \delta$ is to say that x satisfies one of the two inequalities $\xi - \delta < x < \xi$ or $\xi < x < \xi + \delta$.

With the help of the last remark it is easy to prove the following result.

8.4   Proposition Let f be defined on an interval (a, b) except possibly at a point $\xi \in (a, b)$. Then $f(x) \to l$ as $x \to \xi$ if and only if $f(x) \to l$ as $x \to \xi-$ *and* $f(x) \to l$ as $x \to \xi+$.

8.5   *Example* Let f be the function from $\mathbb{R}$ to itself defined by

$$f(x) = \begin{cases} 1-x & (x \leqslant 1) \\ 2x & (x > 1). \end{cases}$$

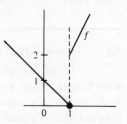

Then

    (i) $\displaystyle\lim_{x \to 1^-} f(x) = 0$ (ii) $\displaystyle\lim_{x \to 1^+} f(x) = 2.$

Note that it follows from proposition 8.4 that

$$\lim_{x \to 1} f(x)$$

does not exist even though $f(1)$ is perfectly well defined ($f(1) = 0$).

Proof (i) $f(x) \to 0$ as $x \to 1 -$. Given any $\epsilon > 0$ we must show how to find a $\delta > 0$ such that

$$|f(x) - 0| < \epsilon$$

provided that $1 - \delta < x < 1$. Since we are only concerned with values of x satisfying $x < 1$, we can replace $f(x)$ by $1 - x$. The condition $|f(x) - 0| < \epsilon$ then becomes $|1 - x| < \epsilon$ which is equivalent to $-\epsilon < x - 1 < \epsilon$. Adding 1 throughout, we see that $|f(x) - 0| < \epsilon$ is the same as $1 - \epsilon < x < 1 + \epsilon$.

The problem is now reduced to the following. Given any $\epsilon > 0$, find a $\delta > 0$ such that

$$1 - \epsilon < x < 1 + \epsilon$$

provided that $1 - \delta < x < 1$.

Obviously, the choice $\delta = \epsilon$ suffices to make this true and this completes the proof that $f(x) \to 0$ as $x \to 1 -$.

(ii) $f(x) \to 2$ as $x \to 1 +$. Given any $\epsilon > 0$, we must show how to find a $\delta > 0$ such that

$$|f(x) - 2| < \epsilon$$

provided that $1 < x < 1 + \delta$. Since we are only concerned with values of x satisfying $x > 1$, we can replace $f(x)$ by $2x$. The condition $|f(x) - 2| < \epsilon$ then becomes $|2x - 2| < \epsilon$ which is equivalent to $-\epsilon < 2(x - 1) < \epsilon$. Thus $|f(x) - 2| < \epsilon$ is the same as $1 - \frac{1}{2}\epsilon < x < 1 + \frac{1}{2}\epsilon$.

The problem is now reduced to the following. Given any $\epsilon > 0$, find a $\delta > 0$ such that

$$1 - \tfrac{1}{2}\epsilon < x < 1 + \tfrac{1}{2}\epsilon$$

provided that $1 < x < 1 + \delta$.

Obviously, the choice $\delta = \frac{1}{2}\epsilon$ suffices to make this true and this completes the proof that $f(x) \to 2$ as $x \to 1 +$.

8.6   Continuity at a point

In the definition of $f(x) \to l$ as $x \to \xi$ given in §8.3, consideration of what happens when x actually *equals* ξ was carefully excluded. When taking limits we are only interested in the behaviour of $f(x)$ as x *tends* to ξ not in what happens when x *equals* ξ.

Thus it is quite possible for $f(x)$ to tend to a limit l as x tends to ξ even though f is not defined at the point ξ (see exercise 8.15(3) below). And, even if $f(\xi)$ *is* defined, it is not necessarily true that $l = f(\xi)$ (see exercise 8.15(2) below).

Having said this, we can now turn to the definition of continuity at a point.

Suppose that f is defined on an interval (a, b) and $\xi \in (a, b)$. Then we say that f is *continuous at the point* ξ if and only if

$$f(x) \to f(\xi) \text{ as } x \to \xi.$$

Roughly speaking, to say that f is continuous at the point ξ means that the graph of f does not have a 'break' at the point ξ.

If f is defined on an interval $(a, b]$ and $f(x) \to f(b)$ as $x \to b -$, we say that f is continuous *on the left* at the point b. If f is defined on an interval $[a, b)$ and $f(x) \to f(a)$ as $x \to a +$, then we say that f is continuous *on the right* at the point a.

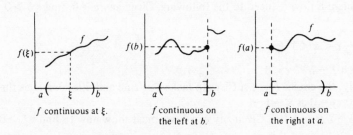

f continuous at ξ.   f continuous on   f continuous on
                       the left at b.   the right at a.

8.7   *Example* For any real numbers α and β, the function $f: \mathbb{R} \to \mathbb{R}$ defined by

$$f(x) = \alpha x + \beta$$

is continuous at every real number ξ.

Proof Assume $\alpha \neq 0$. (If $\alpha = 0$, the proof is even easier.) Let $\epsilon > 0$ be given. Choose $\delta = \epsilon / |\alpha|$. Then, provided that $|x - \xi| < \delta$,

$$|f(x) - f(\xi)| = |\alpha(x - \xi)| = |\alpha| \cdot |x - \xi| < |\alpha| \cdot \delta = \epsilon.$$

Given any $\epsilon > 0$, we have found a $\delta > 0$ such that $|f(x) - f(\xi)| < \epsilon$ provided that $|x - \xi| < \delta$. Hence $f(x) \to f(\xi)$ as $x \to \xi$ and so f is continuous at the point ξ.

Note that we showed that $|f(x) - f(\xi)| < \epsilon$ 'provided that $|x - \xi| < \delta$' instead of 'provided that $0 < |x - \xi| < \delta$' as appears in the definition of $f(x) \to l$ as $x \to \xi$. This is because, when $x = \xi$, it is *automatically* true that $|f(x) - f(\xi)| < \epsilon$ because then $|f(x) - f(\xi)| = 0$. It is only when we are considering a limit $l \neq f(\xi)$ that it is important to exclude the possibility that $x = \xi$.

8.8   Connexion with convergent sequences

8.9   *Theorem* Let f be defined on (a, b) except possibly for $\xi \in (a, b)$. Then $f(x) \to l$ as $x \to \xi$ if and only if, for each sequence $\langle x_n \rangle$ of points of (a, b) such that $x_n \neq \xi$ $(n = 1, 2, \ldots)$ and $x_n \to \xi$ as $n \to \infty$, it is true that $f(x_n) \to l$ as $n \to \infty$.

Proof (i) Suppose first that $f(x) \to l$ as $x \to \xi$. Let $\epsilon > 0$ be given. Then we can find a $\delta > 0$ such that $|f(x) - l| < \epsilon$ provided that $0 < |x - \xi| < \delta$. Now suppose that $\langle x_n \rangle$ is a sequence of points of (a, b) such that $x_n \neq \xi$ $(n = 1, 2, \ldots)$ and $x_n \to \xi$ as $n \to \infty$. Since $\delta > 0$ we can find an N such that for any $n > N$, $|x_n - \xi| < \delta$. But $x_n \neq \xi$ $(n = 1, 2, \ldots)$ and so $0 < |x_n - \xi| < \delta$. But this implies that

$$|f(x_n) - l| < \epsilon.$$

Given any $\epsilon > 0$, we have found an N such that, for any $n > N$, $|f(x_n) - l| < \epsilon$. Thus $f(x_n) \to l$ as $n \to \infty$.

(ii) Now suppose that, for each sequence $\langle x_n \rangle$ of points of (a, b) such that $x_n \neq \xi$ $(n = 1, 2, \ldots)$ and $x_n \to \xi$ as $n \to \infty$, it is true that $f(x_n) \to l$ as $n \to \infty$. We now suppose that it is *not* true that $f(x) \to l$ as $x \to \xi$ and seek a contradiction.

If it is *not* true that $f(x) \to l$ as $x \to \xi$ then, for *some* $\epsilon > 0$, it must be true that for *each* value of $\delta > 0$ we can find an x satisfying $0 < |x - \xi| < \delta$ such that

$$|f(x) - l| \geqslant \epsilon.$$

In particular, if $\delta = 1/n$, then we can find an x_n satisfying $0 < |x_n - \xi| < 1/n$ such that

$$|f(x_n) - l| \geqslant \epsilon.$$

But then $\langle x_n \rangle$ is a sequence of points of (a, b) such that $x_n \neq \xi$ $(n = 1, 2, \ldots)$ and $x_n \to \xi$ as $n \to \infty$ but for which it is *not* true that $f(x_n) \to l$ as $n \to \infty$. This contradicts our assumption above.

Similar results to theorem 8.9 hold for convergence from the left or right. For example, $f(x) \to l$ as $x \to b-$ if and only if, for each sequence $\langle x_n \rangle$ of points of (a, b) such that $x_n \to b$ as $n \to \infty$, it is true that $f(x_n) \to l$ as $n \to \infty$.

8.10   Example We show how theorem 8.9 can be used to prove that the function $f: \mathbb{R} \to \mathbb{R}$ defined by $f(x) = \alpha x + \beta$ is continuous at every point ξ (see example 8.7).

Let $\langle x_n \rangle$ be *any* sequence of real numbers such that $x_n \to \xi$ as $n \to \infty$. By proposition 4.8, $\alpha x_n + \beta \to \alpha \xi + \beta$ as $n \to \infty$. From theorem 8.9 it follows that $\alpha x + \beta \to \alpha \xi + \beta$ as $x \to \xi$ and this concludes the proof.

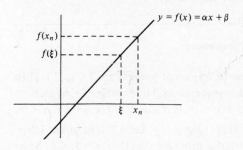

8.11   Properties of limits

The following propositions are analogues of the combination theorem and the sandwich theorem of chapter 4. They are easily deduced from these results with the help of theorem 8.9.

8.12   Proposition Let f and g be defined on an interval (a, b) except possibly at $\xi \in (a, b)$. Suppose that $f(x) \to l$ as $x \to \xi$ and $g(x) \to m$ as $x \to \xi$ and suppose that λ and μ are any real numbers. Then

    (i) $\lambda f(x) + \mu g(x) \to \lambda l + \mu m$ as $x \to \xi$

    (ii) $f(x)g(x) \to lm$ as $x \to \xi$

    (iii) $f(x)/g(x) \to l/m$ as $x \to \xi$ (provided $m \neq 0$).

An important consequence of proposition 8.12 is the following result which we quote as a theorem.

8.13   Theorem A polynomial is continuous at every point. A rational function is continuous at every point at which it is defined.

Proof Example 8.7 shows that $x \to \xi$ as $x \to \xi$. Repeated application of proposition 8.12(ii) then shows that $x^k \to \xi^k$ as $x \to \xi$ for each $k \in \mathbb{N}$. Hence, if

$$P(x) = a_n x^n + a_{n-1} x^{n-1} + \ldots + a_1 x + a_0,$$

then a repeated application of proposition 8.12(i) yields $P(x) \to P(\xi)$ as $x \to \xi$, i.e. P is continuous at ξ.

To show that a rational function is continuous wherever it is defined we appeal to proposition 8.12(iii).

8.14 *Proposition (sandwich theorem)* Let f, g and h be defined on (a, b) except possibly at $\xi \in (a, b)$. Suppose that $g(x) \to l$ as $x \to \xi$, $h(x) \to l$ as $x \to \xi$ and that

$$g(x) \leqslant f(x) \leqslant h(x)$$

except possibly when $x = \xi$. Then $f(x) \to l$ as $x \to \xi$.

Note The analogues of propositions 8.12 and 8.14 for convergence from the left or right are both true.

8.15 *Exercise*

(1) Calculate the following limits

(i) $\lim\limits_{x \to 1} \left\{ \dfrac{x^2 + 4}{x^2 - 4} \right\}$ (ii) $\lim\limits_{x \to 0} \left\{ \dfrac{x^{73} + 5x^{42} + 9}{3x^{23} + 7} \right\}$

(2) Let $f: \mathbb{R} \to \mathbb{R}$ be defined by

$$f(x) = \begin{cases} 3 - x & (x > 1) \\ 1 & (x = 1) \\ 2x & (x < 1). \end{cases}$$

Show that $f(x) \to 2$ as $x \to 1-$ and $f(x) \to 2$ as $x \to 1+$ using only the definitions of §8.1 and §8.2. Deduce that the limit

$$\lim_{x \to 1} f(x)$$

exists but is *not* equal to $f(1)$. Draw a graph.

(3) Let f be defined for all x *except* $x = 0$ by

$$f(x) = \frac{(1 + x)^2 - 1}{x} \quad (x \neq 0).$$

Prove that $f(x) \to 2$ as $x \to 0$ even though $f(0)$ is *not* defined.

(4) Use the definition to show that $|x - \xi| \to 0$ as $x \to \xi$. A function f is defined on an open interval which contains ξ and, for each x in this interval, $f(x)$ lies between ξ and x. Prove that $f(x) \to \xi$ as $x \to \xi$.

(5) Let $n \in \mathbb{N}$. Prove that the function $f: [0, \infty) \to \mathbb{R}$ defined by

$$f(x) = x^{1/n} \quad (x \geqslant 0)$$

is continuous at each $\xi > 0$ and continuous on the right at 0. [Hint: Use the sandwich theorem and exercise 3.11(6) to prove that f is continuous at $\xi > 0$. For continuity on the right at 0, appeal to the definition of $f(x) \to 0$ as $x \to 0+$.]

(6) Suppose that $f(x) \to l$ as $x \to \xi$. If $l > 0$, show that, for some $h > 0$,

$$f(x) > 0$$

provided that $\xi - h < x < \xi + h$ and $x \neq \xi$. What happens if (i) $l < 0$, (ii) $l = 0$?

8.16   Limits of composite functions

Suppose that $f(y) \to l$ as $y \to \eta$ and that $g(x) \to \eta$ as $x \to \xi$. It is tempting to write $y = g(x)$ and conclude that

$$f(g(x)) \to l \text{ as } x \to \xi.$$

But some care is necessary. Consider, for example, the functions $f\colon \mathbb{R} \to \mathbb{R}$ and $g\colon \mathbb{R} \to \mathbb{R}$ defined by

$$f(y) = \begin{cases} 3 & (y = 1) \\ 2 & (y \neq 1) \end{cases} ; \qquad g(x) = 1.$$

Then $f(y) \to 2$ as $y \to 1$ and $g(x) \to 1$ as $x \to 0$. But it is *not* true that $f(g(x)) \to 2$ as $x \to 0$ because, for all values of x, $f(g(x)) = f(1) = 3$.

8.17   Theorem Suppose that $f(y) \to l$ as $y \to \eta$ and that $g(x) \to \eta$ as $x \to \xi$. Then either of the two conditions below is sufficient to ensure that

$$f(g(x)) \to l \text{ as } x \to \xi.$$

(i) f is continuous at η (i.e. $l = f(\eta)$).

(ii) For some open interval I containing ξ, it is true that $g(x) \neq \eta$ for any $x \in I$, except possibly $x = \xi$.

Proof Let $\epsilon > 0$ be given. Since $f(y) \to l$ as $y \to \eta$, we can find a $\Delta > 0$ such that $|f(y) - l| < \epsilon$ provided that $0 < |y - \eta| < \Delta$.

Writing $y = g(x)$, we obtain

$$|f(g(x)) - l| < \epsilon \tag{1}$$

provided that

$$0 < |g(x) - \eta| < \Delta \tag{2}$$

But $g(x) \to \eta$ as $x \to \xi$ and $\Delta > 0$. Hence we can find a $\delta > 0$ such that

$$|g(x) - \eta| < \Delta \tag{3}$$

provided that

$$0 < |x - \xi| < \delta. \tag{4}$$

We would now *like* to say that (4) implies (3), which implies (2), which in turn, implies (1). This would complete the proof. But notice that (3) does not imply (2) in general. We therefore need to use one of the hypotheses (i) or (ii) of the theorem.

If we assume that f is continuous at η, then $l = f(\eta)$ and so $|f(y) - l| < \epsilon$ even when $y = \eta$ (see the remarks at the end of example 8.7). Thus, in this case, we may replace condition (2) by condition (3) and the argument described above goes through.

If instead we assume hypothesis (ii), then we can be sure that $g(x) \neq \eta$ provided that x satisfies (4) for a sufficiently small value of $\delta > 0$. But then condition (3) can be replaced by condition (2) and the argument goes through again.

8.18   Divergence

We write down two sample definitions. The reader will have little difficulty in supplying the definitions in other cases.

We say that $f(x) \to +\infty$ as $x \to \xi +$ if, given any $H > 0$, we can find a $\delta > 0$ such that

$$f(x) > H$$

provided that $\xi < x < \xi + \delta$.

We say that $f(x) \to l$ as $x \to -\infty$ if, given any $\epsilon > 0$, we can find an X such that

$$|f(x) - l| < \epsilon$$

provided that $x < X$.

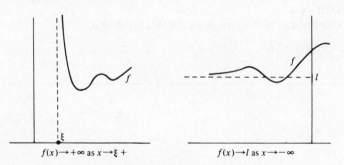

$f(x) \to +\infty$ as $x \to \xi +$ $f(x) \to l$ as $x \to -\infty$

8.19   Example Prove that $x^{-1} \to +\infty$ as $x \to 0+$.

Proof Let $H > 0$ be given. We have to find a $\delta > 0$ such that

$$\frac{1}{x} > H$$

provided that $0 < x < \delta$. Obviously the choice $\delta = H^{-1}$ satisfies the requirement.

8.20   Exercise

(1) Prove the following

(i) $x^{-1} \to -\infty$ as $x \to 0-$

(ii) $x^2 \to +\infty$ as $x \to +\infty$.

(2) Suppose that $f(y) \to l$ as $y \to +\infty$ and $g(x) \to +\infty$ as $x \to +\infty$. Prove that $f(g(x)) \to l$ as $x \to +\infty$. Explain why the difficulties encountered in theorem 8.17 do not occur in this case.

(3) Suppose that $f(y) \to l$ as $y \to 0+$. Prove that $f(x^{-1}) \to l$ as $x \to +\infty$.

(4) Suppose that f is defined on (a, b) except possibly for $\xi \in (a, b)$. Let $\langle x_n \rangle$ be a sequence of points of (a, b) such that $x_n \neq \xi$ $(n = 1, 2, \ldots)$ and $x_n \to \xi$ as $n \to \infty$. Let $\langle y_n \rangle$ be a sequence with the same properties. If $l \neq m$ and $f(x_n) \to l$ as $n \to \infty$ but $f(y_n) \to m$ as $n \to \infty$, show that

$$\lim_{x \to \xi} f(x)$$

does not exist. [Hint: theorem 8.9]

(5) Let $f: \mathbb{R} \to \mathbb{R}$ be defined by

$$f(x) = \begin{cases} 1 & (x \text{ rational}) \\ 0 & (x \text{ irrational}). \end{cases}$$

Show that

$$\lim_{x \to 0} f(x)$$

does not exist.

(6) If f is defined as in the previous question, prove that

$$\lim_{x \to 0} \{xf(x)\} = 0.$$

9 CONTINUITY

9.1 Continuity on an interval

A function f is continuous at a point ξ if and only if $f(x) \to f(\xi)$ as $x \to \xi$ (see §8.6). Geometrically, this means that the graph of f does not have a 'break' or 'jump' above the point ξ.

To say that f is continuous on an interval I should mean that its graph is unbroken above the interval I, i.e. that we can draw the part of the graph which lies above I without lifting our pencil from the paper.

A precise mathematical definition is required. This depends on the type of interval in question.

> A function f is continuous on an *open* interval I if and only if it is continuous at each point of I.

Where an endpoint of an interval belongs to the interval, a slightly more complicated definition is required. The most important case is that of a *compact* interval $[a, b]$.

> A function f is continuous on a *compact* interval $[a, b]$ if and only if it is continuous at each point of (a, b) and continuous on the right at a and on the left at b.

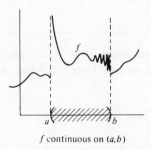

f continuous on (a,b)

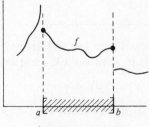

f continuous on $[a,b]$

Note that, in both diagrams, the graph of *f* is unbroken above the interval in question.

In general, a function *f* is continuous on an interval *I* provided that:

(i) it is continuous at each point of *I* which is not an endpoint;

(ii) it is continuous on the right at the left hand endpoint of *I*, *if* this exists and belongs to *I*;

(iii) it is continuous on the left at the right hand endpoint of *I*, *if* this exists and belongs to *I*.

9.2   Examples

(i) All polynomials are continuous on every interval. All rational functions are continuous on any interval not containing a zero of the denominator (theorem 8.13).

(ii) Let $f: (0, \infty) \to \mathbb{R}$ be defined by $f(x) = 1/x$. Then *f* is continuous on the interval $(0, \infty)$ because it is continuous at each point ξ satisfying $\xi > 0$.

(iii) Let $f: \mathbb{R} \to \mathbb{R}$ be defined by

$$f(x) = \begin{cases} x^2 & (x \leqslant 1) \\ 3/2 & (x > 1). \end{cases}$$

Then *f* is continuous at every point *except x* = 1. But *f* is continuous on the *left* at *x* = 1 and thus *f* is continuous on the interval [0, 1]. It is *not* continuous on [1, 2].

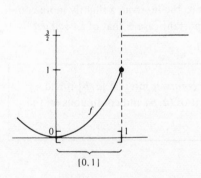

9.3   Proposition Let *f* be defined on an interval *I*. Then *f* is continuous on *I* if and only if, given any $x \in I$ and any $\epsilon > 0$, we can find a $\delta > 0$ such that

$$|f(x) - f(y)| < \epsilon$$

provided that $y \in I$ and satisfies $|x - y| < \delta$.

9.4 *Proposition* Let λ and μ be real numbers and suppose that the functions f and g are continuous on an interval I. Then so are the functions

(i) $\lambda f + \mu g$

(ii) fg

(iii) f/g (provided $g(x) \neq 0$ for any $x \in I$).

9.5 *Proposition* Let $g : I \rightarrow J$ be continuous on the interval I and let $f : J \rightarrow \mathbb{R}$ be continuous on the interval J. Then $f \circ g$ is continuous on I.

9.6 *Proposition* Let f be continuous on the interval I. If $\xi \in I$ and $\langle x_n \rangle$ is a sequence of points of I such that $x_n \rightarrow \xi$ as $n \rightarrow \infty$, then

$$f(x_n) \rightarrow f(\xi) \text{ as } n \rightarrow \infty.$$

These four propositions are easily proved. The first is proved by an appeal to the limit definitions of § §8.1, 8.2 and 8.3. The second follows from proposition 8.2, the third from theorem 8.17 and the fourth from theorem 8.9. The last of these results can be remembered in the form

$$\lim_{n \rightarrow \infty} f(x_n) = f(\lim_{n \rightarrow \infty} x_n)$$

i.e. for a *continuous* function the limit and function symbols can be validly exchanged. (We say that the symbols 'commute'.)

9.7 Continuity property

We now come to a theorem which may seem so obvious as to be hardly worth mentioning. It is, however, of a fundamental importance comparable to that of the Continuum Property of §2.2 and will serve as a foundation stone in our development of the calculus in the next few chapters.

9.8 *Theorem* (*continuity property*)

Let f be continuous on a compact interval $[a, b]$. Then the image of $[a, b]$ under f is also a compact interval.

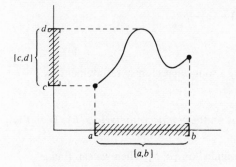

We split the content of the continuity property into three lesser theorems and prove these separately. Given that f is continuous on $I = [a, b]$, it follows from theorem 9.9 that $J = f(I)$ is an interval. From theorem 9.11 it follows that the interval J is bounded and from theorem 9.12 that J includes its endpoints.

9.9   *Theorem* Let f be continuous on an interval I. Then the image of I under f is also an interval.

Proof To show that $J = f(I) = \{f(x) : x \in I\}$ is an interval, we need to show that, if y_1 and y_2 are elements of J and $y_1 \leqslant \lambda \leqslant y_2$, then $\lambda \in J$ (see §2.9).

Since $y_1 \in J$ and $y_2 \in J$, the subsets of I defined by

$$S = \{x : f(x) \leqslant \lambda\}; \qquad T = \{x : f(x) \geqslant \lambda\}$$

are non-empty. Also, every point of the interval I belongs to one or other of these sets. It follows that a point of one of the sets is at zero distance from the other (exercise 2.13(6)). Suppose that $s \in S$ is at zero distance from T. Then a sequence $\langle t_n \rangle$ of points of T can be found such that $t_n \to s$ as $n \to \infty$ (exercise 4.29(6)). Since f is continuous on I, it follows that $f(t_n) \to f(s)$ as $n \to \infty$ (proposition 9.6). But

$$f(t_n) \geqslant \lambda \quad (n = 1, 2, \ldots)$$

and therefore $f(s) \geqslant \lambda$ (theorem 4.23). We already know that $f(s) \leqslant \lambda$ and so it follows that $f(s) = \lambda$. Hence $\lambda \in J$.

A similar argument applies if a point of T is at zero distance from S.

9.10   *Corollary* (*intermediate value theorem*) Let f be continuous on an interval I containing a and b. If λ lies between $f(a)$ and $f(b)$, then we can find a ξ between a and b such that $\lambda = f(\xi)$.

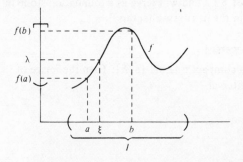

Proof The corollary is simply a somewhat clumsy restatement of theorem 9.9.

9.11   *Theorem* Let f be continuous on the *compact* interval $[a, b]$. Then f is bounded on $[a, b]$.

Proof Suppose that f is unbounded on $[a, b]$. Then we can find a sequence $\langle x_n \rangle$ of points of $[a, b]$ such that

$$|f(x_n)| \to +\infty \text{ as } n \to \infty \tag{1}$$

(see exercise 4.29(6)). Since $[a, b]$ is compact, there exists a subsequence $\langle x_{n_r} \rangle$ which converges to a point $\xi \in [a, b]$ by the Bolzano–Weierstrass theorem (see exercise 5.21(4)). Because f is continuous on $[a, b]$, $f(x_{n_r}) \to f(\xi)$ as $r \to \infty$ (proposition 9.6). But this contradicts (1).

9.12 *Theorem* Let f be continuous on the compact interval $[a, b]$. Then f achieves a maximum value d and a minimum value c on $[a, b]$.

Proof We know from the previous theorem that f is bounded on $[a, b]$. Let d be the supremum of f on $[a, b]$ and consider a sequence $\langle x_n \rangle$ of points of $[a, b]$ such that $f(x_n) \to d$ as $n \to \infty$ (see exercise 4.29(6)). Since $[a, b]$ is compact, $\langle x_n \rangle$ contains a subsequence $\langle x_{n_r} \rangle$ which converges to a point $\xi \in [a, b]$. Because f is continuous on $[a, b]$, it follows that $f(x_{n_r}) \to f(\xi)$ as $r \to \infty$. Hence $f(\xi) = d$ and thus the supremum d is actually a maximum.

A similar argument shows the infimum to be a minimum.

9.13 *Example* The function $f : \mathbb{R} \to \mathbb{R}$ defined by

$$f(x) = \begin{cases} 5 & (x \geq 1) \\ 2 & (x < 1) \end{cases}$$

is not continuous on $[0, 2]$. Observe that $2 = f(0) < 4 < f(2) = 5$ but there is no value of ξ between 0 and 2 such that $f(\xi) = 4$. Note that, for λ satisfying $2 < \lambda < 5$,

$$\{x : f(x) \leq \lambda\} = \{2\}; \qquad \{x : f(x) \geq \lambda\} = \{5\}.$$

9.14 *Example* The function $f : (0, \infty) \to \mathbb{R}$ defined by $f(x) = 1/x$ is continuous on the *open* interval $(0, 1)$ but is *not* bounded on $(0, 1)$. The function $g : \mathbb{R} \to \mathbb{R}$ defined by $g(x) = x$ is continuous everywhere and happens to be bounded on the *open* interval $(0, 1)$. But g does *not* attain a maximum value or a minimum value on $(0, 1)$.

9.15 *Example* Show that the equation

$$17x^7 - 19x^5 - 1 = 0$$

has a solution ξ which satisfies $-1 < \xi < 0$.

Proof The function $f : \mathbb{R} \to \mathbb{R}$ defined by $f(x) = 17x^7 - 19x^5 - 1$ is a polynomial and hence is continuous everywhere. In particular it is continuous on $[-1, 0]$. Now $f(-1) = 1$ and $f(0) = -1$. Since $-1 < 0 < 1$, it follows from

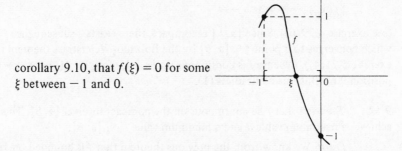

corollary 9.10, that $f(\xi) = 0$ for some
ξ between -1 and 0.

9.16   Example Let $f : [a, b] \to [a, b]$ be continuous on $[a, b]$. Then, for
some $\xi \in [a, b]$,

$$f(\xi) = \xi.$$

Thus a continuous function from a compact interval to itself 'fixes' some
point of the interval. (This is the one-dimensional version of Brouwer's fixed
point theorem.)

Proof The image of $[a, b]$ is a subset of $[a, b]$. Thus $f(a) \geqslant a$ and
$f(b) \leqslant b$. The function $g : [a, b] \to \mathbb{R}$ defined by $g(x) = f(x) - x$ is continuous
on $[a, b]$ by proposition 9.4. But $g(a) \geqslant 0$ and $g(b) \leqslant 0$. By corollary 9.10, for
some $\xi \in [a, b]$ it is true that $g(\xi) = 0$. Thus $f(\xi) = \xi$.

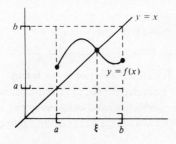

9.17   Exercise

(1) Each of the following expressions defines a function on $(-2, 2)$.
Decide in each case whether or not the function is continuous on (*a*)
$(-2, 2)$ and (*b*) $[0, 1]$. In each case draw a graph.

(i) $f(x) = \dfrac{2x + 3}{2x - 5}$ (ii) $f(x) = |x - 1|$

(iii) $f(x) = \begin{cases} 1 & (x < 1) \\ 2 & (x \geqslant 1) \end{cases}$ (iv) $f(x) = \begin{cases} 1 & (0 \leqslant x \leqslant 1) \\ 0 & \text{(otherwise)} \end{cases}$

(v) $f(x) = \dfrac{x^2 + 4}{x^2 - 4}$ (vi) $f(x) = \begin{cases} x(x - 1) & (0 < x < 1) \\ 0 & \text{(otherwise)} \end{cases}$

Which of these functions are bounded on $(-2, 2)$? Which attain a maximum on $(-2, 2)$? Which attain a minimum on $(-2, 2)$?

(2) A function f is continuous on an interval I and for each rational number $r \in I$ it is true that

$$f(r) = r^2.$$

Prove that $f(x) = x^2$ for any $x \in I$. [Hint: proposition 9.6.]

(3) Show that all polynomials of *odd* degree have at least one (real) root.

(4) Suppose that f is continuous at every point and that $f(x) \to 0$ as $x \to +\infty$ and $f(x) \to 0$ as $x \to -\infty$. Prove that f attains a maximum value or a minimum value on the set $\mathbb{R}$.

(5) Let f be continuous on the *compact* interval I. Suppose that, for each $x \in I$, there exists a $y \in I$ such that

$$|f(y)| \leqslant \tfrac{1}{2}|f(x)|.$$

Prove the existence of a $\xi \in I$ for which $f(\xi) = 0$. [Hint: theorem 9.12.]

(6) Let I be a closed interval and let $0 < \alpha < 1$. Let $f : I \to I$ satisfy the inequality

$$|f(x) - f(y)| \leqslant \alpha |x - y|$$

for each $x \in I$ and $y \in I$. (Such a function is called a contraction mapping). Prove that f is continuous on I.

Let $x_1 \in I$ and define $x_{n+1} = f(x_n)$ $(n = 1, 2, \ldots)$. Prove that the sequence $\langle x_n \rangle$ converges and that its limit l satisfies $l = f(l)$. [Hint: exercise 5.21(1).]

10 DIFFERENTIATION

10.1 Derivatives

Suppose that f is defined on an open interval I containing the point ξ. Then f is said to be *differentiable* at the point ξ if and only if the limit

$$\lim_{x \to \xi} \frac{f(x) - f(\xi)}{x - \xi}$$

exists. If the limit exists, it is called the *derivative* of f at the point ξ and denoted by $f'(\xi)$ or $Df(\xi)$.

For a function f which is differentiable at ξ we therefore have

$$f'(\xi) = \lim_{x \to \xi} \frac{f(x) - f(\xi)}{x - \xi}.$$

Equivalently, we may write

$$f'(\xi) = \lim_{h \to 0} \frac{f(\xi + h) - f(\xi)}{h}. \tag{1}$$

From an intuitive point of view, to say that a function f is differentiable at ξ means that a tangent can be drawn to the curve $y = f(x)$ where $x = \xi$. The slope of this tangent is then equal to $f'(\xi)$. This geometric interpretation is based on the diagram below.

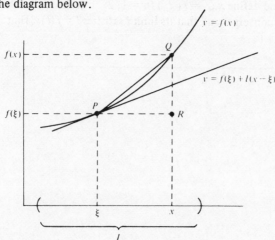

In order that the line $y = f(\xi) + l(x - \xi)$ be tangent to $y = f(x)$ where $x = \xi$, we require that the slope of the chord PQ approach the slope l of this line as $x \to \xi$. The slope of the chord PQ is

$$\frac{QR}{PR} = \frac{f(x) - f(\xi)}{x - \xi}$$

and hence our requirement reduces to

$$\lim_{x \to \xi} \frac{f(x) - f(\xi)}{x - \xi} = l$$

as in the definition of differentiability.

(Note: It follows immediately from our definition that a function f is differentiable at ξ if and only if there exists an l such that

$$f(\xi + h) - f(\xi) - lh = o(h) \quad (h \to 0). \tag{2}$$

Here $o(h)$ denotes a quantity which tends to zero when divided by h. This form of the definition lends itself more readily to generalisation.)

10.2 Higher derivatives

If f is differentiable at each point of an open interval I, we say that f is differentiable on I. In this case it is natural to define f' or Df to be the function from I to $\mathbb{R}$ whose value at each point $x \in I$ is $f'(x)$. We can then define the second derivative $f''(\xi)$ or $D^2f(\xi)$ at the point ξ by

$$f''(\xi) = \lim_{x \to \xi} \frac{f'(x) - f'(\xi)}{x - \xi}$$

provided that the limit exists. Similarly for third derivatives and so on.

10.3 Examples

(i) Let $f: \mathbb{R} \to \mathbb{R}$ be defined by $f(x) = x^2$. Then

$$f'(x) = Dx^2 = 2x$$

for all values of x.

Proof We have

$$\frac{f(x + h) - f(x)}{h} = \frac{(x + h)^2 - x^2}{h} = \frac{2xh + h^2}{h} = 2x + h \to 2x \text{ as } h \to 0.$$

(Note that we studiously ignore what happens when $h = 0$ – see §8.4. It is as well that this is permissible, for the expression

$$\frac{f(x + h) - f(x)}{h}$$

is quite meaningless when $h = 0$.)

(ii) Let $f: (0, \infty) \to \mathbb{R}$ be defined by $f(x) = 1/x$ $(x > 0)$. Then

$$f'(x) = D\frac{1}{x} = -\frac{1}{x^2}$$

for all values of $x > 0$.

Proof We have

$$\frac{f(x+h) - f(x)}{h} = \frac{1}{h}\left\{\frac{1}{x+h} - \frac{1}{x}\right\} = \frac{1}{h}\left\{\frac{-h}{x(x+h)}\right\}$$

$$= -\frac{1}{x(x+h)} \to \frac{-1}{x^2} \text{ as } h \to 0.$$

10.4   More notation

We shall avoid the notation introduced in this section because, when using this notation, it is difficult to make statements with a precision adequate for our purposes. In many applications, however, it is unnecessary to maintain this level of precision. In such cases, the notation described below can be a very useful and powerful tool and it would be pedantic to deny oneself the use of it.

Replace ξ by x and h by δx in formula (1) of §10.1. Then

$$f'(x) = \lim_{\delta x \to 0} \frac{f(x + \delta x) - f(x)}{\delta x} = \lim_{\delta x \to 0} \frac{\delta y}{\delta x}. \tag{3}$$

If $y = f(x)$, the quantity $\delta y = f(x + \delta x) - f(x)$ is usually said to be the 'small change in y consequent on the small change δx in x'.

From formula (3) it is only a short step to the much used (and commonly abused) notation

$$f'(x) = \frac{dy}{dx}.$$

This notation has the drawback that it uses the symbol x ambiguously. It is used simultaneously as the point at which the derivative is to be evaluated *and* the variable with respect to which one is differentiating. In applications where this distinction is unimportant, the ambiguity creates no problems. But, for the material of this book, the distinction is important.

Further confusion is possible when 'differentials' are introduced. The *differential df* of a function f may be regarded as the function of *two* variables given by

$$df(x; h) = f'(x)h. \tag{4}$$

Formula (2) can then be rewritten as

$$f(x + h) - f(x) - df(x; h) = o(h) \quad (h \to 0) \tag{5}$$

provided that f is differentiable at x.

For a fixed value of x, the equation

$$k = df(x; h) = f'(x)h \tag{6}$$

is the equation of a straight line. If the h and k axes are drawn as below, (6) is the equation of the tangent to the graph of the function f at the point x.

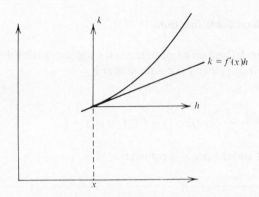

The reader may justifiably feel that little or nothing has been added to our understanding by the introduction of the idea of a differential, the idea being more useful when functions of several variables are being considered. Our only reason for discussing differentials is to explain the meaning of the equation

$$dy = f'(x) \, dx.$$

This is simply equation (6) with the variable h replaced by dx and the variable k replaced by dy. Thus dx and dy should be thought of as variables. The reason one chooses this notation is to make the formula

$$dy = \frac{dy}{dx} dx \tag{7}$$

a valid one.

It is *false* that dy denotes 'the small change in y consequent on a small change dx in x'. Formula (5) tells us that this is *nearly* true for small values of dx but it will only be *exactly* true for functions f with a straight line graph. Should it be necessary to discuss 'small changes in x and y', the notation δx and δy is available and one can say that δy is *approximately* equal to $f'(x)\delta x$.

Even more *false* (if that is possible) is the idea that dy and dx are somehow 'infinitesimal quantities' obtained by allowing δx and δy to tend to zero. One would then regard

$$\frac{dy}{dx}$$

as the quotient of these quantities and abandon all the apparatus of limits. This metaphysical notion is admittedly an attractive one and it is on this idea that Newton based his 'Theory of Fluxions'. Unfortunately it cannot stand up to close logical scrutiny.

(Note: A common error is to confuse the terms derivative and differential. A derivative is sometimes referred to as a 'differential coefficient' (because of (7) above) but should not be called a differential.)

10.5   Properties of differentiable functions

10.6   Theorem Let *f* be defined on an open interval *I* which contains the point ξ. If *f* is differentiable at ξ, then *f* is continuous at ξ.

Proof We have

$$f(x) - f(\xi) = \frac{f(x) - f(\xi)}{x - \xi} \cdot (x - \xi) \to f'(\xi) \cdot 0 \text{ as } x \to \xi.$$

Hence $f(x) \to f(\xi)$ as $x \to \xi$ and the proof is complete.

10.7   Example Let $f: \mathbb{R} \to \mathbb{R}$ be defined by

$$f(x) = \begin{cases} 1 & (x \geqslant 1) \\ -1 & (x < 1). \end{cases}$$

We can see immediately that *f* is not differentiable at the point 1 because it is not continuous there.

10.8   Example Let $f: \mathbb{R} \to \mathbb{R}$ be defined by

$$f(x) = \begin{cases} x & (x \geqslant 1) \\ x^2 & (x < 1). \end{cases}$$

Then *f* is continuous at 1 but *not* differentiable (The graph has a 'corner'). We have

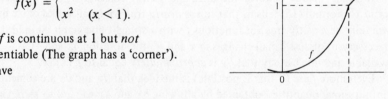

$$\lim_{h \to 0+} \frac{f(1+h) - f(1)}{h} = \lim_{h \to 0+} \frac{(1+h) - 1}{h} = 1$$

$$\lim_{h \to 0-} \frac{f(1+h) - f(1)}{h} = \lim_{h \to 0-} \frac{(1+h)^2 - 1}{h} = 2.$$

Thus the converse of theorem 10.6 is false.

10.9   *Theorem* Suppose that f and g are defined on an open interval I containing the point ξ. Let λ and μ be any real numbers. Then, if f and g are differentiable at ξ,

(i) $D\{\lambda f + \mu g\} = \lambda Df + \mu Dg$

(ii) $D\{fg\} = fDg + gDf$

(iii) $D\{f/g\} = \dfrac{gDf - fDg}{g^2}$ (provided $g(\xi) \neq 0$)

at the point ξ.

Proof

(i) $\dfrac{1}{h}\{\lambda f(\xi + h) + \mu g(\xi + h) - \lambda f(\xi) - \mu g(\xi)\}$

$$= \lambda \left\{\frac{f(\xi + h) - f(\xi)}{h}\right\} + \mu \left\{\frac{g(\xi + h) - g(\xi)}{h}\right\}$$

$$\to \lambda f'(\xi) + \mu g'(\xi) \text{ as } h \to 0.$$

(ii) $\dfrac{f(\xi + h)g(\xi + h) - f(\xi)g(\xi)}{h}$

$$= \frac{1}{h}\{f(\xi + h)g(\xi + h) - f(\xi)g(\xi + h) + f(\xi)g(\xi + h) - f(\xi)g(\xi)\}$$

$$= g(\xi + h)\left\{\frac{f(\xi + h) - f(\xi)}{h}\right\} + f(\xi)\left\{\frac{g(\xi + h) - g(\xi)}{h}\right\}$$

$$\to g(\xi)f'(\xi) + f(\xi)g'(\xi) \text{ as } h \to 0.$$

(Note that $g(\xi + h) \to g(\xi)$ as $h \to 0$ because, by theorem 10.6, g is continuous at ξ.)

(iii) $\dfrac{1}{h}\left\{\dfrac{f(\xi + h)}{g(\xi + h)} - \dfrac{f(\xi)}{g(\xi)}\right\} = \dfrac{f(\xi + h)g(\xi) - g(\xi + h)f(\xi)}{hg(\xi + h)g(\xi)}$

$$= \frac{1}{g(\xi + h)g(\xi)}\left\{g(\xi)\frac{f(\xi + h) - f(\xi)}{h} - f(\xi)\frac{g(\xi + h) - g(\xi)}{h}\right\}$$

$$\to \frac{1}{\{g(\xi)\}^2}\{g(\xi)f'(\xi) - f(\xi)g'(\xi)\} \text{ as } h \to 0.$$

10.10   *Example* If n is a natural number, then, for any x,

$$Dx^n = nx^{n-1}.$$

Proof This is obvious when $n = 1$ since

$$\frac{(x + h) - x}{h} = 1 \to 1 \text{ as } h \to 0.$$

Now assume its truth when $n = k$. Thus $Dx^k = kx^{k-1}$. Using theorem 10.9(ii),

$$D(x^{k+1}) = D(x^k.x) = x^k.1 + x.kx^{k-1} = (k+1)x^k$$

and the result follows by induction.

10.11   *Exercise*

(1) Prove that, for any x,

$$D\left\{\frac{1}{1+x^2}\right\} = \frac{-2x}{(1+x^2)^2}.$$

(2) If n is a negative integer and $x \neq 0$, prove that $Dx^n = nx^{n-1}$.

(3) Let $f: \mathbb{R} \to \mathbb{R}$ be defined by

$$f(x) = \begin{cases} 2x & (x \geq 1) \\ x^2 + 1 & (x < 1). \end{cases}$$

Prove that f is differentiable at the point 1 and has derivative 2.
Let $g: \mathbf{R} \to \mathbf{R}$ be defined by $g(x) = |x|$. Prove that g is not differentiable at the point 0. [Hint: Consider the right and left hand limits separately.]

(4) A polynomial P of degree n has the property that $P(\xi) = 0$ and $P'(\xi) = 0$. Prove that

$$P(x) = (x - \xi)^2 Q(x)$$

for all x, where $Q(x)$ is a polynomial of degree $n - 2$. (See exercise 3.11(3).)

(5) Let $f: \mathbb{R} \to \mathbb{R}$ be n times differentiable at the point ξ. Show that the 'Taylor polynomial' P defined by

$$P(x) = f(\xi) + \frac{(x-\xi)}{1!}f'(\xi) + \frac{(x-\xi)^2}{2!}f''(\xi) + \ldots$$

$$+ \frac{(x-\xi)^{n-1}}{(n-1)!}f^{(n-1)}(\xi)$$

has the property $P^{(k)}(\xi) = f^{(k)}(\xi)$ $(k = 0, 1, 2, \ldots, n-1)$. (Note that $f^{(k)}(\xi)$ denotes the derivative of order k at the point ξ.)

†(6) Let f and g be n times differentiable at the point ξ. Prove 'Leibniz's rule',

i.e. $$D^n fg = \sum_{j=0}^{n} \binom{n}{j} D^j f D^{n-j} g$$

at the point ξ. [Hint: Recall the proof of the binomial theorem.]

10.12 Composite functions

The next theorem is usually remembered in the form

$$\frac{dz}{dx} = \frac{dz}{dy} \cdot \frac{dy}{dx}.$$

10.13 *Theorem* Suppose that g is differentiable at the point x and that f is differentiable at the point $y = g(x)$. Then

$$(f \circ g)'(x) = f'(y)g'(x).$$

Proof Recall that $f \circ g(x) = f(g(x))$. We therefore have to consider

$$\frac{f(g(x + h)) - f(g(x))}{h}.$$

Write $k = g(x + h) - g(x)$. Since g is differentiable at x, it is continuous there and so $k \to 0$ as $h \to 0$. Suppose that $k \neq 0$. Then

$$\frac{f(g(x+h)) - f(g(x))}{h} = \frac{f(g(x + h)) - f(g(x))}{g(x + h) - g(x)} \cdot \frac{g(x + h) - g(x)}{h}$$

$$= \frac{f(y + k) - f(y)}{k} \cdot \frac{g(x + h) - g(x)}{h}.$$

It is tempting to deduce the conclusion of the theorem immediately. But care is necessary. What happens if, for some values of h, $k = g(x + h) - g(x) = 0$? To get round this difficulty we introduce the function

$$F(k) = \begin{cases} \dfrac{f(y + k) - f(y)}{k} & (k \neq 0) \\ \\ f'(y) & (k = 0). \end{cases}$$

Since f is differentiable at y, $F(k) \to f'(y)$ as $k \to 0$. Because $F(0) = f'(y)$, it follows that F is continuous at the point 0. From theorem 8.17(i) we deduce that

$$F(k) \to f'(y) \text{ as } h \to 0$$

where $k = g(x + h) - g(x)$.

For $k \neq 0$ we have shown that

$$\frac{f(g(x + h)) - f(g(x))}{h} = F(k) \cdot \frac{g(x + h) - g(x)}{h}.$$

But this equation is also true when $k = 0$ since then both sides are zero. It follows that

$$\frac{f(g(x + h)) - f(g(x))}{h} \to f'(y)g'(x) \text{ as } h \to 0$$

and this is what had to be proved.

10.14   *Example* $D\{1 + x^{97}\}^{100} = 100(1 + x^{97})^{99}.97x^{96}$.

Proof Set $f(y) = y^{100}$ and $g(x) = 1 + x^{97}$. By theorem 10.13,

$$
\begin{aligned}
D\{1 + x^{97}\}^{100} &= Df(g(x)) = f'(g(x))g'(x) \\
&= 100(g(x))^{99}.97x^{96} \\
&= 100\{1 + x^{97}\}^{99}.97x^{96}
\end{aligned}
$$

10.15   *Exercise*

(1) If $x > 0$ and $n \in \mathbb{N}$, prove that

$$D\{x^{1/n}\} = \frac{1}{n}x^{1/n}x^{-1}.$$

[Hint: exercise 3.11(6). Recall that the continuity of the nth root function was considered in exercise 8.15(5).]

(2) Use the previous question and the rule for differentiating a composite function (theorem 10.13) to show that, for any rational number r,

$$D\{x^{r}\} = rx^{r-1}$$

provided that $x > 0$.

(3) Evaluate the following derivatives for those values of x for which they exist.

(i) $D\{1 + x^{1/27}\}^{3/5}$ (ii) $D\sqrt{(x + \sqrt{(x + \sqrt{x})})}$.

†(4) Suppose that

$$\frac{d}{dx}\{f(x^{2})\} = \frac{d}{dx}\{f(x)\}^{2}$$

when $x = 1$. Prove that $f'(1) = 0$ or $f(1) = 1$.

(5) Let $f: \mathbb{R} \to \mathbb{R}$ be differentiable at x. Suppose that f admits an inverse function $f^{-1}: \mathbb{R} \to \mathbb{R}$ which is differentiable at $y = f(x)$. Prove that

$$Df^{-1}(y) = \frac{1}{Df(x)}.$$

[Hint: $f^{-1} \circ f(x) = x$]. The function $g: \mathbb{R} \to \mathbb{R}$ defined by $g(x) = x^{3}$ admits an inverse function $g^{-1}: \mathbb{R} \to \mathbb{R}$. Is g^{-1} differentiable at the point 0?

†(6) A function $f: \mathbb{R} \to \mathbb{R}$ is defined by

$$f(x) = \begin{cases} x & (x \text{ rational}) \\ -x & (x \text{ irrational}). \end{cases}$$

Show that $f \circ f(x) = x$ for all values of x. What may be deduced from theorem 10.13?

11 MEAN VALUE THEOREMS

11.1 Local maxima and minima

Let f be defined on an open interval (a, b) and let $\xi \in (a, b)$. We say that f has a *local maximum* at ξ if

$$f(x) \leqslant f(\xi)$$

for all values of x in *some* open interval I which contains ξ. Similarly for a local minimum.

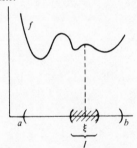

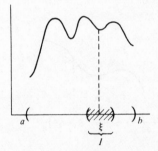

f has *local* max at ξ. f has *local* min at ξ.

Very roughly, f has a local maximum at ξ if its graph has a 'little hill' above the point ξ. Similarly, f has a local minimum at ξ if its graph has a 'little valley' above the point ξ.

If $f(\xi)$ is the maximum value of f on the whole interval (a, b), then obviously f has a local maximum at ξ. But the converse need not be true (see the diagrams above).

11.2   Theorem Suppose that f is differentiable on (a, b) and that $\xi \in (a, b)$. If f has a local maximum or minimum at ξ, then

$$f'(\xi) = 0.$$

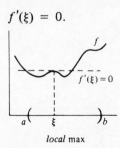

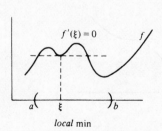

local max   *local* min

Proof By definition

$$\frac{f(x) - f(\xi)}{x - \xi} \to f'(\xi) \text{ as } x \to \xi.$$

Suppose that $f'(\xi) > 0$. From Exercise 8.15(6), it follows that for some open interval $I = (\xi - h, \xi + h)$

$$\frac{f(x) - f(\xi)}{x - \xi} > 0 \qquad\qquad (1)$$

provided that $x \in I$ and $x \neq \xi$.

Let x_1 be *any* number in the interval $(\xi - h, \xi)$. Then $x_1 - \xi < 0$ and hence it follows from (1) that $f(x_1) < f(\xi)$. Thus f cannot have a local minimum at ξ. Let x_2 be *any* number in the interval $(\xi, \xi + h)$. Then $x_2 - \xi > 0$ and so it follows from (1) that $f(x_2) > f(\xi)$. Thus f cannot have a local maximum at ξ.

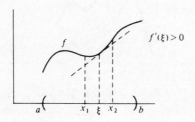

A similar argument applied to $-f$ deals with the case when $f'(\xi) < 0$. The only remaining possibility is $f'(\xi) = 0$.

11.3   Stationary points

A point ξ at which $f'(\xi) = 0$ is called a *stationary point* of f. Not all stationary points give rise to a local maximum or local minimum. The reader will be familiar with the case when $f'(\xi) = 0$ and f has a *point of inflexion* at ξ as in the diagrams below.

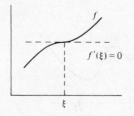
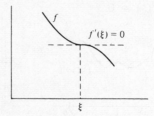

But worse behaviour than this is possible. The diagram below illustrates a case where $f'(\xi) = 0$ but f has no local maximum, no local minimum, and no point of inflexion at ξ. (See the solution to exercise 16.5(6).)

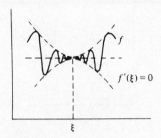

11.4   Theorem (Rolle's theorem) Suppose that f is continuous on $[a, b]$ and differentiable on (a, b). If $f(a) = f(b)$, then, for some $\xi \in (a, b)$,

$$f'(\xi) = 0.$$

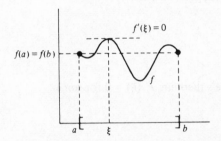

Proof Since f is continuous on the compact interval $[a, b]$, it follows from the continuity property (theorem 9.12) that f attains a maximum M at some point ξ_1 in the interval $[a, b]$ and attains a minimum m at some point ξ_2 in the interval $[a, b]$.

Suppose ξ_1 and ξ_2 are both endpoints of $[a, b]$. Because $f(a) = f(b)$ it then follows that $m = M$ and hence f is constant on $[a, b]$. But then $f'(\xi) = 0$ for all $\xi \in (a, b)$.

Suppose that ξ_1 is not an endpoint of $[a, b]$. Then $\xi_1 \in (a, b)$ and f has a local maximum at ξ_1. Thus, by theorem 11.2, $f'(\xi_1) = 0$. Similarly if ξ_2 is not an endpoint of $[a, b]$.

[Note the importance of the continuity property in this proof.]

11.5   Mean value theorem

11.6   Theorem (mean value theorem) Suppose that f is continuous on $[a, b]$ and differentiable on (a, b). Then, for some $\xi \in (a, b)$,

$$f'(\xi) = \frac{f(b) - f(a)}{b - a}.$$

In the diagram, the slope of the chord PQ is

$$\frac{f(b) - f(a)}{b - a}.$$

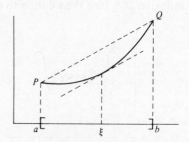

Thus, for some $\xi \in (a, b)$, the tangent to f at ξ is parallel to PQ.

Proof Let F be defined on $[a, b]$ by

$$F(x) = f(x) + hx$$

where h is a constant. Then F is continuous on $[a, b]$ and differentiable on (a, b). We choose the constant h so that $F(a) = F(b)$. Then

$$f(a) + ha = f(b) + hb$$

and so

$$h = - \left\{ \frac{f(b) - f(a)}{b - a} \right\}.$$

Since F satisfies the conditions of Rolle's theorem, $F'(\xi) = 0$ for some $\xi \in (a, b)$. But then

$$F'(\xi) = f'(\xi) + h = 0$$

and the theorem follows.

11.7   Theorem Suppose that f is continuous on $[a, b]$ and differentiable on (a, b). If $f'(x) = 0$ for *each* $x \in (a, b)$, then f is constant on $[a, b]$.

Proof Let $y \in (a, b]$. Then f satisfies the conditions of the mean value theorem on $[a, y]$. Hence, for some $\xi \in (a, y)$,

$$f'(\xi) = \frac{f(y) - f(a)}{y - a}.$$

But $f'(\xi) = 0$ and so $f(y) = f(a)$ for any $y \in [a, b]$.

(Note: This theorem will be invaluable in chapter 13, about integration. It may seem intuitively obvious but, without the mean value theorem (which depends ultimately on the continuity property), it would be very hard to prove.)

11.8   Exercise

(1) Find the stationary points of the function $f: \mathbb{R} \to \mathbb{R}$ defined by

$f(x) = x(x - 1)(x - 2)$. Determine the maximum and minimum values of f on the interval $[0, 3]$.

(2) Suppose that f and g are continuous on $[a, b]$, differentiable on (a, b) and that $g'(x) \neq 0$ for any $x \in (a, b)$. Prove that, for some $\xi \in (a, b)$,

$$\frac{f'(\xi)}{g'(\xi)} = \frac{f(b) - f(a)}{g(b) - g(a)}.$$

This is the Cauchy mean value theorem. [Hint: consider $F = f + hg$ where h is a constant.]

(3) In addition to the hypotheses of question 2, suppose that $f(a) = g(a) = 0$. Deduce that

$$\lim_{x \to a+} \frac{f(x)}{g(x)} = \lim_{x \to a+} \frac{f'(x)}{g'(x)}$$

provided that the second limit exists. This is L'Hôpital's rule. [Hint: exercise 8.15(4).]

Use L'Hôpital's rule to evaluate

$$\lim_{y \to +\infty} \{y - \sqrt{(1 + y^2)}\}$$

[Hint: put $y = x^{-1}$.]

(4) Suppose that $f: \mathbb{R} \to \mathbb{R}$ is differentiable at every point and that

$$f'(x) = x^2$$

for all x. *Prove* that $f(x) = \frac{1}{3}x^3 + c$, where c is a constant.

*(5) Let $f: \mathbb{R} \to \mathbb{R}$ be n times differentiable at every point. Let P be a polynomial of degree $n - 1$ for which

$$f(\xi_j) = P(\xi_j) \quad (j = 0, 1, \ldots n)$$

where $\xi_0, \xi_1, \ldots, \xi_n$ are all distinct points. Prove that, for some ξ,

$$f^{(n)}(\xi) = 0.$$

*(6) Let $g: \mathbb{R} \to \mathbb{R}$ satisfy $g(0) = g(1) = 0$ and suppose that $f: \mathbb{R} \to \mathbb{R}$ is differentiable at every point and $y = f(x)$ is a solution of the differential equation

$$g(x)\frac{dy}{dx} + y = 1.$$

Prove that $f(x) = 1$ for any x satisfying $0 \leqslant x \leqslant 1$. [Hint: Use Rolle's theorem to show that any open subinterval of $(0, 1)$ contains a point ξ at which $f(\xi) = 1$. Then appeal to the continuity of f in the manner of exercise 9.17(2).]

11.9   Taylor's theorem

Suppose that f is
differentiable at the point ξ. Then
the equation of the tangent to
$y = f(x)$ at the point ξ is

$$y = f(\xi) + (x - \xi)f'(\xi).$$

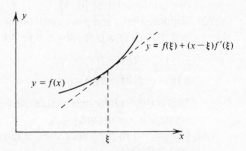

Thus, for values of x which are 'close' to ξ, it is reasonable to expect that
$f(\xi) + (x - \xi)f'(\xi)$ is a 'good' approximation to $f(x)$. But how large may the
error in this approximation be?

Suppose now that f is $n - 1$ times differentiable at the point ξ. Then the first
$n - 1$ derivatives at the point ξ of the polynomial

$$P(x) = f(\xi) + \frac{1}{1!}(x - \xi)f'(\xi) + \frac{1}{2!}(x - \xi)^2 f''(\xi) + \ldots$$

$$+ \frac{1}{(n-1)!}(x - \xi)^{n-1}f^{(n-1)}(\xi)$$

are the *same* as those of f. (See exercise 10.11(5).)

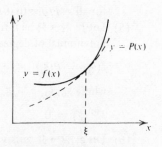

We might therefore reasonably expect that
$P(x)$ will be a 'very good' approximation to
$f(x)$ for values of x which are 'close' to ξ. But
again, how large may the error be in this
approximation?

These questions are answered to some extent by the following version of
Taylor's theorem. If $n = 1$, the theorem reduces to theorem 11.6 (the mean
value theorem). One may therefore regard Taylor's theorem as the nth order
mean value theorem.

11.10   Theorem (Taylor's theorem) Suppose that f is n times differentiable on
an open interval I which contains the point ξ. Given any $x \in I$,

$$f(x) = f(\xi) + \frac{1}{1!}(x - \xi)f'(\xi) + \frac{1}{2!}(x - \xi)^2 f''(\xi) + \ldots$$

$$+ \frac{1}{(n-1)!}(x - \xi)^{n-1}f^{(n-1)}(\xi) + E_n$$

where the error E_n satisfies

$$E_n = \frac{1}{n!}(x - \xi)^n f^{(n)}(\eta)$$

for *some* value of η between ξ and x.

Proof If y lies between x and ξ, put

$$F(y) = f(x) - f(y) - (x - y)f'(y) - \ldots - \frac{(x - y)^{n-1}}{(n - 1)!} f^{(n-1)}(y).$$

Taylor's theorem asserts that, for some η between x and ξ,

$$F(\xi) = \frac{1}{n!}(x - \xi)^n f^{(n)}(\eta) \tag{2}$$

The function F has the simplifying property that

$$F'(y) = \frac{-(x - y)^{n-1}}{(n - 1)!} f^{(n)}(y).$$

The function

$$G(y) = F(y) - \left(\frac{x - y}{x - \xi}\right)^n F(\xi)$$

has the property that $G(\xi) = G(x) = 0$. It follows from Rolle's theorem that there exists an η between x and ξ for which

$$0 = G'(\eta) = F'(\eta) + \frac{n(x - \eta)^{n-1}}{(x - \xi)^n} F(\xi)$$

$$= -\frac{(x - \eta)^{n-1}}{(n - 1)!} f^{(n)}(\eta) + \frac{n(x - \eta)^{n-1}}{((x - \xi)^n} F(\xi).$$

Identity (2) follows.

11.11 Exercise

(1) If $n \in \mathbb{N}$ and $x \in \mathbb{R}$, use Taylor's theorem to prove the binomial theorem in the form

$$(1 + x)^n = 1 + nx + \frac{n(n - 1)}{2!} x^2 + \ldots + x^n.$$

(2) Use Taylor's theorem with $n = 3$ and $\xi = 4$ to obtain an approximation to $\sqrt{5}$ for which the error is at most 2^{-9}.

(3) Let f be n times differentiable on an open interval I which contains the point ξ and suppose that $D^n f$ is continuous at the point ξ. You are given

$$f'(\xi) = f''(\xi) = \ldots = f^{(n-1)}(\xi) = 0$$

but $f^{(n)}(\xi) \neq 0$. In each of the following cases decide whether or not f has a local maximum or a local minimum at ξ. In those cases where f has neither a local maximum nor a local minimum discuss its behaviour close to the point ξ.

(i) $f^{(n)}(\xi) > 0$ and n even.

(ii) $f^{(n)}(\xi) > 0$ and n odd.

(iii) $f^{(n)}(\xi) < 0$ and n even.

(iv) $f^{(n)}(\xi) < 0$ and n odd.

[Hint: exercise 8.15(6).]

Discuss the behaviour of the functions given below at the point 0.

(v) $f(x) = x^{93}$ (vi) $g(x) = x^{94}$ (vii) $h(x) = -x^{96}$.

12 MONOTONE FUNCTIONS

12.1 Definitions

Let f be defined on a set S. We say that f *increases* on the set S if and only if, for each $x \in S$ and $y \in S$ with $x < y$, it is true that

$$f(x) \leqslant f(y).$$

If strict inequality always holds, we say that f is *strictly increasing* on the set S.
Similar definitions hold for decreasing and strictly decreasing.

A function which is either increasing or decreasing is called *monotone*. A function which is *both* increasing and decreasing must be a constant.

12.2 Example

The function $f \colon \mathbb{R} \to \mathbb{R}$ defined by $f(x) = x^3$ is strictly increasing on $\mathbb{R}$. The function $f \colon (0, \infty) \to \mathbb{R}$ defined by $f(x) = 1/x$ is strictly decreasing on $(0, \infty)$.

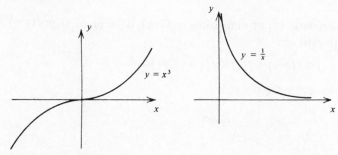

12.3 Limits of monotone functions

We introduce the notation

$$f(\xi -) = \lim_{x \to \xi-} f(x); \quad f(\xi +) = \lim_{x \to \xi+} f(x)$$

provided that these limits exist. Do *not* confuse $f(\xi -)$ and $f(\xi +)$ with $f(\xi)$, which is the value of the function f at ξ. In example 8.5, $f(1 -) = 0 = f(1)$ *but* $f(1 +) = 2$.

12.4   Theorem

(i) Let f be increasing and bounded above on (a, b) with smallest upper bound L. Then $f(x) \to L$ as $x \to b\, -$.

(ii) Let f be increasing and bounded below on (a, b) with largest lower bound l. Then $f(x) \to l$ as $x \to a\, +$.

Proof We only prove (i). Let $\epsilon > 0$ be given. We have to find a $\delta > 0$ such that, given any x satisfying $b - \delta < x < b$,

$$|f(x) - L| < \epsilon$$

i.e.
$$L - \epsilon < f(x) < L + \epsilon.$$

The inequality $f(x) < L + \epsilon$ is automatically satisfied because L is an upper bound for f on (a, b). Since $L - \epsilon$ is *not* an upper bound for f on (a, b), there exists a $y \in (a, b)$ such that $f(y) > L - \epsilon$. But f increases on (a, b). Therefore, for any x satisfying $y < x < b$,

$$L - \epsilon < f(y) \leqslant f(x).$$

The choice $\delta = b - y$ then completes the proof.

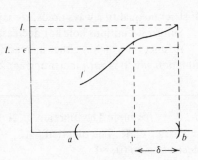

12.5   Corollary

Let f be increasing on (a, b). If $\xi \in (a, b)$, then $f(\xi\, -)$ and $f(\xi\, +)$ both exist and

$$f(x) \leqslant f(\xi\, -) \leqslant f(\xi) \leqslant f(\xi\, +) \leqslant f(y)$$

provided that $a < x < \xi < y < b$.

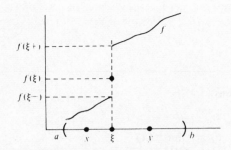

Proof The function f is bounded above on the interval (a, ξ) by $f(\xi)$. By theorem 12.4, the *smallest* upper bound is $f(\xi\, -)$. It follows that, for any $x \in (a, \xi)$,

$$f(x) \leqslant f(\xi -) \leqslant f(\xi).$$

A similar argument for the interval (ξ, b) yields the other inequalities.

12.6   Differentiable monotone functions

12.7   Theorem Suppose that f is continuous on $[a, b]$ and differentiable on (a, b).

(i) If $f'(x) \geqslant 0$ for each $x \in (a, b)$, then f is increasing on $[a, b]$. If $f'(x) > 0$ for each $x \in (a, b)$, then f is strictly increasing on $[a, b]$.

(ii) If $f'(x) \leqslant 0$ for each $x \in (a, b)$, then f is decreasing on $[a, b]$. If $f'(x) < 0$ for each $x \in (a, b)$, then f is strictly decreasing on $[a, b]$.

Proof Let c and d be *any* numbers in $[a, b]$ which satisfy $c < d$. Then f satisfies the conditions of the mean value theorem on $[c, d]$ and hence, for some $\xi \in (c, d)$,

$$f'(\xi) = \frac{f(d) - f(c)}{d - c}.$$

If $f'(x) \geqslant 0$ for each $x \in (a, b)$, then $f'(\xi) \geqslant 0$ and hence $f(d) \geqslant f(c)$. Thus f increases on $[a, b]$.

If $f'(x) > 0$ for each $x \in (a, b)$, then $f'(\xi) > 0$ and hence $f(d) > f(c)$. Thus f is strictly increasing on $[a, b]$.

Similarly, for the other cases.

12.8   Example Consider the function $f \colon \mathbb{R} \to \mathbb{R}$ defined by $f(x) = x(1 - x)$. We have $f'(x) = 1 - 2x$. Hence $f'(x) \geqslant 0$ when $x \leqslant \frac{1}{2}$ and $f'(x) \leqslant 0$ when $x \geqslant \frac{1}{2}$. It follows that f increases on $(-\infty, \frac{1}{2}]$ and decreases on $[\frac{1}{2}, \infty)$.

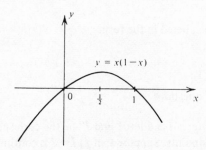

12.9   Inverse functions

A strictly increasing function f defines a 1:1 correspondence between its domain I and its range J. Hence f always has an inverse function f^{-1} and this is strictly increasing on J. (Proof?)

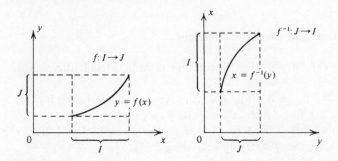

In the case when I is an interval and f is continuous on I, then $J = f(I)$ is also an interval (theorem 9.9). Observe that $f^{-1}: J \to I$ is continuous on J. This is a simple consequence of corollary 12.5. (If f^{-1} had a discontinuity at $\xi \in J$, then we could find a $\lambda \in I$ for which $f(\lambda)$ would be undefined.)

f^{-1} discontinuous at ξ

$f^{-1}(\xi -) < \lambda < f^{-1}(\xi +)$

$f(\lambda)$ undefined

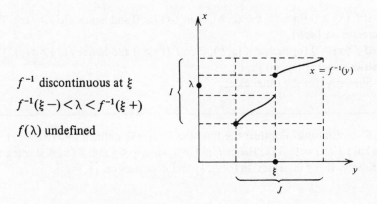

The next theorem is usually remembered in the form

$$\frac{dx}{dy} = \left(\frac{dy}{dx}\right)^{-1}$$

although this notation begs a lot of questions.

12.10   Theorem Let I and J be intervals and let I^o and J^o be the corresponding open intervals with the same endpoints. Suppose that $f: I \to J$ is continuous on I and that $J = f(I)$.
     If f is differentiable on I^o and

$$Df(x) > 0 \quad (x \in I^o)$$

then $f^{-1}: J \to I$ exists and is continuous on J. Also f^{-1} is differentiable on J^o and

$$Df^{-1}(y) = \frac{1}{Df(x)} \quad (y \in J^{\circ})$$

provided that $y = f(x)$.

Proof It follows from theorem 12.7 that f strictly increasing on I and hence $f^{-1}: J \to I$ exists. The continuity of f^{-1} has already been discussed. There remains the consideration of its derivative.

Let $y \in J^{\circ}$. Then $x = f^{-1}(y) \in I^{\circ}$. Put $k = f^{-1}(y + h) - f^{-1}(y)$. Then $f^{-1}(y + h) = f^{-1}(y) + k = x + k$. Thus $y + h = f(x + k)$ and it follows that

$$h = f(x + k) - y = f(x + k) - f(x).$$

Since f^{-1} is continuous on J, $k \to 0$ as $h \to 0$. Also, f^{-1} is strictly increasing and so $k \neq 0$ unless $h = 0$.

We may therefore appeal to theorem 8.17(ii) to obtain

$$\frac{f^{-1}(y + h) - f^{-1}(y)}{h} = \frac{k}{f(x + k) - f(x)} \to \frac{1}{f'(x)} \text{ as } h \to 0.$$

Note that all the above results apply equally well to strictly *decreasing* functions.

12.11 Roots

As an application of our observations on inverse functions, we now discuss the theory of nth roots.

Let n be a natural number and consider the function f defined on $[0, \infty)$ by

$$f(x) = x^n.$$

Since $f(0) = 0$ and $f(x) \to + \infty$ as $x \to + \infty$, it follows from the continuity property (theorem 9.9) that the range of f is also $[0, \infty)$.

Observe that

$$f'(x) = nx^{n-1} > 0 \quad (x > 0).$$

Thus f is strictly increasing on $[0, \infty)$ and, as in theorem 12.10, it has an inverse function f^{-1}. We write

$$f^{-1}(y) = y^{1/n}.$$

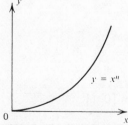

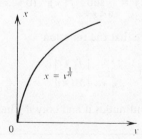

This argument shows that, given any $y \geqslant 0$ there exists exactly one $x \geqslant 0$ (namely $x = y^{1/n}$) such that $y = x^n$. Our basic assumption about nth roots made in §1.9 has therefore been justified.

In exercise 10.15(1), we calculated the derivative of the nth root function. We can therefore check the validity of the formula of theorem 12.10 in this case.

$$Dy^{1/n} = \frac{1}{Dx^n} = \frac{1}{nx^{n-1}} = \frac{1}{n(y^{1/n})^{n-1}} = \frac{1}{n} y^{1/n} y^{-1}.$$

If m is an arbitrary integer, the rule for differentiating a composite function (theorem 10.13) then yields

$$D(y^{m/n}) = D(y^m)^{1/n} = \frac{1}{n}(y^m)^{1/n} y^{-m} m y^{m-1} = \frac{m}{n} y^{m/n} y^{-1}.$$

12.12   *Exercise*

(1) Suppose that f increases on the compact interval $[a, b]$. Prove that f attains a maximum at b and a minimum at a.

(2) Let $n \in \mathbb{N}$. Prove that the function $f : [0, \infty) \to \mathbb{R}$ defined by

$$f(x) = (x + 1)^{1/n} - x^{1/n}$$

decreases on $[0, \infty)$.

(3) Suppose that f is differentiable and increasing on an open interval I. Prove that $f'(x) \geqslant 0$ for each $x \in I$. If f is strictly increasing on I, does it follow that $f'(x) > 0$ for each $x \in I$? Justify your answer.

(4) Let $f : \mathbb{R} \to \mathbb{R}$ be defined by $f(x) = 1 + x + x^3$. Show that f has an inverse function $f^{-1} : \mathbb{R} \to \mathbb{R}$. Calculate the value of $Df^{-1}(y)$ when $y = -1$.

†(5) A function f increases on the interval I and for each a and b in I it is true that, if λ lies between $f(a)$ and $f(b)$, then a $\xi \in I$ can be found such that $f(\xi) = \lambda$ (see corollary 9.10). Prove that f is continuous on I. Does the same conclusion hold if the hypothesis that f increases is abandoned?

†(6) Let $f : [0, 1] \to (0, \infty)$ be continuous on $[0, 1]$. Let $M : [0, 1] \to (0, \infty)$ be defined by

$$M(x) = \sup_{0 \leqslant y \leqslant x} f(y) \quad (0 \leqslant x \leqslant 1).$$

Prove that the function

$$\phi(x) = \lim_{n \to \infty} \left\{ \frac{f(x)}{M(x)} \right\}^n$$

is continuous if and only if f increases on $[0, 1]$.

12.13 Convex functions

Let f be defined on an interval I. We say that f is *convex* on I if it is true that, for each $\alpha > 0$ and $\beta > 0$ with $\alpha + \beta = 1$,

$$f(\alpha x + \beta y) \leqslant \alpha f(x) + \beta f(y) \qquad (1)$$

whenever $x \in I$ and $y \in I$. If the inequality sign is reversed, we say that f is *concave* on I. It is obvious that f is concave on I if and only if $-f$ is convex on I.

The geometric interpretation is that any point on a chord drawn to the graph of a convex function lies on or above the graph.

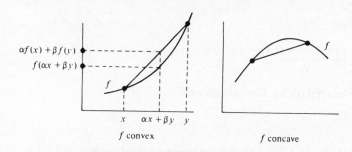

f convex f concave

It is sometimes useful to rewrite the definition of a convex function in the following way. A function f is convex on an interval I if and only if, for all points x_1, x_2 and x_3 on the interval I satisfying $x_1 < x_2 < x_3$,

$$\frac{f(x_2) - f(x_1)}{x_2 - x_1} \leqslant \frac{f(x_3) - f(x_2)}{x_3 - x_2}. \qquad (2)$$

Equivalently, a function f is convex on an interval I if and only if for all points x_1, x_2 and x_3 in the interval satisfying $x_1 < x_2 < x_3$,

$$\frac{f(x_2) - f(x_1)}{x_2 - x_1} \leqslant \frac{f(x_3) - f(x_1)}{x_3 - x_1}. \qquad (3)$$

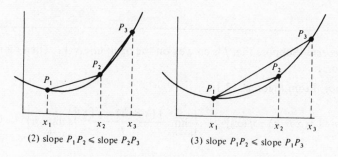

(2) slope $P_1 P_2 \leqslant$ slope $P_2 P_3$ (3) slope $P_1 P_2 \leqslant$ slope $P_1 P_3$

That (2) and (3) have the same content as (1) is easily seen by making the substitutions $x_1 = x$, $x_2 = \alpha x + \beta y$ and $x_3 = y$ (see exercise 12.21(2)).

12.14   Theorem Suppose that f is convex on the *open* interval I. Then the limits

$$\lim_{h \to 0-} \frac{f(x+h)-f(x)}{h}; \quad \lim_{h \to 0+} \frac{f(x+h)-f(x)}{h} \tag{4}$$

both exist for each $x \in I$.

Proof Suppose that $0 < h_1 < h_2$. Put $x = x_1, x_2 = x + h_1$ and $x_3 = x + h_2$ in (3). Then

$$\frac{f(x+h_1)-f(x)}{h_1} \leqslant \frac{f(x+h_2)-f(x)}{h_2}.$$

Hence the function

$$F(h) = \frac{f(x+h)-f(x)}{h}$$

increases in some interval $(0, \delta)$. The existence of

$$\lim_{h \to 0+} F(h)$$

then follows from theorem 12.4. A similar argument establishes the existence of the left hand limit.

12.15   Example The function $f: \mathbb{R} \to \mathbb{R}$ defined by $f(x) = |x|$ is convex on $\mathbb{R}$. (The chords lie above the graph).
Note however that the two limits of
(4) are *not* equal when $x = 0$.

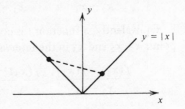

12.16   Theorem Suppose that f is convex on the open interval I. Then f is continuous on I.

Proof From theorem 12.14,

$$\lim_{h \to 0-} \{f(x+h)-f(x)\} = \left\{ \lim_{h \to 0-} \frac{f(x+h)-f(x)}{h} \right\} \left\{ \lim_{h \to 0-} h \right\} = 0.$$

Similarly,

$$\lim_{h \to 0+} \{f(x+h)-f(x)\} = \left\{ \lim_{h \to 0+} \frac{f(x+h)-f(x)}{h} \right\} \left\{ \lim_{h \to 0+} h \right\} = 0.$$

12.17 *Example* The function $f: \mathbb{R} \to \mathbb{R}$ defined by

$$f(x) = \begin{cases} 1 & (|x| < 1) \\ 2 & (|x| \geq 1) \end{cases}$$

is convex on the *compact* interval
$[-1, 1]$. Observe that it is *not*
continuous on $[-1, 1]$.

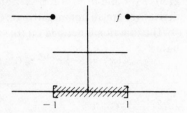

Example 12.15 shows that a function f which is convex on an open interval I need not be differentiable on I. However, if f *is* differentiable on I, then the following theorem provides a useful characterisation of convexity.

12.18 *Theorem* Let I be an open interval and suppose that f is differentiable on I. Then f is convex on I if and only if Df increases on I.

Proof (i) Suppose first that Df increases on I. Let $x_1 < x_2 < x_3$. By the mean value theorem

$$\frac{f(x_2) - f(x_1)}{x_2 - x_1} = f'(\xi); \qquad \frac{f(x_3) - f(x_2)}{x_3 - x_2} = f'(\eta)$$

where $x_1 < \xi < x_2 < \eta < x_3$. Since Df increases, $f'(\xi) \leq f'(\eta)$ and therefore inequality (2) holds. Thus f is convex on I.

(ii) Suppose next that f is convex on I. Let $x_1 < x_2 < x_3 < x_4$. By inequality (2),

$$\frac{f(x_2) - f(x_1)}{x_2 - x_1} \leq \frac{f(x_3) - f(x_2)}{x_3 - x_2} \leq \frac{f(x_4) - f(x_3)}{x_4 - x_3}.$$

Ignore the middle term in this inequality and let $x_2 \to x_1 +$ and $x_3 \to x_4 -$. We obtain $f'(x_1) \leq f'(x_4)$ and it follows that Df increases on I.

If f is twice differentiable on I, an even simpler characterisation of convexity results.

12.19 *Theorem* Let I be an open interval and suppose that f is twice differentiable on I. Then f is convex on I if and only if

$$f''(x) \geq 0$$

for all $x \in I$.

Proof The theorem follows from theorems 12.7 and 12.18.

12.20   Examples The functions f, g and h defined on $(0, \infty)$ by $f(x) = x^2$, $g(x) = 1/x$ and $h(x) = x + x^{-1}$ are all convex on $(0, \infty)$ because their second derivatives are all non-negative on $(0, \infty)$.

The function ϕ defined on $(0, \infty)$ by $\phi(x) = \sqrt{x}$ is concave on $(0, \infty)$. We have

$$\phi''(x) = -\tfrac{1}{4}x^{-3/2} \leqslant 0 \quad (x > 0).$$

12.21   Exercise

(1) Let $f: \mathbb{R} \to \mathbb{R}$ be defined by $f(x) = (x-1)(x-2)(x-3)$. Find the ranges of values of x for which f is (a) increasing, (b) decreasing, (c) convex, (d) concave. Draw a graph.

(2) Let f be convex on the interval I and let x_1, x_2 and x_3 be points of I which satisfy $x_1 < x_2 < x_3$. Prove that

$$\frac{f(x_2) - f(x_1)}{x_2 - x_1} \leqslant \frac{f(x_3) - f(x_1)}{x_3 - x_1} \leqslant \frac{f(x_3) - f(x_2)}{x_3 - x_2}.$$

(3) Let f be convex on the open interval I and let $J = f(I)$. If f is strictly increasing on I, show that f^{-1} is concave on J. What happens if f is strictly decreasing on I?

*(4) Suppose that f is convex and differentiable on the open interval I. If $\xi \in I$, prove that

$$f(x) - f(\xi) \geqslant f'(\xi)(x - \xi) \quad (x \in I).$$

Interpret this result geometrically. What result is obtained if f is concave on I?

Let $\phi: \mathbb{R} \to \mathbb{R}$ be a strictly increasing, convex, differentiable function on $\mathbb{R}$ and suppose that $\phi(\xi) = 0$. If $x_1 > \xi$ and

$$x_{n+1} = x_n - \frac{\phi(x_n)}{\phi'(x_n)} \quad (n = 1, 2, \ldots)$$

prove that $x_n \to \xi$ as $n \to \infty$. (This method of obtaining an approximation to the zero of ϕ is called the Newton–Raphson process.)

†(5) Let f be differentiable, convex and bounded on $\mathbb{R}$. Show that f is a constant.

†(6) Let f be continuous on an interval I and satisfy

$$f\left(\frac{x+y}{2}\right) \leqslant \frac{f(x) + f(y)}{2}$$

for each $x \in I$ and $y \in I$. Prove that, for any $x_1, x_2, \ldots, x_n$ in the interval I,

$$f\left(\frac{x_1 + x_2 + \ldots + x_n}{n}\right) \leqslant \frac{1}{n}\{f(x_1) + f(x_2) + \ldots + f(x_n)\}.$$

[Hint: recall the induction argument of example 3.10.] Deduce that f is convex on I. [Hint: establish inequality (1) of §12.13. Begin with the case when α and β are rational.]

13 INTEGRATION

13.1 Area

Suppose that f is continuous on the compact interval $[a, b]$.

Does it make sense to discuss the 'area' under the graph of f? And, if so, how can we compute its value?

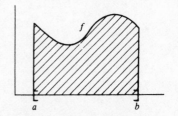

If it makes sense at all to talk about the area under the graph, then presumably this area must be at least as big as the areas S_1 and S_2 which have been shaded in the diagrams below.

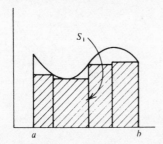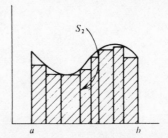

Let $\mathcal{S}$ denote the set of *all* numbers S which can be obtained as the sums of the areas of little rectangles as in the diagrams above. Then the 'area' under the graph of f must be at least as big as *every* element of $\mathcal{S}$. It seems reasonable to identify the 'area' under the graph of f with the smallest number larger than every element of $\mathcal{S}$, i.e. with sup $\mathcal{S}$.

These remarks are intended to motivate the formal mathematical definition given in the next section.

13.2   The integral

Suppose that f is continuous on the compact interval $[a, b]$. We propose to give a definition of the integral

$$\int_a^b f(x)\, dx$$

of f over the interval $[a, b]$.

We begin by defining a *partition P* of the interval $[a, b]$. This is a finite set $P = \{y_0, y_1, \ldots, y_n\}$ of real numbers with the property that

$$a = y_0 < y_1 < y_2 < \ldots < y_{n-1} < y_n = b.$$

Since f is continuous on $[a, b]$, it is bounded on $[a, b]$. It therefore makes sense, given a partition $P = \{y_0, y_1, \ldots, y_n\}$ of $[a, b]$, to define

$$m_1 = \inf_{y_0 \leqslant x \leqslant y_1} f(x)$$

$$m_2 = \inf_{y_1 \leqslant x \leqslant y_2} f(x)$$

and so on.

For each partition P, we can then form the sum

$$S(P) = \sum_{k=1}^n m_k(y_k - y_{k-1}).$$

The diagram below illustrates the case $n = 4$. The value of $S(P)$ is the area of the shaded region.

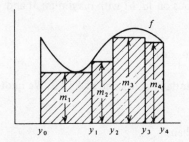

Suppose that f is bounded above on $[a, b]$ by H. Thus $f(t) \leqslant H$ for any $t \in [a, b]$. Then

$$S(P) = \sum_{k=1}^n m_k(y_k - y_{k-1})$$

$$\leqslant H \sum_{k=1}^n (y_k - y_{k-1}) = H\{(y_1 - y_0) + \ldots + (y_n - y_{n-1})\}$$

$$= H(y_n - y_0) = H(b - a).$$

It follows that the set of all numbers $S(P)$, where P is a partition of $[a, b]$, is bounded above by $H(b-a)$. Hence it has a *smallest* upper bound and we may define

$$\int_a^b f(x)\, dx = \sup_P S(P) \tag{1}$$

where the supremum extends over *all* partitions P of $[a, b]$.

We wish to think of

$$\int_a^b f(x)\, dx$$

as the 'area under the graph of f'. But notice that any area below the x-axis has to be counted as *negative* for this interpretation.

Note also that, if $b > a$, we define

$$\int_b^a f(x)\, dx = -\int_a^b f(x)\, dx.$$

13.3   Some properties of the integral

Much of the power of the idea of integration lies in its connexion with differentiation. For example, it would be very painful indeed if all integrals had to be worked out directly from the definition. Our first priority is therefore to prove the theorems which link integration with differentiation. Before we can do this, however, we need to derive a few basic results directly from the definition.

13.4   Proposition Let f be continuous on $[a, b]$ with maximum M and minimum m. Then

$$m(b-a) \le \int_a^b f(x)\, dx \le M(b-a).$$

The proof of this proposition is essentially contained in §13.2. We quote two simple corollaries.

13.5   Corollary If c is a constant, then

$$\int_a^b c\, dx = c(b-a).$$

13.6   Corollary Let f be continuous on $[a, b]$ and satisfy $|f(t)| < \kappa$ for any $t \in [a, b]$. Then, for any ξ and x in the interval $[a, b]$,

$$\left| \int_\xi^x f(t)\, dt \right| < \kappa |x - \xi| \quad (x \ne \xi).$$

The next proposition is also very simple, provided that one recalls the properties of sup and inf. (See exercise 2.13(2) and exercise 7.16(6i).)

13.7 *Proposition* Let f be continuous on $[a, b]$ and let c be a constant. Then

$$\int_a^b \{f(t) + c\}\, dt = \int_a^b f(t)\, dt + c(b - a).$$

The next and final result of this section is not quite so easy and so we give the proof.

13.8 *Theorem* Let f be continuous on $[a, b]$ and let $a \leqslant c \leqslant b$. Then

$$\int_a^b f(t)\, dt = \int_a^c f(t)\, dt + \int_c^b f(t)\, dt.$$

Proof We use the notation of §13.2. Suppose, in the first place, that P_1 and P_2 are *any* partitions of $[a, c]$ and $[c, b]$ respectively. Then the set P of points in at least one of the sets P_1 and P_2 is a partition of $[a, b]$.

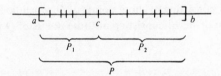

It is obvious that $S(P) = S(P_1) + S(P_2)$. Moreover,

$$S(P) \leqslant \int_a^b f(t)\, dt$$

because of definition (1) of §13.2. Thus, given any partition P_1 of $[a, c]$ and any partition P_2 of $[c, b]$,

$$S(P_1) + S(P_2) \leqslant \int_a^b f(t)\, dt$$

$$S(P_1) \leqslant \int_a^b f(t)\, dt - S(P_2). \tag{2}$$

Hence, given any partition P_2 of $[c, b]$, the right hand side of (2) is an upper bound for the set of all numbers of the form $S(P_1)$ where P_1 is a partition of $[a, c]$. Therefore

$$\sup_{P_1} S(P_1) \leqslant \int_a^b f(t)\, dt - S(P_2)$$

where the supremum extends over all partitions P_1 of $[a, c]$. Recalling the definition of an integral, we obtain

$$\int_a^c f(t)\,dt \leqslant \int_a^b f(t)\,dt - S(P_2)$$

and so

$$S(P_2) \leqslant \int_a^b f(t)\,dt - \int_a^c f(t)\,dt.$$

A similar argument now yields

$$\int_c^b f(t)\,dt \leqslant \int_a^b f(t)\,dt - \int_a^c f(t)\,dt$$

and therefore

$$\int_a^b f(t)\,dt \geqslant \int_a^c f(t)\,dt + \int_a^b f(t)\,dt. \tag{3}$$

Now let P be *any* partition of $[a, b]$ and let Q be the partition obtained from P by inserting the extra point c (if it does not already belong to P). It is easily seen that

$$S(P) \leqslant S(Q).$$

Let P_1 be the partition of $[a, b]$ consisting of those points of Q which lie in $[a, c]$ and let P_2 be the partition of $[c, b]$ consisting of those points of Q which lie in $[c, b]$.

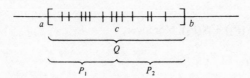

We have

$$S(P) \leqslant S(Q) = S(P_1) + S(P_2)$$

$$\leqslant \int_a^c f(t)\,dt + \int_c^b f(t)\,dt.$$

The right hand side of this inequality is therefore an upper bound for the set of all numbers $S(P)$ where P is a partition of $[a, b]$. Hence

$$\sup_P S(P) \leqslant \int_a^c f(t)\,dt + \int_c^b f(t)\,dt$$

i.e. $$\int_a^b f(t)\,dt \leqslant \int_a^c f(t)\,dt + \int_c^b f(t)\,dt. \tag{4}$$

Combining (3) and (4), we obtain the conclusion of the theorem.

13.9   Differentiation and integration

Let f be defined on (a, b). Suppose that F is continuous on $[a, b]$ and differentiable on (a, b) and satisfies

$$F'(x) = f(x)$$

for each $x \in (a, b)$. Then we shall say that F is a *primitive* (or anti-derivative) of f on $[a, b]$.

13.10   Example If $f(x) = x^3$, then the function F defined by $F(x) = \frac{1}{4}x^4 + 3$ is a primitive for f.

13.11   Theorem Let F be a primitive for f on $[a, b]$ and let G be defined on $[a, b]$. Then G is a primitive for f on $[a, b]$ if and only if, for some constant c,

$$G(x) = F(x) + c$$

for each $x \in [a, b]$.

Proof Obviously $F(x) + c$ is a primitive for f on $[a, b]$. Suppose therefore that G is a primitive for f on $[a, b]$. Then $F - G$ is continuous on $[a, b]$, differentiable on (a, b) and, for each $x \in (a, b)$,

$$D\{F(x) - G(x)\} = F'(x) - G'(x) = f(x) - f(x) = 0.$$

From theorem 11.7 it follows that $F - G$ is constant on $[a, b]$. Hence the result.

What have primitives to do with integration? Suppose that f is continuous on $[a, b]$. Then we may define a function F on $[a, b]$ by

$$F(x) = \int_a^x f(t)\, dt \quad (a \leqslant x \leqslant b).$$

In the next theorem, we shall show that F is a primitive of f on $[a, b]$. But first we give a rough, intuitive argument to indicate why one might suppose this to be true in the first place. Given $x \in (a, b)$, we seek to explain why one might suppose that $F'(x) = f(x)$.

The number $F(x)$ may be interpreted geometrically as the 'area under the graph of f between a and x'. The shaded region in the diagram below should therefore have area $F(x + h) - F(x)$.

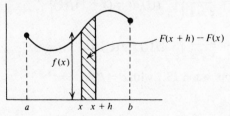

Since f is continuous, one feels that, if h is very small, the shaded region will 'nearly' be a rectangle and hence its area will be approximately $hf(x)$. Thus

$$F(x + h) - F(x) \doteq hf(x)$$

and

$$\frac{F(x + h) - F(x)}{h} \doteq f(x). \tag{5}$$

This equation is only approximate, but the smaller h becomes the better the approximation should be. Allowing $h \to 0$, we might therefore expect that $F'(x) = f(x)$.

This argument is dubious on a number of counts – particularly at the stage where we divide by the small number h to obtain (5). In the next theorem we provide a precise proof of the result.

13.12   Theorem Suppose that f is continuous on $[a, b]$ and that F is defined on $[a, b]$ by

$$F(x) = \int_a^x f(t) \, dt \quad (a \leqslant x \leqslant b).$$

Then F is a primitive of f on $[a, b]$, i.e. F is continuous on $[a, b]$, differentiable on (a, b) and $F'(x) = f(x)$ for each $x \in (a, b)$.

Proof Since f is continuous on $[a, b]$, it is bounded on $[a, b]$. Suppose that $|f(t)| < \kappa \ (a \leqslant t \leqslant b)$.

By theorem 13.8, for any x and ξ in $[a, b]$,

$$F(x) - F(\xi) = \int_\xi^x f(t) \, dt$$

and therefore, by corollary 13.6,

$$|F(x) - F(\xi)| < \kappa |x - \xi|.$$

And, from this inequality, it follows that F is continuous on $[a, b]$.

A more subtle argument is needed to show that F is differentiable on (a, b). If x and ξ are in (a, b) and $x \neq \xi$, then

$$\frac{F(x) - F(\xi)}{x - \xi} - f(\xi) = \frac{1}{x - \xi} \{F(x) - F(\xi) - (x - \xi)f(\xi)\}$$

$$= \frac{1}{x - \xi} \left\{ \int_\xi^x f(t) \, dt - (x - \xi)f(\xi) \right\}$$

$$= \frac{1}{x - \xi} \int_\xi^x \{f(t) - f(\xi)\} \, dt.$$

The last step is justified by proposition 13.7 with $c = f(\xi)$.

Let $\epsilon > 0$ be given. If $\xi \in (a, b)$, then f is continuous at the point ξ and so, for some $\delta > 0$,

$$|f(t) - f(\xi)| < \epsilon$$

provided that $|t - \xi| < \delta$. Hence, provided that $|x - \xi| < \delta$, it follows that $|f(t) - f(\xi)| < \epsilon$ for every value of t in the compact interval with endpoints ξ and x. From corollary 13.6, we conclude that

$$\left| \frac{F(x) - F(\xi)}{x - \xi} - f(\xi) \right| = \frac{1}{|x - \xi|} \left| \int_{\xi}^{x} \{f(t) - f(\xi)\} \, dt \right|$$

$$< \frac{1}{|x - \xi|} \cdot \epsilon |x - \xi|$$

$$= \epsilon$$

provided that $0 < |x - \xi| < \delta$. But this is what is meant by the assertion

$$\frac{F(x) - F(\xi)}{x - \xi} \to f(\xi) \text{ as } x \to \xi.$$

We have therefore shown that F is differentiable on (a, b) and $F'(\xi) = f(\xi)$ for each $\xi \in (a, b)$.

13.13   Example

$$D \left\{ \int_{0}^{x} \frac{dt}{1 + t^{100}} \right\} = \frac{1}{1 + x^{100}}.$$

13.14   Theorem Any function f which is continuous on $[a, b]$ has a primitive on $[a, b]$. If G is any primitive of f, then

$$\int_{a}^{b} f(t) \, dt = G(b) - G(a) = [G(t)]_{a}^{b}.$$

Proof Let F be defined on $[a, b]$ by

$$F(x) = \int_{a}^{x} f(t) \, dt \quad (a \leqslant x \leqslant b).$$

Then

$$\int_{a}^{b} f(t) \, dt = F(b) = F(b) - F(a).$$

By theorem 13.12, F is a primitive of f on $[a, b]$ and, by theorem 13.11, any other primitive G satisfies

$$G(x) = F(x) + c \quad (a \leqslant x \leqslant b)$$

for some constant c. Hence

$$\int_a^b f(t)\,dt = \{F(b) + c\} - \{F(a) + c\}$$

$$= G(b) - G(a).$$

13.15   *Example* Since

$$D\left\{-\frac{1}{2(1+t^2)}\right\} = \frac{t}{(1+t^2)^2},$$

it follows from theorem 13.14 that

$$\int_0^1 \frac{t}{(1+t^2)^2}\,dt = \left[-\frac{1}{2(1+t^2)}\right]_0^1 = -\frac{1}{4} + \frac{1}{2} = \frac{1}{4}.$$

13.16   Riemann integral

The integral

$$\int_a^b f(x)\,dx \tag{6}$$

was defined as the *supremum* of all areas like that shaded in the left hand diagram below. It would be equally sensible to define the integral as the *infimum* of all areas like that shaded in the right hand diagram below.

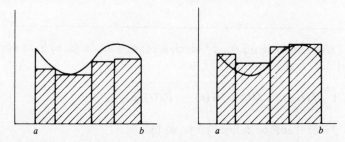

If f is continuous on $[a, b]$, these two definitions yield the same result. The reason is simple. The use of the second definition would yield the result

$$-\int_a^b -f(x)\,dx. \tag{7}$$

But (6) and (7) are equal because F is a primitive for f on $[a, b]$ if and only if $-F$ is a primitive for $-f$.

If f is *not* continuous on $[a, b]$, the two definitions do not necessarily yield the same result. But, *if* they do yield the same result, we say that f is *Riemann integrable* and call the common number obtained from the two definitions the *Riemann integral* of f on $[a, b]$.

13.17 Example We know that

$$\int_0^1 x^2 \, dx = [\tfrac{1}{3}x^3]_0^1 = \tfrac{1}{3}.$$

To illustrate the idea discussed in §13.16, we evaluate the integral without using the theory of differentiation.

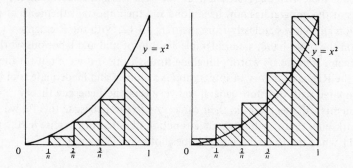

Introduce the partition $P_n = \{0, 1/n, 2/n, 3/n, \ldots, 1\}$ of $[0, 1]$. The shaded area of the left hand diagram is smaller than the integral and the shaded area of the right hand diagram is larger. Thus

$$\sum_{k=1}^n \left(\frac{k-1}{n}\right)^2 \frac{1}{n} \leqslant \int_0^1 x^2 \, dx \leqslant \sum_{k=1}^n \left(\frac{k}{n}\right)^2 \frac{1}{n}$$

$$\frac{1}{n^3}\{0^2 + 1^2 + \ldots + (n-1)^2\} \leqslant \int_0^1 x^2 \, dx \leqslant \frac{1}{n^3}\{1^2 + 2^2 + \ldots + n^2\}$$

$$\frac{1}{n^3} \cdot \frac{1}{6}(n-1)n(2n-1) \leqslant \int_0^1 x^2 \, dx \leqslant \frac{1}{n^3} \cdot \frac{1}{6}n(n+1)(2n+1)$$

(exercise 3.11(1i))

$$\frac{1}{3}\left(1 - \frac{1}{n}\right)\left(1 - \frac{1}{2n}\right) \leqslant \int_0^1 x^2 \, dx \leqslant \frac{1}{3}\left(1 + \frac{1}{n}\right)\left(1 + \frac{1}{2n}\right).$$

Considering the limit as $n \to \infty$ yields

$$\int_0^1 x^2 \, dx = \tfrac{1}{3}.$$

It is important to note that the definition of an integral given in §13.2 is valid *only* for functions f which are continuous on $[a, b]$ while the definition of the Riemann integral given above is valid for a wider class of functions. It is easily proved, for example, that any function which is bounded and has only a finite number of discontinuities on $[a, b]$ is Riemann integrable on $[a, b]$. Even easier is to prove that a function f which is monotone on $[a, b]$ is Riemann integrable although such a function may have an infinite number of discontinuities. (Proofs?) Moreover, the results of §13.3 (notably theorem 13.8) remain true for Riemann integrable functions and hence the vital theorem 13.12 is also valid for Riemann integrable functions provided that it is only claimed that $F'(x) = f(x)$ for those values of $x \in (a, b)$ at which f is continuous. Why then have we chosen to restrict our definition of the integral to functions f which are continuous on $[a, b]$ rather than to use the more general Riemann integral? The reason is that, with our restriction, all the hard work in establishing the properties of the integral has now been done and the important theorems of the next section all follow painlessly from theorem 13.12. With the Riemann integral, on the other hand, a considerable amount of dull and laborious work would remain. It might be worth ploughing through this if it were not for the fact that the Riemann theory of integration is a clumsy and inadequate tool when compared with the more general and very versatile Lebesgue theory which students will meet later in their career. As to the Lebesgue theory, we shall remark only that it will integrate a function like that of exercise 8.20(5) over $[0, 1]$ while the Riemann integral does not exist. (Why not?)

13.18   Exercise

(1) Prove that, given any natural number n,

$$\int_a^b x^n \, dx = \left[\frac{1}{n+1} x^{n+1} \right]_a^b$$

Show that the same result holds if n is replaced by any rational number $\alpha \neq -1$ provided that $0 < a < b$.

(2) Since the integrand is a square, the following result must be wrong. Explain why.

$$\int_0^2 \frac{dx}{(x-1)^2} = \left[-\frac{1}{x-1} \right]_0^2 = -1 - 1 = -2.$$

(3) If α is a positive rational number, prove that

$$\lim_{n \to \infty} \frac{1}{n^{\alpha+1}} \{ 1^\alpha + 2^\alpha + 3^\alpha + \ldots + n^\alpha \} = \frac{1}{\alpha+1}.$$

13.19   More properties of the integral

As mentioned above, the remaining important properties of the integral of a continuous function now follow painlessly from the theorems of §13.9 which connect integration and differentiation.

13.20 Theorem Suppose that f and g are continuous on $[a, b]$ and that λ and μ are any real numbers. Then

$$\int_a^b \{\lambda f(t) + \mu g(t)\}\, dt = \lambda \int_a^b f(t)\, dt + \mu \int_a^b g(t)\, dt.$$

Proof Let F and G be primitives of f and g respectively on $[a, b]$. By theorem 10.9(i) the function $H = \lambda F + \mu G$ is a primitive of $\lambda f + \mu g$ on $[a, b]$. Hence, by theorem 13.14,

$$\int_a^b \{\lambda f(t) + \mu g(t)\}\, dt = [\lambda F(t) + \mu G(t)]_a^b$$

$$= \lambda [F(t)]_a^b + \mu [G(t)]_a^b$$

$$= \lambda \int_a^b f(t)\, dt + \mu \int_a^b g(t)\, dt.$$

13.21 Theorem (integration by parts) Suppose that f and g are continuous on $[a, b]$ and have primitives F and G respectively on $[a, b]$. Then

$$\int_a^b f(t)G(t)\, dt = [F(t)G(t)]_a^b - \int_a^b F(t)g(t)\, dt.$$

Proof By theorem 10.9(ii),

$$D(FG) = fG + Fg$$

and thus FG is a primitive of $fG + Fg$ on $[a, b]$. Hence

$$\int_a^b \{f(t)G(t) + F(t)g(t)\}\, dt = [F(t)G(t)]_a^b$$

and the theorem follows.

As is clear from its proof, the formula for integrating by parts is just the integral analogue of the formula for differentiating a product. The next result is the integral analogue of the formula for differentiating a composite function, i.e. a 'function of a function'. It is the rule for changing the variable in an integral and is usually remembered in the form

$$dt = \frac{dt}{du} \cdot du.$$

13.22 Theorem Suppose that ϕ has a derivative which is continuous on $[a, b]$ and that f is continuous on an open interval which contains the image of $[a, b]$ under ϕ. Then

$$\int_{\phi(a)}^{\phi(b)} f(t)\, dt = \int_a^b f(\phi(u))\phi'(u)\, du.$$

Proof Let

$$F(x) = \int_{\phi(a)}^{x} f(t)\, dt.$$

From theorem 10.13 and theorem 13.12,

$$\frac{d}{du} F(\phi(u)) = F'(\phi(u))\phi'(u) = f(\phi(u))\phi'(u).$$

It follows from theorem 13.14 that

$$\int_{a}^{b} f(\phi(u))\phi'(u)\, du = [F(\phi(u))]_{a}^{b}$$

and this is what had to be proved.

13.23   Theorem Suppose that f and g are continuous on $[a, b]$ and that, for any $t \in [a, b]$, $f(t) \le g(t)$. Then

$$\int_{a}^{b} f(t)\, dt \le \int_{a}^{b} g(t)\, dt.$$

Proof Let H be a primitive of $g - f$ on $[a, b]$. Then $DH(t) = g(t) - f(t) \ge 0 \ (a < t < b)$. By theorem 12.7, H therefore increases on $[a, b]$. Thus $H(b) \ge H(a)$ and hence

$$\int_{a}^{b} \{g(t) - f(t)\}\, dt = H(b) - H(a) \ge 0.$$

13.24   Theorem Suppose that f is continuous on $[a, b]$. Then

$$\left| \int_{a}^{b} f(t)\, dt \right| \le \int_{a}^{b} |f(t)|\, dt.$$

Proof Since $- |f(t)| \le f(t) \le |f(t)| \ (a \le t \le b)$, it follows that

$$-\int_{a}^{b} |f(t)|\, dt \le \int_{a}^{b} f(t)\, dt \le \int_{a}^{b} |f(t)|\, dt$$

and hence the result. (Note that $|f| = \{f^2\}^{1/2}$ and is therefore continuous.)

13.25   Theorem *(Cauchy–Schwarz inequality)* Suppose that f and g are continuous on $[a, b]$. Then

$$\left\{ \int_{a}^{b} f(t)g(t)\, dt \right\}^2 \le \int_{a}^{b} \{f(t)\}^2\, dt \int_{a}^{b} \{g(t)\}^2\, dt.$$

Proof This is much the same as that of example 1.11. For any x,

$$0 \le \int_{a}^{b} \{xf(t) + g(t)\}^2\, dt$$

$$= x^2 \int_a^b \{f(t)\}^2 \, dt + 2x \int_a^b f(t)g(t) \, dt + \int_a^b \{g(t)\}^2 \, dt$$

$$= Ax^2 + 2Bx + C.$$

Thus the quadratic equation $Ax^2 + 2Bx + C = 0$ cannot have two (distinct) roots. Hence

$$B^2 \leqslant AC$$

as required.

13.26 Exercise

(1) Suppose that g is continuous on $[a, b]$ and that $g(t) \geqslant 0 \; (a \leqslant t \leqslant b)$. Prove that

$$\int_a^b g(t) \, dt = 0$$

if and only if $g(t) = 0$ for each $t \in [a, b]$.

(2) Suppose that f is twice differentiable on $[a, b]$ and that f'' is continuous on $[a, b]$. Prove that

$$\int_a^b x f''(x) \, dx = \{bf'(b) - f(b)\} - \{af'(a) - f(a)\}.$$

*(3) Let f be positive and continuous on $[1, \infty)$. Suppose that

$$F(x) = \int_1^x f(t) \, dt \leqslant \{f(x)\}^2 \quad (x \geqslant 1).$$

Prove that $f(x) \geqslant \frac{1}{2}(x - 1) \quad (x \geqslant 1)$. [Hint: The integral $\int_1^x \{F(t)\}^{-1/2} F'(t) \, dt$ is relevant.]

*(4) Suppose that f and g are continuous on $[a, b]$ and that $g(t) \geqslant 0$ $(a \leqslant t \leqslant b)$. Prove that

$$\int_a^b f(t)g(t) \, dt = f(\xi) \int_a^b g(t) \, dt$$

for some $\xi \in [a, b]$. [Hint: continuity property.] What has this to do with theorem 11.6?

*(5) Suppose that g is continuous on $[a, b]$ and that f is differentiable on $[a, b]$ and its derivative is continuous on $[a, b]$ with $f'(t) \geqslant 0$ $(a \leqslant t \leqslant b)$. Prove that

$$\int_a^b f(t)g(t) \, dt = f(a) \int_a^\xi g(t) \, dt + f(b) \int_\xi^b g(t) \, dt$$

for some $\xi \in [a, b]$. What happens if $f'(t) \leqslant 0$ $(a \leqslant t \leqslant b)$? [Hint: integrate by parts.]

*(6) By integrating many times by parts, show that the error term in Taylor's theorem (theorem 11.10) can be expressed in the form

$$E_n = \frac{1}{(n-1)!} \int_\xi^x (x-t)^{n-1} f^{(n)}(t)\, dt$$

provided that $f^{(n)}$ is continuous on I. Show how the form of the error term given in theorem 11.10 can be obtained from that above.

13.27   Improper integrals

We write down some sample definitions. Suppose that f is continuous on $[0, \infty)$. Then we define

$$\int_0^{\to\infty} f(x)\, dx = \lim_{X \to \infty} \int_0^X f(x)\, dx$$

provided that the limit exists. Similarly, if f is continuous on $(a, b]$, then we define

$$\int_{\to a}^b f(x)\, dx = \lim_{y \to a+} \int_y^b f(x)\, dx$$

provided that the limit exists. Note that, if f is continuous everywhere, we define

$$\int_{\to -\infty}^{\to\infty} f(x)\, dx = \int_{\to -\infty}^0 f(x)\, dx + \int_0^{\to\infty} f(x)\, dx$$

provided that the appropriate limits exist, i.e. we allow the upper and lower limits to recede to $+\infty$ and $-\infty$ respectively *independently*. We do *not* define the improper integral by

$$\lim_{X \to \infty} \int_{-X}^X f(x)\, dx.$$

13.28   *Examples*

(i) $\displaystyle\int_1^{\to\infty} \frac{dx}{x^2} = \lim_{X \to \infty} \int_1^X \frac{dx}{x^2} = \lim_{X \to \infty} \left[-\frac{1}{x} \right]_1^X = \lim_{X \to \infty} \left(1 - \frac{1}{X} \right) = 1.$

On the other hand, if $0 < y < 1$,

$$\int_y^1 \frac{dx}{x^2} = \left[-\frac{1}{x} \right]_y^1 = \frac{1}{y} - 1 \to +\infty \text{ as } y \to 0+$$

and hence the improper integral $\displaystyle\int_{\to 0}^1 \frac{dx}{x^2}$ does not exist.

(ii) $\int_{\to 0}^{1} \frac{dx}{\sqrt{x}} = \lim_{y \to 0+} \int_{y}^{1} \frac{dx}{\sqrt{x}} = \lim_{y \to 0+} [2x^{1/2}]_{y}^{1} = \lim_{y \to 0+} (2 - 2\sqrt{y}) = 2.$

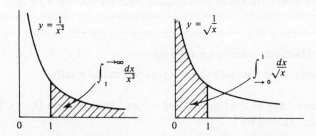

(iii) Note finally the following example. Observe that $\int_{0}^{\to \infty} x \, dx$ does not exist (why not?). Hence $\int_{\to -\infty}^{\to \infty} x \, dx$ cannot exist even though

$$\int_{-X}^{X} x \, dx = \tfrac{1}{2}X^{2} - \tfrac{1}{2}X^{2} \to 0 \text{ as } X \to +\infty.$$

The integrals defined in this section are called *improper* because they are not given by the definition of §13.2. It is perhaps pedantic to include the little arrows in the notation for improper integrals and most authors would simply delete them. But for those who propose to study integration more deeply, it is probably better not to do without the little arrows since this will avoid confusion with the differing definitions of the powerful Lebesque theory of integration.

13.29 *Proposition* Suppose that ϕ is a function which is continuous and non-negative on the open interval I. Suppose also that f is continuous on I and satisfies

$$|f(x)| \leqslant \phi(x) \quad (x \in I).$$

If the improper integral of ϕ over the interval I exists, then so does that of f.

This is, of course, an analogue of the comparison test (theorem 6.15), for series and may be proved in much the same way. If f happens to be non-negative on I a simpler proof may be based on theorem 12.4 concerning the convergence of monotone functions.

13.30 *Example* To prove the existence of the integral

$$\int_{1}^{\to \infty} \frac{dx}{1 + x^{2}}$$

we simply have to observe that, for any $x \geqslant 1$,

$$\frac{1}{1+x^2} \leqslant \frac{1}{x^2}.$$

Since $\int_1^{\to \infty} x^{-2} dx$ exists, the result then follows from proposition 13.29.

13.31   Euler–Maclaurin summation formula

We discuss only a simplified version of this useful result.

13.32   Theorem Let f be continuous, positive and decreasing on $[1, \infty)$. Then the sequence $\langle \Delta_n \rangle$ defined by

$$\Delta_n = \sum_{k=1}^{n} f(k) - \int_1^n f(x) \, dx$$

is decreasing and bounded below by zero. Hence it converges.

Proof Since f decreases, for any $k = 1, 2, \ldots,$

$$f(k+1) \leqslant \int_k^{k+1} f(x) \, dx \leqslant f(k)$$

Thus

$$\Delta_{n+1} - \Delta_n = \left\{ \sum_{k=1}^{n+1} f(k) - \int_1^{n+1} f(x) \, dx \right\} - \left\{ \sum_{k=1}^{n} f(k) - \int_1^{n} f(x) \, dx \right\}$$

$$= f(n+1) - \int_n^{n+1} f(x) \, dx \leqslant f(n+1) - f(n+1) = 0$$

and so $\langle \Delta_n \rangle$ decreases. Also

$$\Delta_n = \sum_{k=1}^{n} f(k) - \sum_{k=1}^{n-1} \int_k^{k+1} f(x) \, dx$$

$$\geqslant \sum_{k=1}^{n} f(k) - \sum_{k=1}^{n-1} f(k) = f(n) > 0$$

and so $\langle \Delta_n \rangle$ is bounded below.

From theorem 13.32 it follows that, if f is positive, continuous and decreasing on $[1, \infty)$, then the series and the improper integral

$$\sum_{n=1}^{\infty} f(n) \text{ and } \int_1^{\to \infty} f(x) \, dx$$

either both converge or else both diverge. The theorem therefore provides a criterion for the convergence of a series or an improper integral and for this reason is often referred to as the *integral test*.

13.33 *Example* Take $f(x) = x^{-1}$ in theorem 13.32. We know from theorem 6.5, that the series $\sum_{n=1}^{\infty} n^{-1}$ diverges to $+\infty$. From theorem 13.32, we conclude that

$$\int_1^n \frac{dx}{x} \to +\infty \text{ as } n \to \infty.$$

In particular, the improper integral $\int_1^{\to\infty} x^{-1}dx$ does not exist.

13.34 *Exercise*

(1) Discuss the existence of the following improper integrals.

(i) $\displaystyle\int_0^{\to 1} \frac{dx}{\sqrt{(1-x)}}$ (ii) $\displaystyle\int_0^{\to\infty} \frac{x^2\,dx}{(1+x^3)^2}$ (iii) $\displaystyle\int_0^{\to\infty} \frac{1+x+x^3}{1+x+x^5}\,dx$

(iv) $\displaystyle\int_0^{\to 1} \frac{dx}{(1-x)^{3/2}}$ (v) $\displaystyle\int_0^{\to 1} \frac{dx}{(1-x)}$ (vi) $\displaystyle\int_{\to 0}^1 \frac{dx}{x}$.

[Hint: for (v) and (vi) recall example 13.33.]

(2) Prove that, for $0 < y < \frac{1}{2}$,

$$\int_{(1/2)-y}^{(1/2)+y} \frac{2x-1}{x(1-x)}\,dx = 0.$$

Discuss the existence of the improper integral $\displaystyle\int_{\to 0}^{\to 1} \frac{2x-1}{x(x-1)}\,dx$.

(3) Let α be a rational number with $\alpha > 1$. In theorem 6.6 it was shown that the series

$$\sum_{n=1}^{\infty} \frac{1}{n^\alpha}$$

converges. Base another proof of this result on theorem 13.32.

---

14 EXPONENTIAL AND LOGARITHM

14.1 Logarithm

We define the *logarithm* for each $x > 0$ by

$$\log x = \int_1^x \frac{dt}{t}.$$

It follows immediately from the definition that $\log 1 = 0$ and that, for each $x > 0$,

$$D \log x = \frac{1}{x}.$$

Also,

$$D^2 \log x = -\frac{1}{x^2}.$$

We conclude that the logarithm is strictly increasing and concave on $(0, \infty)$. Further, in view of example 13.33 and exercise 13.34 (1vi),

$$\log x \to +\infty \text{ as } x \to +\infty$$

and

$$\log x \to -\infty \text{ as } x \to 0+.$$

(For an alternative proof of these results, see exercise 14.3(1) below.) With this information the graph $y = \log x$ can easily be drawn.

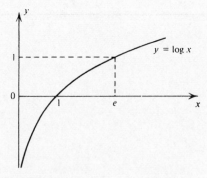

The number e is defined by the equation

$$\log e = 1.$$

The value of e is approximately $2 \cdot 718$ (see exercise 15.6(4)).

14.2 *Theorem* Suppose that $x > 0, y > 0$ and r is rational. Then
  (i) $\log xy = \log x + \log y$
  (ii) $\log (x^r) = r \log x$.

  Proof (i) For a fixed value of $y > 0$, consider the function

$$f(x) = \log xy - \log x.$$

We have

$$f'(x) = \frac{1}{xy} \cdot y - \frac{1}{x} = \frac{1}{x} - \frac{1}{x} = 0$$

for each $x > 0$. Thus f is constant (theorem 11.7) and

$$\log xy - \log x = c \quad (x > 0).$$

To obtain the value of c, put $x = 1$. Then

$$c = \log y - \log 1 = \log y.$$

  (ii) Consider the function

$$f(x) = \log (x^r) - r \log x.$$

We have

$$f'(x) = \frac{1}{x^r} \cdot rx^{r-1} - \frac{r}{x} = 0.$$

Hence f is a constant and

$$\log (x^r) - r \log x = c.$$

To obtain the value of c, put $x = 1$. Then

$$c = \log 1 - r \log 1 = 0.$$

Note The logarithm we are discussing here is the '*natural logarithm*'. This is sometimes stressed by writing 'ln' rather than 'log'. In pure mathematics, little occasion arises for the use of 'logarithms to the base 10' as found in the familiar logarithmic tables.

14.3 *Exercise*
  (1) Prove that $\log 2 > 0$. By considering $\log 2^n$ and $\log 2^{-n}$, show that $\log x \to +\infty$ as $x \to +\infty$ and that $\log x \to -\infty$ as $x \to 0+$.
  *(2) Prove that, for each $y > 0$, $\log y \leqslant y - 1$. If s is a positive rational number and $x > 1$, deduce that

$$\log x \leqslant \frac{x^s}{s}.$$

Given a positive rational number r, prove that

(i) $x^{-r} \log x \to 0$ as $x \to +\infty$

(ii) $y^r \log y \to 0$ as $y \to 0 +$.

(These results are usually remembered by saying 'powers drown logarithms'.) [Hint: take $0 < s < r$ when proving (i).]

*(3) Show that $F(x) = x \log x - x$ is a primitive for $\log x$ on $(0, \infty)$. Deduce that

$$\int_0^1 \log (1 + x)\, dx = 2 \log 2 - 1.$$

Hence show that

$$\log \frac{1}{n} \left\{ \frac{(2n)!}{n!} \right\}^{1/n} \to \log \frac{4}{e} \text{ as } n \to \infty.$$

[Hint: recall example 13.17.]

*(4) Show that the equation $\log x = \alpha x$ has solutions if and only if $\alpha e \leqslant 1$. Show that

$$e \log x \leqslant x \quad (x > 0)$$

with equality if and only if $x = e$.

Suppose that $x_1 > e$ and $x_{n+1} = e \log x_n$ $(n = 1, 2, \ldots)$. Prove that $x_n \to e$ as $n \to \infty$.

*(5) Prove the existence of a real number γ (Euler's constant) with the property that

$$1 + \frac{1}{2} + \frac{1}{3} + \ldots + \frac{1}{n} - \log n \to \gamma \text{ as } n \to \infty.$$

[Hint: theorem 13.32.] Hence show that

$$\sum_{n=1}^{\infty} \frac{(-1)^{n-1}}{n} = 1 - \frac{1}{2} + \frac{1}{3} - \frac{1}{4} + \ldots = \log 2.$$

(6) Evaluate

(i) $D \{(\log x)^s\}$ $(x > 0$ and s rational$)$

(ii) $D \{\log \log x\}$ $(x > 1)$.

Hence show that the series

$$\sum_{n=1}^{\infty} \frac{1}{n(\log n)^r}$$

converges if $r > 1$ and diverges if $r \leqslant 1$. (See theorems 6.5 and 6.6 for comparison.) [Hint: theorem 13.32 again.]

14.4   Exponential

From theorem 12.10 we may deduce the existence of an inverse function to the logarithm. We call this inverse function the exponential function and write

$$y = \exp x \text{ if and only if } x = \log y.$$

Since the logarithm function has domain $(0, \infty)$ and range $\mathbb{R}$, the exponential function has domain $\mathbb{R}$ and range $(0, \infty)$.

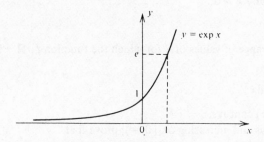

Observe that the exponential function is strictly increasing on $\mathbb{R}$ and that

$$\exp x \to +\infty \text{ as } x \to +\infty$$

$$\exp x \to 0 \text{ as } x \to -\infty.$$

Also

$$D \exp x = \frac{1}{D \log y} = \frac{1}{1/y} = y = \exp x.$$

Thus the exponential function is its own derivative.

14.5   *Exercise*

(1) If x and y are real numbers and r is rational, prove that
   (i) $\exp(x + y) = (\exp x)(\exp y)$
   (ii) $\exp(rx) = (\exp x)^r$.
   [Hint: These may be deduced from theorem 14.2. Alternatively, they may be proved directly. In the case of (i), for example, by differentiating

$$\exp(x + y)/(\exp x)$$

with respect to x.]

(2) Prove that, for any rational number r,
   (i) $x^{-r} \exp x \to +\infty$ as $x \to +\infty$
   (ii) $x^r \exp x \to 0$ as $x \to -\infty$.
   (Roughly speaking, 'exponentials drown powers'.)

(3) Use L'Hôpital's rule (exercise 11.8(3)) to prove that

(i) $\lim_{x \to 0} \dfrac{\exp x - 1}{x} = 1$ (ii) $\lim_{x \to 0} \dfrac{\log(1 + x)}{x} = 1$.

(4) Suppose that h is positive and continuous on $[0, \infty)$ and that $h(x) \geqslant H(x)(x \geqslant 0)$, where

$$H(x) = 1 + \int_0^x h(t)\,dt.$$

Prove that, for any $x > 0$,

$$h(x) \geqslant \exp x.$$

(5) Determine the range of values of x for which the function $f: \mathbb{R} \to \mathbb{R}$ defined by

$$f(x) = \exp\{18x^3\}$$

is (a) convex, (b) concave.

(6) If f is continuous and increasing on $[0, \infty)$, prove that

$$\int_0^n f(x)\,dx \leqslant \sum_{k=1}^n f(k) \leqslant \int_1^{n+1} f(x)\,dx.$$

Deduce that $n \log n - n \leqslant \log n! \leqslant (n + 1) \log(n + 1) - n$ and hence show that

$$\frac{n^n}{n!} \leqslant \exp n \leqslant \frac{(n + 1)^{n+1}}{n!}.$$

14.6   Powers

If $a > 0$ and r is rational, then we have defined the expression a^r. We propose to extend this definition to include non-rational exponents.

If $a > 0$ and x is real, we define

$$a^x = \exp(x \log a).$$

But does this definition agree with our old definition in the case when x is rational? It follows from theorem 14.2(ii) that, if r is rational, then

$$\exp\{r \log a\} = \exp\{\log(a^r)\} = a^r.$$

Recall that e is defined by $\log e = 1$. Equivalently, $e = \exp 1$. Observe that

$$e^x = \exp\{x \log e\} = \exp x$$

which explains the familiar notation for the exponential function.

14.7   Exercise

(1) If a and b are positive, prove the following.

  (i) $a^{x+y} = a^x a^y$ (ii) $(ab)^x = a^x b^x$

  (iii) $a^{-x} = \dfrac{1}{a^x}$ (iv) $(a^x)^y = a^{xy}$.

(2) Prove that the derivative of a^x with respect to x is $(\log a)a^x$.

(3) Write

$$\left(1 + \frac{x}{n}\right)^n = \exp\left\{n \log\left(1 + \frac{x}{n}\right)\right\}$$

and hence show that

$$\lim_{n \to \infty} \left(1 + \frac{x}{n}\right)^n = e^x$$

(see example 4.19). [Hint: theorem 8.9 and exercise 14.5(3ii).]

(4) Write

$$n^{1/n} = \exp\left\{\frac{1}{n} \log n\right\}$$

and hence show that $n^{1/n} \to 1$ as $n \to \infty$ (see exercises 4.20(6) and 5.7(1)).

(5) Discuss the existence of the following improper integrals.

  (i) $\displaystyle\int_{\to -\infty}^{\to \infty} e^{-x^2/2}\, dx$ (ii) $\displaystyle\int_{\to 0}^{\to \infty} x^{y-1} e^{-x} dx$ (iii) $\displaystyle\int_{\to -\infty}^{\to \infty} \frac{x}{1+x^2}\, dx$.

  [Hint: proposition 13.29.]

†(6) Obtain all solutions of the functional equations

  (i) $f(x+y) = f(x) + f(y)$ $(x \in \mathbb{R}, y \in \mathbb{R})$

  (ii) $f(x+y) = f(x)f(y)$ $(x \in \mathbb{R}, y \in \mathbb{R})$

  (iii) $f(xy) = f(x) + f(y)$ $(x > 0, y > 0)$

  (iv) $f(xy) = f(x)f(y)$ $(x > 0, y > 0)$

for which the derivative of f exists and is continuous wherever f is defined. For (ii) and (iv) you may assume that f never takes the value zero.

15 POWER SERIES

15.1 Interval of convergence

A *power series* about the point ξ is an expression of the form

$$\sum_{n=0}^{\infty} a_n(x - \xi)^n$$

in which x is a variable.

15.2 **Theorem** The set of values of x for which a power series

$$\sum_{n=0}^{\infty} a_n(x - \xi)^n$$

converges is an interval with midpoint ξ.

Proof We shall show that, if the power series converges when $x = y$, then it converges for all x satisfying $|x - \xi| < |y - \xi|$. The theorem then follows.

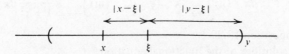

If the power series converges when $x = y$, then

$$a_n(y - \xi)^n \to 0 \text{ as } n \to \infty \quad \text{(theorem 6.9)}.$$

Hence the sequence $\langle a_n(y - \xi)^n \rangle$ is bounded (theorem 4.25). Thus, for some H,

$$|a_n(y - \xi)^n| \leqslant H \quad (n = 1, 2, \ldots).$$

Now suppose that $|x - \xi| < |y - \xi|$. Then

$$\rho = \frac{|x - \xi|}{|y - \xi|} < 1.$$

Hence

$$|a_n(x - \xi)^n| = |a_n(y - \xi)^n| \cdot \rho^n \leqslant H\rho^n \quad (n = 1, 2, \ldots).$$

The convergence of the power series

$$\sum_{n=0}^{\infty} a_n(x - \xi)^n$$

now follows from the comparison test (theorem 6.15) since $\Sigma_{n=0}^{\infty} \rho^n$ converges.

We call the set of values of x for which a power series converges its *interval of convergence*. If the endpoints of the interval of convergence are $\xi - R$ and $\xi + R$, we call R the *radius of convergence* of the power series. (If the interval of convergence is the set $\mathbb{R}$ of all real numbers, we say that the radius of convergence is infinite.)

The proof of theorem 15.2 shows that a power series converges *absolutely* at all points of its interval of convergence with the possible exception of the endpoints. At the endpoints the series may converge absolutely or it may converge conditionally or it may diverge.

It is sometimes useful to observe that the radius of convergence R is given by

and
$$\left. \begin{aligned} \frac{1}{R} &= \limsup_{n \to \infty} |a_n|^{1/n} \\ \frac{1}{R} &= \lim_{n \to \infty} \left| \frac{a_{n+1}}{a_n} \right|. \end{aligned} \right\}$$

The formulae are justified by propositions 6.17 and 6.18. In both cases, appropriate conventions must be adopted if the right hand side of the formula happens to be 0 or $+\infty$. In the second case, of course, the sequence $\langle |a_{n+1}|/|a_n| \rangle$ may oscillate. If this happens, the limit does not exist and so the second formula becomes useless.

15.3   *Example* We list some power series about the point 0, together with their intervals of convergence. In each case, the radius of convergence can be calculated using the second of the formulae above. In the last three cases one is then left with the problem of whether or not the endpoints of the interval of convergence belong to the interval of convergence.

Power series	*Interval of convergence*
$\displaystyle\sum_{n=0}^{\infty} \frac{x^n}{n!}$	$\mathbb{R}$
$\displaystyle\sum_{n=0}^{\infty} (nx)^n$	$\{0\}$
$\displaystyle\sum_{n=0}^{\infty} x^n$	$(-1, 1)$

$$\sum_{n=1}^{\infty} \frac{x^n}{n} \qquad [-1, 1)$$

$$\sum_{n=1}^{\infty} \frac{x^n}{n^2} \qquad [-1, 1]$$

15.4   Taylor series

Suppose that f can be differentiated as often as we choose in some open interval I containing the point ξ. Then its *Taylor series* expansion about the point ξ is

$$\sum_{n=0}^{\infty} \frac{(x - \xi)^n}{n!} f^{(n)}(\xi) = f(\xi) + \frac{(x - \xi)}{1!} f'(\xi) + \frac{(x - \xi)^2}{2!} f''(\xi) + \dots$$

It should *not* be automatically assumed that this power series converges with sum $f(x)$. Indeed there is no reason, in general, to suppose that it converges at all (except when $x = \xi$) and, even if it does converge, its sum need not be $f(x)$ (see exercise 15.6(6) below). A function f which *is* the sum of its Taylor series expansion in some open interval containing ξ is said to be *analytic* at the point ξ.

The most natural way of showing that a function is analytic is to prove that the error term in Taylor's theorem (theorem 11.10) tends to zero as $n \to \infty$.

15.5   *Example*

It is particularly easy to write down the Taylor series expansion of the exponential function about the point 0. If $f(x) = \exp x$, then $f^{(n)}(x) = \exp x$. Since $\exp 0 = 1$, the expansion is

$$\sum_{n=0}^{\infty} \frac{x^n}{n!}.$$

From examples 15.3 we know that this power series converges for all x. Is its sum e^x? From Taylor's theorem we know that

$$e^x = 1 + \frac{x}{1!} + \frac{x^2}{2!} + \dots + \frac{x^{n-1}}{(n-1)!} + \frac{x^n}{n!} e^{\eta},$$

where η lies between 0 and x. Hence

$$\left| e^x - \left\{ 1 + \frac{x}{1!} + \frac{x^2}{2!} + \dots + \frac{x^{n-1}}{(n-1)!} \right\} \right| = \left| \frac{x^n}{n!} e^{\eta} \right|$$

$$\leqslant \frac{|x|^n}{n!} e^{|x|}$$

$$\to 0 \text{ as } n \to \infty$$

(by exercise 4.20(4) or by theorem 6.9). The partial sums of the power series therefore converge to e^x, and thus, for all values of x,

$$e^x = 1 + \frac{x}{1!} + \frac{x^2}{2!} + \frac{x^3}{3!} + \dots$$

15.6   **Exercise**

(1) Determine the intervals of convergence of the following power series

(i) $\sum_{n=1}^{\infty} \left(\frac{x}{n}\right)^n$ (ii) $\sum_{n=0}^{\infty} n! x^n$ (iii) $\sum_{n=0}^{\infty} (-1)^n \frac{x^{2n}}{(2n)!}$

(iv) $\sum_{n=0}^{\infty} (-1)^n \frac{x^{2n+1}}{(2n+1)!}$ (v) $\sum_{n=1}^{\infty} \frac{(n!)^2}{(2n)!} x^n$ (vi) $\sum_{n=1}^{\infty} \frac{x^n}{\sqrt{n}}$.

(2) Prove that the Taylor series expansion of $\log(1+x)$ about the point 0 is

$$\sum_{n=1}^{\infty} (-1)^{n-1} \frac{x^n}{n} = x - \frac{x^2}{2} + \frac{x^3}{3} - \frac{x^4}{4} + \dots$$

Using Taylor's theorem in the form given by exercise 13.26(6) show that this power series converges to $\log(1+x)$ for $-1 < x \leqslant 1$. Hence give another proof of the identity

$$\log 2 = 1 - \tfrac{1}{2} + \tfrac{1}{3} - \tfrac{1}{4} + \dots$$

(see exercise 14.3(5)).

(3) Prove that the application of Taylor's theorem in the form of theorem 11.10 (with $\xi = 0$) to the function $\log(1+x)$ yields a remainder term of the form

$$E_n = \frac{(-1)^{n-1}}{n} \left(\frac{x}{1+\eta}\right)^n$$

where η lies between 0 and x. For what values of x is it possible to show that $E_n \to 0$ as $n \to \infty$ without further information about the manner in which η depends on x and n?

(4) Prove that

$$\frac{1}{(n+1)!} + \frac{1}{(n+2)!} + \dots < \frac{1}{(n+1)!} \cdot \frac{n+2}{n+1}.$$

Deduce that

$$0 < e - \left\{1 + \frac{1}{1!} + \frac{1}{2!} + \frac{1}{3!} + \frac{1}{4!}\right\} < \frac{1}{100}$$

and hence show that $2{\cdot}7083 < e < 2{\cdot}7184$.

*(5) Prove that e is irrational. [Hint: assume that $e = m/n$, where m and n are natural numbers and seek a contradiction using the inequality of question 4.]

†(6) Let $f\colon \mathbb{R} \to \mathbb{R}$ be defined by

$$f(x) = \begin{cases} e^{-1/x^2} & (x \neq 0) \\ 0 & (x = 0). \end{cases}$$

Show that, for any $x \neq 0$,

$$f^{(n)}(x) = P_n\left(\frac{1}{x}\right) e^{-1/x^2}$$

where P is a polynomial of degree $3n$. [Hint: use induction.] Using exercise 14.5(2), deduce that, for each natural number n,

$$\frac{f^{(n-1)}(x)}{x} \to 0 \text{ as } x \to 0 +.$$

Hence show that f can be differentiated as many times as we choose at the point 0 and $f^{(n)}(0) = 0$ $(n = 0, 1, 2, \ldots)$.

Write down the Taylor series expansion of f about the point 0. For what values of x does this converge to $f(x)$?

15.7   Continuity and differentiation

The following proposition is useful. Its proof has been placed in the appendix since it is somewhat involved. The reader will encounter the proof again at a later stage when studying 'uniform convergence'.

15.8   *Proposition* Suppose that the power series

$$f(x) = \sum_{n=0}^{\infty} a_n(x - \xi)^n$$

has interval of convergence I. Then its sum is continuous on I and differentiable on I (except at the endpoints). Moreover

$$f'(x) = \sum_{n=1}^{\infty} na_n(x - \xi)^{n-1}.$$

15.9   Example Proposition 15.8 allows us to tackle exercise 15.6(2) in an alternative way.

We know that the power series

$$\sum_{n=1}^{\infty} (-1)^{n-1} \frac{x^n}{n}$$

has interval of convergence $(-1, 1]$ (see Example 15.3). Hence, by proposition 15.8, it is differentiable on $(-1, 1)$ and

$$D\left\{\sum_{n=1}^{\infty} (-1)^{n-1} \frac{x^n}{n}\right\} = \sum_{n=1}^{\infty} (-1)^{n-1} x^{n-1}$$

$$= 1 - x + x^2 - x^3 + \ldots$$

$$= \frac{1}{1+x} \quad (-1 < x < 1).$$

It follows that, for each $x \in (-1, 1)$,

$$D\left\{\log(1+x) - \sum_{n=1}^{\infty} (-1)^{n-1} \frac{x^n}{n}\right\} = \frac{1}{1+x} - \frac{1}{1+x} = 0$$

and therefore that

$$\log(1+x) = \sum_{n=1}^{\infty} (-1)^{n-1} \frac{x^n}{n} + c \quad (-1 < x < 1)$$

where c is a constant (theorem 11.7). By substituting $x = 0$, we see that the constant $c = 0$.

Since the power series is continuous on $(-1, 1]$,

$$\log 2 = \lim_{x \to 1-} \left\{\log(1+x)\right\} = \lim_{x \to 1-} \left\{\sum_{n=1}^{\infty} (-1)^{n-1} \frac{x^n}{n}\right\}$$

$$= \sum_{n=1}^{\infty} \frac{(-1)^{n-1}}{n}.$$

15.10   Exercise

(1) Let α be a real number. By considering

$$D\left\{(1+x)^{-\alpha} \sum_{n=0}^{\infty} \frac{\alpha(\alpha-1)\ldots(\alpha-n+1)}{n!} x^n\right\}$$

for $|x| < 1$, show that

$$(1+x)^\alpha = 1 + \alpha x + \frac{\alpha(\alpha-1)}{2!} x^2 + \ldots$$

provided that $|x| < 1$ (general binomial theorem).

(2) Prove that the power series

$$\sum_{n=0}^{\infty} n^2 x^n$$

converges for $|x| < 1$ and determine its sum.

(3) Suppose that the power series

$$f(x) = \sum_{n=0}^{\infty} a_n x^n$$

has interval of convergence I. If $y \in I$ but is not an endpoint of I, prove that

$$\int_0^y f(x)\,dx = \sum_{n=0}^{\infty} \frac{a_n}{n+1} y^{n+1}.$$

(4) Prove that

$$\sum_{n=0}^{\infty} \frac{1}{n^2} = -\int_0^{\to 1} \frac{\log(1-x)}{x}\,dx.$$

(5) Suppose that the two power series

$$\sum_{n=0}^{\infty} a_n(x-\xi)^n, \qquad \sum_{n=0}^{\infty} b_n(x-\xi)^n$$

converge in some open interval I containing ξ. Prove that their sums are equal for each $x \in I$ if and only if

$$a_n = b_n \,(n = 0, 1, 2, \ldots).$$

(6) Suppose that the power series

$$y = f(x) = \sum_{n=0}^{\infty} a_n x^n$$

converges for all real x and satisfies the differential equation

$$\frac{dy}{dx} = y.$$

Show that $a_{n+1} = \dfrac{1}{n+1} a_n \,(n = 0, 1, 2, \ldots)$ and deduce that

$$f(x) = a_0 e^x.$$

16 TRIGONOMETRIC FUNCTIONS

16.1 Introduction

We based our definitions of the exponential and logarithm functions on the formula

$$\log x = \int_1^x \frac{dt}{t}.$$

We could equally well have begun with the differential equation

$$\frac{dy}{dx} = y$$

and defined the exponential function as the sum of a power series as indicated in exercise 15.10(6).

In a similar way, the definitions of the trigonometric functions can be based on the formula

$$\arctan x = \int_0^x \frac{dt}{1+t^2}.$$

We prefer, however, to base them on the differential equation

$$\frac{d^2y}{dx^2} + y = 0.$$

We already have some intuitive ideas about the sine and cosine functions from elementary geometry and trigonometry. It is therefore sensible to begin by indicating why the sine and cosine functions, as conceived of in trigonometry, should be expected to satisfy this differential equation.

In the diagram, the angle x is measured in radians and $y = \sin x$ and $z = \cos x$.

By the sort of argument considered adequate in elementary trigonometry, we obtain

$$z = \cos x = \frac{dy}{dx}; \qquad y = \sin x = -\frac{dz}{dx}$$

from which it follows that

$$\frac{d^2y}{dx^2} = \frac{dz}{dx} = -y; \qquad \frac{d^2z}{dx^2} = -\frac{dy}{dx} = -z.$$

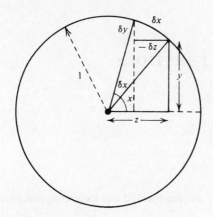

16.2   Sine and cosine

Suppose that the power series

$$f(x) = \sum_{n=0}^{\infty} a_n x^n$$

converges for all real values of x and that

$$f''(x) + f(x) = 0.$$

By proposition 15.8,

$$f(x) = a_0 + a_1 x + a_2 x^2 + \ldots + a_n x^n + \ldots$$
$$f'(x) = a_1 + 2a_2 x + 3a_3 x^2 + \ldots + (n+1)a_{n+1} x^n + \ldots$$
$$f''(x) = 2a_2 + 3.2a_3 x + 4.3a_4 x^2 + \ldots + (n+2)(n+1)a_{n+2} x^n + \ldots$$

Thus, to satisfy the differential equation, a_0 and a_1 may be chosen in any way, but then the remaining coefficients must be chosen to satisfy

$$(n+2)(n+1)a_{n+2} + a_n = 0 \quad (n = 0, 1, 2, \ldots).$$

This recurrence relation is readily solved and we obtain

$$f(x) = a_0 \sum_{n=0}^{\infty} (-1)^n \frac{x^{2n}}{(2n)!} + a_1 \sum_{n=0}^{\infty} (-1)^n \frac{x^{2n+1}}{(2n+1)!}.$$

We want $\sin 0 = 0$ and $D \sin 0 = \cos 0 = 1$. We therefore *define*

$$\sin x = \sum_{n=0}^{\infty} (-1)^n \frac{x^{2n+1}}{(2n+1)!} = x - \frac{x^3}{3!} + \ldots$$

Similarly

$$\cos x = \sum_{n=0}^{\infty} (-1)^n \frac{x^{2n}}{(2n)!} = 1 - \frac{x^2}{2!} + \ldots$$

Both these power series converge for *all* values of x by comparison with the power series for e^x.

Note that we take these formulae as *definitions*. The definitions of elementary trigonometry concerning the 'opposite', the 'adjacent' and the 'hypotenuse' are not precise enough for our purposes in this book. The remarks of §16.1 are simply intended to indicate *why* we chose to define sine and cosine as we did.

16.3   *Exercise*

(1) Prove the following.

    (i) $\cos 0 = 1$ (ii) $\sin 0 = 0$

    (iii) $\cos(-x) = \cos x$ (iv) $\sin(-x) = -\sin x$.

    Show that, for all values of x,

    (v) $D \cos x = -\sin x$ (vi) $D \sin x = \cos x$.

(2) Let y be any real number and define

$$g(x) = \sin(x+y) - \sin x \cos y - \cos x \sin y$$

$$h(x) = \cos(x+y) - \cos x \cos y + \sin x \sin y.$$

    Differentiate $\{g(x)\}^2 + \{h(x)\}^2$ with respect to x and hence show that, for all values of x and y,

$$\sin(x+y) = \sin x \cos y + \cos x \sin y$$

$$\cos(x+y) = \cos x \cos y - \sin x \sin y.$$

(3) Use the previous question to prove the following.

    (i) $\cos^2 x + \sin^2 x = 1$ (ii) $|\cos x| \leqslant 1$ and $|\sin x| \leqslant 1$

    (iii) $\sin 2x = 2 \sin x \cos x$ (iv) $\cos 2x = \cos^2 x - \sin^2 x$.

(4) Use L'Hôpital's rule to prove that

    (i) $\dfrac{\sin x}{x} \to 1$ as $x \to 0$

    (ii) $\dfrac{1 - \cos x}{x^2} \to \dfrac{1}{2}$ as $x \to 0$.

†(5) Use the mean value theorem to show that, for all values of x

$$\left| \frac{\sin x}{x} \right| \leqslant 1.$$

Hence show that the series $\sum_{n=1}^{\infty} \sin(1/n^2)$ converges. Discuss the convergence of $\sum_{n=1}^{\infty} \sin(1/n)$.

*(6) Use exercise 13.26(5) to prove that

$$\left| \int_m^n \frac{\sin t}{t} \, dt \right| < \frac{4}{m}$$

provided that $n > m > 0$. Explain why it is possible to deduce the existence of the limit

$$\lim_{n \to \infty} \int_1^n \frac{\sin t}{t} \, dt.$$

16.4   Periodicity

Since $\cos 0 = 1$ and $\cos(-x) = \cos x$ and the cosine function is continuous, there exists a $\xi > 0$ such that $\cos x > 0$ for $x \in (-\xi, \xi)$.

But the cosine function is not positive all the time. If this were the case, then it would follow from the formula $D^2 \cos x = -\cos x$ that the cosine function was concave. But a bounded, differentiable function cannot be concave unless it is constant (exercise 12.21(5)).

It follows that there exists a smallest positive number ξ for which $\cos \xi = 0$. We *define* the real number π by

$$\tfrac{1}{2}\pi = \xi.$$

By definition, $\cos \tfrac{1}{2}\pi = \cos(-\tfrac{1}{2}\pi) = 0$ and $\cos x > 0$ for $-\tfrac{1}{2}\pi < x < \tfrac{1}{2}\pi$. We show that $\sin \tfrac{1}{2}\pi = 1$. It follows from the formula $\cos^2 x + \sin^2 x = 1$ that $\sin^2 \tfrac{1}{2}\pi = 1$. But $D \sin x = \cos x$ and so the sine function increases on $[-\tfrac{1}{2}\pi, \tfrac{1}{2}\pi]$. Since $\sin 0 = 0$ it follows that $\sin \tfrac{1}{2}\pi > 0$.

We now appeal to the formulae of exercise 16.3(2). We have

$$\sin(x + \tfrac{1}{2}\pi) = \sin x \cos \tfrac{1}{2}\pi + \cos x \sin \tfrac{1}{2}\pi = \cos x$$

$$\cos(x + \tfrac{1}{2}\pi) = \cos x \cos \tfrac{1}{2}\pi - \sin x \sin \tfrac{1}{2}\pi = -\sin x$$

from which it follows, in turn, that

$$\sin(x + 2\pi) = \sin x$$

$$\cos(x + 2\pi) = \cos x.$$

Because of these last formulae we say that the sine and cosine functions are *periodic* with *period* 2π,

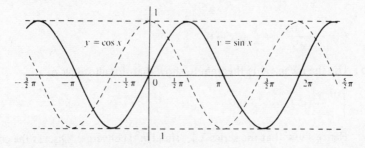

16.5   *Exercise*

(1) Prove that the cosine function decreases on $[0, \pi]$ and increases on $[\pi, 2\pi]$. Prove that it is concave on $[-\frac{1}{2}\pi, \frac{1}{2}\pi]$ and convex on $[\frac{1}{2}\pi, \frac{3}{2}\pi]$.

(2) We define the tangent function by

$$\tan x = \frac{\sin x}{\cos x} \quad (x \neq (n + \tfrac{1}{2})\pi: n = 0, \pm 1, \pm 2, \ldots).$$

Prove that $\tan (x + \pi) = \tan x$ provided that $x \neq (n + \frac{1}{2})\pi$ where $n = 0$, $\pm 1, \pm 2, \ldots$ Show that the tangent function is strictly increasing on $(-\frac{1}{2}\pi, \frac{1}{2}\pi)$ and that $\tan x \to +\infty$ as $x \to \frac{1}{2}\pi -$ and $\tan x \to -\infty$ as $x \to -\frac{1}{2}\pi +$.

(3) Show that the sine function is strictly increasing and continuous on $[-\frac{1}{2}\pi, \frac{1}{2}\pi]$ and that the image of this interval under the sine function is $[-1, 1]$. If we ignore the fact that the sine function is defined outside the interval $[-\frac{1}{2}\pi, \frac{1}{2}\pi]$, the function we obtain therefore admits an inverse function with domain $[-1, 1]$ and range $[-\frac{1}{2}\pi, \frac{1}{2}\pi]$. We call this function the arcsine function. Draw a graph of this function and calculate $D \arcsin x$ for $-1 < x < 1$. (Some authors use the notation $\sin^{-1} x$, but we prefer not to, since the sine function has no inverse. The arcsine function is the inverse of the *restriction* of the sine function to $[-\frac{1}{2}\pi, \frac{1}{2}\pi]$).

Discuss the arccosine function obtained as the inverse of the restriction of the cosine function to $[0, \pi]$. Calculate $D \arccos x$ and explain how the arcsine and arccosine functions are related.

(4) Show that the arctangent function, obtained as the inverse of the restriction of the tangent function to $(-\frac{1}{2}\pi, \frac{1}{2}\pi)$, has domain $\mathbb{R}$ and range $(-\frac{1}{2}\pi, \frac{1}{2}\pi)$. Draw a graph and calculate $D \arctan x$. Hence show that

$$\int_0^{\to \infty} \frac{dx}{1 + x^2} = \frac{\pi}{2}.$$

†(5) Suppose that f has a continuous derivative on $\mathbb{R}$ and that

$$f(x) + f(y) = f\left(\frac{x + y}{1 - xy}\right)$$

for each x and y such that $xy < 1$. Prove that, for some constant C, $f(x) = C \arctan x$. By writing $x = y$ in the formula and considering what happens when $x \to 1 -$, prove that $\arctan 1 = \frac{1}{4}\pi$.

Show that, for $-1 < x \leqslant 1$,

$$\arctan x = x - \frac{x^3}{3} + \frac{x^5}{5} - \frac{x^7}{7} + \ldots$$

and hence obtain the identity

$$\tfrac{1}{4}\pi = 1 - \tfrac{1}{3} + \tfrac{1}{5} - \tfrac{1}{7} + \ldots$$

†(6)   Let $f: \mathbb{R} \to \mathbb{R}$ be defined by

$$f(x) = \begin{cases} \sin \dfrac{1}{x} & (x \neq 0) \\[2mm] 0 & (x = 0). \end{cases}$$

Let $g: \mathbb{R} \to \mathbb{R}$ and $h: \mathbb{R} \to \mathbb{R}$ be defined by $g(x) = xf(x)$ and $h(x) = x^2 f(x)$. Draw graphs and prove the following.

(i) f is *not* continuous at the point 0.

(ii) g *is* continuous at the point 0 but *not* differentiable there.

(iii) h *is* differentiable at 0 and $h'(0) = 0$.

17 THE GAMMA FUNCTION

17.1 Introduction

In this chapter we consider Stirling's formula and the gamma function. These have important uses in mathematical physics, statistics and elsewhere. As in our development of the more familiar exponential, logarithmic and trigonometric functions, the treatment provides an opportunity to apply the theoretical work of earlier chapters and so consists chiefly of a linked sequence of exercises. In particular, the properties of the gamma function are obtained using the convexity ideas of chapter 12 (rather than the usual complex analysis approach). The exercises on this topic will be found more demanding than those set previously.

17.2 Stirling's formula

If $\langle a_n \rangle$ and $\langle b_n \rangle$ are sequences of real numbers, the notation $a_n \sim b_n$ means that

$$\frac{a_n}{b_n} \to 1 \text{ as } n \to \infty.$$

17.3 Proposition (Stirling's formula)

$$n! \sim \sqrt{(2\pi)} n^n n^{1/2} e^{-n}.$$

Proof Consider the sequence $\langle d_n \rangle$ defined by

$$d_n = \log n! - (n + \tfrac{1}{2}) \log n + n.$$

We seek to show that $\langle d_n \rangle$ decreases and so we examine the sign of

$$d_n - d_{n+1} = -\log(n+1) - (n + \tfrac{1}{2}) \log n + (n + \tfrac{3}{2}) \log(n+1) - 1$$

$$= \left(n + \frac{1}{2}\right) \log\left(\frac{n+1}{n}\right) - 1$$

$$= \frac{2n+1}{2} \log\left\{\frac{1 + (2n+1)^{-1}}{1 - (2n+1)^{-1}}\right\} - 1.$$

But, for $|x| < 1$,

$$f(x) = \frac{1}{2x} \log \left| \frac{1+x}{1-x} \right| - 1$$

$$= \frac{1}{2x} \left\{ \left(x - \frac{x^2}{2} + \frac{x^3}{3} - \ldots \right) - \left(-x - \frac{x^2}{2} - \frac{x^3}{3} - \ldots \right) \right\} - 1$$

$$= \frac{x^2}{3} + \frac{x^4}{5} + \frac{x^6}{7} + \ldots \tag{1}$$

In particular, $f(x) \geqslant 0$ for $|x| < 1$ and thus $d_n - d_{n+1} \geqslant 0$ $(n = 1, 2, \ldots)$. Therefore $\langle d_n \rangle$ decreases.

From (1) it also follows that, for $|x| < 1$,

$$f(x) \leqslant \frac{x^2}{3} \{1 + x^2 + x^4 + \ldots\}$$

$$= \frac{x^2}{3(1 - x^2)}.$$

Thus

$$d_n - d_{n+1} \leqslant \frac{1}{3\{(2n+1)^2 - 1\}} = \frac{1}{12n} - \frac{1}{12(n+1)}.$$

It follows that the sequence $\langle d_n - (12n)^{-1} \rangle$ is increasing. In particular,

$$d_n - (12n)^{-1} \geqslant d_1 - \tfrac{1}{12} \quad (n = 1, 2, \ldots)$$

and hence the sequence $\langle d_n \rangle$ is bounded below. Since $\langle d_n \rangle$ decreases, it follows that $\langle d_n \rangle$ converges.

Suppose that $d_n \to d$ as $n \to \infty$. Since the exponential function is continuous at every point, $\exp d_n \to \exp d$ as $n \to \infty$. Let $C = \exp d$ then

$$\frac{n!}{n^n n^{1/2} e^{-n}} \to C \text{ as } n \to \infty.$$

It remains to show that $C = \sqrt{(2\pi)}$. We leave this as an exercise (exercise 17.5(1) and (2)).

17.4[†]   The gamma function

The gamma function is defined on $(0, \infty)$ by the formula

$$\Gamma(x) = \int_{\to 0}^{\to \infty} t^{x-1} e^{-t} dt \quad (x > 0).$$

To justify the existence of the improper integral, we appeal to proposition 13.29. In view of the inequality

$$t^{x-1} e^{-t} \leqslant t^{x-1} \quad (t > 0)$$

the existence of the integral $\int_{\to 0}^1 t^{x-1}e^{-t}dt$ follows from that of $\int_{\to 0}^1 t^{x-1}dt$, provided that $x > 0$. Since $t^{x+1}e^{-t} \to 0$ as $t \to +\infty$, we have, for some $H > 0$,

$$t^{x-1}e^{-t} \leqslant Ht^{-2} \quad (t \geqslant 1)$$

and hence the existence of $\int_1^{\to \infty} t^{x-1}e^{-t}dt$ follows from that of $\int_1^{\to \infty} t^{-2}dt$.

17.5† Exercise

†(1) Let $I_n = \int_0^{\pi/2} \sin^n x\, dx$ $(n = 0, 1, 2, \ldots)$. Prove that $\langle I_n \rangle$ is a decreasing sequence of positive numbers which satisfies $nI_n = (n-1)I_{n-2}$ $(n = 2, 3, \ldots)$.
   Deduce that

$$\frac{I_{2n}}{I_{2n+1}} \to 1 \text{ as } n \to \infty.$$

†(2) With the notation of the previous question, establish the identities

$$I_{2n} = \frac{(2n)!}{(2^n n!)^2}\frac{\pi}{2} \quad (n = 0, 1, \ldots)$$

$$I_{2n+1} = \frac{(2^n n!)^2}{(2n+1)!} \quad (n = 0, 1, \ldots).$$

   Hence show that the constant C obtained in §17.3 satisfies $C = \sqrt{(2\pi)}$.

†(3) Let $d_n = \log n! - (n + \frac{1}{2})\log n + n$. Show that the sequence $\langle d_n - (12n)^{-1} \rangle$ increases but that the sequence $\langle d_n - (12n+1)^{-1} \rangle$ decreases. Deduce that

$$e^{1/(12n+1)} \leqslant \frac{n!}{\sqrt{(2\pi)}n^n n^{1/2}e^{-n}} \leqslant e^{1/12n}.$$

   Hence estimate the error on approximating to 100! by Stirling's formula. Is this error large or small compared with the value of 100!?

†(4) Show that $\Gamma(x+1) = x\Gamma(x)(x > 0)$. Deduce that, for $n = 1, 2, 3, \ldots,$

$$\Gamma(n+1) = n!$$

†(5) Prove that the gamma function is continuous on $(0, \infty)$. [Hint: If $0 < \alpha < a \leqslant x \leqslant y \leqslant b < \beta$, prove that, for some constant H which does not depend on x or y, $|\Gamma(x) - \Gamma(y)| \leqslant H|x - y|\{\Gamma(\alpha) + \Gamma(\beta)\}$.]

†(6) Prove that the logarithm of the gamma function is *convex* on $(0, \infty)$. [Hint: exercise 12.21(6) and theorem 13.25.]

17.6[†]   Properties of the gamma function

The gamma function provides a generalisation of the factorial function (see exercise 17.5 (4)). Is it the only such generalisation?

17.7[†]   *Theorem* Let f be positive and continuous on $(0, \infty)$ and let its logarithm be convex on $(0, \infty)$. If f satisfies the functional equation

$$f(x + 1) = xf(x) \quad (x > 0)$$

and $f(1) = 1$, then

$$f(x) = \Gamma(x) \quad (x > 0).$$

Proof The proof consists of showing that, under the hypotheses of the theorem, for each $x > 0$,

$$f(x) = \lim_{n \to \infty} \frac{n^x n!}{x(x + 1) \ldots (x + n)}. \tag{2}$$

It follows from exercise 17.5 (4, 5 and 6) that the gamma function satisfies the hypotheses of the theorem. Since a sequence can have at most one limit, we can therefore conclude from (2) that $f(x) = \Gamma(x)$ $(x > 0)$.

The proof of (2) uses the convexity of $\log f$. Suppose that $s \leqslant t \leqslant s + 1$. Then we may write $t = \alpha s + \beta (s + 1)$ where $\alpha \geqslant 0, \beta \geqslant 0$ and $\alpha + \beta = 1$. Now $t = (\alpha + \beta)s + \beta = s + \beta$ and so $\beta = t - s$.

From the convexity of $\log f$, it follows that

$$\log f(t) \leqslant \alpha \log f(s) + \beta \log f(s + 1)$$
$$f(t) \leqslant \{f(s)\}^\alpha \{f(s + 1)\}^\beta$$
$$= \{f(s)\}^\alpha \{sf(s)\}^\beta$$
$$= s^\beta f(s) = s^{t-s} f(s). \tag{3}$$

Since $s \leqslant t \leqslant s + 1$, we also have $t - 1 \leqslant s \leqslant t$. Making appropriate substitutions in (3), we obtain

$$f(s) \leqslant (t - 1)^{s-t+1} f(t - 1) = (t - 1)^{s-t} f(t). \tag{4}$$

Combining (3) and (4) yields the inequality

$$(t - 1)^{t-s} f(s) \leqslant f(t) \leqslant s^{t-s} f(s).$$

Now suppose that $0 \leqslant x < 1$ and that n is a natural number. We may take $s = n + 1$ and $t = x + n + 1$. Then

$$(x + n)^x f(n + 1) \leqslant f(x + n + 1) \leqslant (n + 1)^x f(n + 1). \tag{5}$$

From this inequality it follows that

$$(x + n)^x n! \leqslant (x + n)(x + n - 1) \ldots xf(x) \leqslant (n + 1)^x n!$$

$$\left(1 + \frac{x}{n}\right)^x \leqslant \frac{(x+n)(x+n-1)\dots xf(x)}{n^x n!} \leqslant \left(1 + \frac{1}{n}\right)^x.$$

This completes the proof of the formula (2) in the case when $0 < x \leqslant 1$. The general case is easily deduced with the help of the functional equation $f(x+1) = xf(x)$.

17.8† *Exercise*

†(1) If $x > 0$, prove that

$$\Gamma(x) = \int_{\to 0}^{1} \left\{\log \frac{1}{t}\right\}^{x-1} dt.$$

†(2) Prove that $\Gamma(x) \sim \sqrt{(2\pi)} x^x x^{-1/2} e^{-x}$. This is Stirling's formula for the gamma function. [Hint: see theorem 17.7, inequality (5).]

†(3) Use L'Hôpital's rule to show that

$$\lim_{z \to 0} \left\{\frac{\log(1+z) - z}{z^2}\right\} = -\frac{1}{2}.$$

With the help of Stirling's formula for the gamma function, deduce that, for all $x > 0$,

$$(\sqrt{u}) f_u(u + x\sqrt{u}) \to \frac{1}{\sqrt{(2\pi)}} e^{-x^2/2} \text{ as } u \to +\infty$$

where

$$f_u(y) = \frac{1}{\Gamma(u)} y^{u-1} e^{-y}.$$

†(4) The *beta function* is the function of two variables defined for $x > 0$ and $y > 0$ by the formula

$$B(x,y) = \int_{\to 0}^{\to 1} t^{x-1}(1-t)^{y-1} dt.$$

Check that the improper integral exists provided that $x > 0$ and $y > 0$. Prove that, for a given fixed value of $y > 0$, $B(x,y)$ is a positive, continuous function of x on $(0, \infty)$ whose logarithm is convex on $(0, \infty)$.

†(5) With the notation of the previous question, prove that, for a given fixed value of $y > 0$, the function $f: (0, \infty) \to \mathbb{R}$ defined by

$$f(x) = \frac{\Gamma(x+y)}{\Gamma(y)} B(x,y) \quad (x>0)$$

satisfies the conditions of theorem 17.7. Deduce that, for $x > 0$ and $y > 0$,

$$B(x,y) = \frac{\Gamma(x)\Gamma(y)}{\Gamma(x+y)}.$$

†(6) Use the previous question to evaluate $\Gamma(\frac{1}{2})$. Hence show that

$$\int_{\to -\infty}^{\to \infty} e^{-x^2/2}\, dx = \sqrt{2\pi}.$$

18 VECTORS

18.1 Introduction

We have been concerned so far with functions $f: A \to B$ for which A and B are sets of real numbers. But many of the theorems obtained generalise very elegantly to the case when A and B are sets of vectors. In the next chapter, we shall explain how the particularly important idea of a derivative generalises to the vector case. As a preliminary, some simple facts about vectors are explained in the current chapter. The chapter is no substitute for a course in linear algebra and some previous knowledge of vectors and matrices is assumed. On the other hand, the treatment is systematic in so far as it goes with a strong emphasis on the geometric significance of the ideas.

18.2 Vectors

The set $\mathbb{R}^n$ consists of all objects of the form

$$(x_1, x_2, \ldots, x_n)$$

in which $x_1, x_2, \ldots, x_n$ are real numbers. We usually use a single symbol for such an object and write

$$\mathbf{x} = (x_1, x_2, \ldots, x_n).$$

The real numbers $x_1, x_2, \ldots, x_n$ are called the *co-ordinates* or *components* of $\mathbf{x}$.

It is often convenient to refer to an object $\mathbf{x}$ in $\mathbb{R}^n$ as a *vector*. When doing so, ordinary real numbers are called *scalars*. If $\mathbf{x} = (x_1, x_2, \ldots, x_n)$ and $\mathbf{y} = (y_1, y_2, \ldots, y_n)$ are vectors and α is a scalar, we define *vector addition* and *scalar multiplication* by

$$\mathbf{x} + \mathbf{y} = (x_1 + y_1, x_2 + y_2, \ldots, x_n + y_n)$$

$$\alpha \mathbf{x} = (\alpha x_1, \alpha x_2, \ldots, \alpha x_n).$$

These definitions have a simple geometric interpretation which we shall illustrate in the case $n = 2$. An object $\mathbf{x} \in \mathbb{R}^2$ may be thought of as a point in the plane referred to rectangular Cartesian axes. Alternatively, we can think of $\mathbf{x}$ as an arrow with its blunt end at the origin and its sharp end at the point (x_1, x_2).

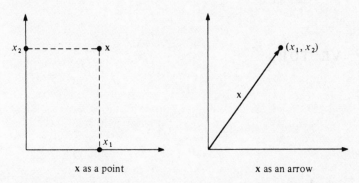

x as a point   x as an arrow

Vector addition and scalar multiplication can then be illustrated as in the diagrams below. For obvious reasons, the rule for adding two vectors is called the *parallelogram law*.

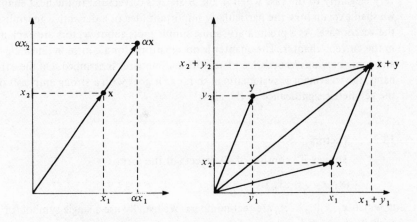

18.3   *Example* Let $\mathbf{x} = (1, 2, 3)$ and $\mathbf{y} = (2, 0, 5)$. Then

$$\mathbf{x} + \mathbf{y} = (1, 2, 3) + (2, 0, 5) = (3, 2, 8)$$
$$2\mathbf{x} = 2(1, 2, 3) = (2, 4, 6).$$

18.4   Length and angle in $\mathbb{R}^n$

The idea of the modulus of a real number is replaced in $\mathbb{R}^n$ by that of the norm of a vector. We define the *norm* of a vector $\mathbf{x}$ in $\mathbb{R}^n$ by

$$\|x\| = \{x_1^2 + x_2^2 + \ldots + x_n^2\}^{1/2}.$$

It is useful to think of $\|x\|$ as the *length* of the vector $\mathbf{x}$. In $\mathbb{R}^2$ this interpretation is justified by Pythagoras' theorem.

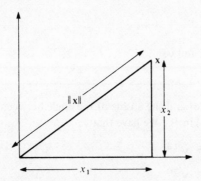

The *inner product* (or *scalar product*) of two vectors $\mathbf{x}$ and $\mathbf{y}$ in $\mathbb{R}^n$ is defined by

$$\langle \mathbf{x}, \mathbf{y} \rangle = x_1 y_1 + x_2 y_2 + \ldots + x_n y_n.$$

It is easy to check the following properties:

(i) $\langle \mathbf{x}, \mathbf{x} \rangle = \|\mathbf{x}\|^2$

(ii) $\langle \mathbf{x}, \mathbf{y} \rangle = \langle \mathbf{y}, \mathbf{x} \rangle$

(iii) $\langle \alpha\mathbf{x} + \beta\mathbf{y}, \mathbf{z} \rangle = \alpha\langle \mathbf{x}, \mathbf{z} \rangle + \beta\langle \mathbf{y}, \mathbf{z} \rangle$.

The geometric significance of the inner product can be discussed using the *cosine rule* (i.e. $c^2 = a^2 + b^2 - 2ab \cos\gamma$) in the diagram below.

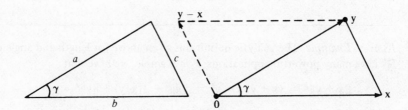

Rewriting the cosine rule in terms of the vectors introduced in the right hand diagram we obtain that

$$\|\mathbf{x} - \mathbf{y}\|^2 = \|\mathbf{x}\|^2 + \|\mathbf{y}\|^2 - 2\|\mathbf{x}\|.\|\mathbf{y}\| \cos\gamma.$$

But

$$\|\mathbf{x} - \mathbf{y}\|^2 = \langle \mathbf{x} - \mathbf{y}, \mathbf{x} - \mathbf{y} \rangle = \langle \mathbf{x}, \mathbf{x} - \mathbf{y} \rangle - \langle \mathbf{y}, \mathbf{x} - \mathbf{y} \rangle$$

$$= \langle \mathbf{x}, \mathbf{x} \rangle - 2\langle \mathbf{x}, \mathbf{y} \rangle + \langle \mathbf{y}, \mathbf{y} \rangle$$

$$= \|\mathbf{x}\|^2 + \|\mathbf{y}\|^2 - 2\langle \mathbf{x}, \mathbf{y} \rangle$$

from which it follows that

$$\langle \mathbf{x}, \mathbf{y} \rangle = \|\mathbf{x}\|.\|\mathbf{y}\| \cos\gamma.$$

Of course, like all arguments based on geometric intuition, this does not *prove* anything. It simply indicates why it is helpful to think of

$$\frac{\langle x, y \rangle}{\|x\|.\|y\|}$$

as the cosine of the *angle* between x and y.

18.5   Example Find the lengths of and the cosine of the angle between the vectos x = (1, 2, 3) and y = (2, 0, 5) in $\mathbb{R}^3$. We have that

$$\|x\| = \{1^2 + 2^2 + 3^2\}^{1/2} = \sqrt{14}$$

$$\|y\| = \{2^2 + 0^2 + 5^2\}^{1/2} = \sqrt{29}$$

$$\frac{\langle x, y \rangle}{\|x\|.\|y\|} = \frac{1.2 + 2.0 + 3.5}{\sqrt{14}.\sqrt{29}} = \frac{17}{\sqrt{14}.\sqrt{29}}.$$

Two vectors x and y are said to be *orthogonal* (or *perpendicular* or *normal*) if and only if

$$\langle x, y \rangle = 0.$$

This accords with our geometric interpretation because cos γ = 0 if and only if γ represents a right angle.

18.6   Example The analytic definitions given above for length and angle in $\mathbb{R}^n$ have many powerful applications. For example, we have that

$$\|x + y\|^2 = \langle x + y, x + y \rangle = \|x\|^2 + 2\langle x, y \rangle + \|y\|^2$$

and hence $\|x + y\|^2 = \|x\|^2 + \|y\|^2$ if and only if $\langle x, y \rangle = 0$. But this last statement is just the analytic version of Pythagoras' theorem.

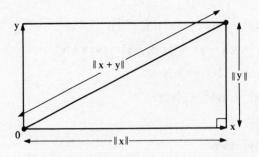

18.7   Note It will be clear from the discussion above that vectors have a certain amount in common with real numbers. It is important, however, to be clear about the ways in which they *differ* from real numbers. Observe that the inner product $\langle \mathbf{x}, \mathbf{y} \rangle$ of two vectors is *not* a vector like $\mathbf{x}$ and $\mathbf{y}$ but is a scalar. Multiplication is therefore peculiar in $\mathbb{R}^n$ while division is, in general, non-existent. Moreover, there is no natural way to define the expression $\mathbf{x} < \mathbf{y}$ which makes sense for all $\mathbf{x}$ and $\mathbf{y}$ in $\mathbb{R}^n$ when $n \geq 2$. Sometimes definitions for $\mathbf{x} < \mathbf{y}$ are introduced which make sense for *some* $\mathbf{x}$ and $\mathbf{y}$ but these are not useful in the context of this chapter and we shall simply regard the expression $\mathbf{x} < \mathbf{y}$ as *meaningless*.

18.8   Inequalities

We met the *Cauchy–Schwarz inequality* in example 1.11. Expressed in terms of inner products and norms, this takes the elegant form

$$|\langle \mathbf{x}, \mathbf{y} \rangle| \leq \|\mathbf{x}\|.\|\mathbf{y}\|.$$

Exercise 1.12(6) was to deduce Minkowski's inequality. This may be expressed in the form

$$\|\mathbf{x} + \mathbf{y}\| \leq \|\mathbf{x}\| + \|\mathbf{y}\|$$

when it becomes instantly recognisable as the vector version of the triangle inequality (theorem 1.17). We give a separate proof.

18.9   Theorem (triangle inequality) For any $\mathbf{x}$ and $\mathbf{y}$ in $\mathbb{R}^n$,

$$\|\mathbf{x} + \mathbf{y}\| \leq \|\mathbf{x}\| + \|\mathbf{y}\|.$$

Proof We have that

$$\begin{aligned}
\|\mathbf{x} + \mathbf{y}\|^2 &= \langle \mathbf{x} + \mathbf{y}, \mathbf{x} + \mathbf{y} \rangle = \langle \mathbf{x}, \mathbf{x} + \mathbf{y} \rangle + \langle \mathbf{y}, \mathbf{x} + \mathbf{y} \rangle \\
&= \langle \mathbf{x}, \mathbf{x} \rangle + 2\langle \mathbf{x}, \mathbf{y} \rangle + \langle \mathbf{y}, \mathbf{y} \rangle \\
&= \|\mathbf{x}\|^2 + 2\langle \mathbf{x}, \mathbf{y} \rangle + \|\mathbf{y}\|^2 \\
&\leq \|\mathbf{x}\|^2 + 2\|\mathbf{x}\|.\|\mathbf{y}\| + \|\mathbf{y}\|^2 \qquad \text{(Cauchy–Schwarz)} \\
&= (\|\mathbf{x}\| + \|\mathbf{y}\|)^2.
\end{aligned}$$

The diagram indicates why the triangle inequality is the analytic equivalent of the geometric proposition that one side of a triangle is shorter than the sum of the other two sides.

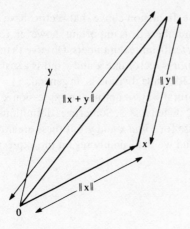

18.10   Distance

The idea of the distance $|x - y|$ between two real numbers x and y is a vital one. The same is true of the distance between two vectors. We define the *distance* between two vectors **x** and **y** in $\mathbb{R}^n$ in the natural way to be the real number

$$\|\mathbf{x} - \mathbf{y}\|.$$

When interpreting this idea in $\mathbb{R}^n$, it is better to think of **x** and **y** as the points at the ends of the arrows rather than as the arrows themselves.

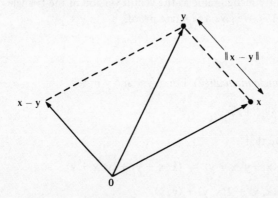

18.11   *Example* The distance between the vectors **x** = (1, 2, 3) and **y** = (2, 0, 5) in $\mathbb{R}^3$ is given by

$$\|\mathbf{y} - \mathbf{x}\| = \{(2 - 1)^2 + (0 - 2)^2 + (5 - 3)^2\}^{1/2} = 3.$$

18.12 Direction

With the exception of the zero vector

$$\mathbf{0} = (0, 0, 0, \ldots, 0)$$

all vectors $\mathbf{x}$ in $\mathbb{R}^n$ determine a *direction*. For example, the two-dimensional vector $(2, -2)$ points in a 'south-easterly' direction.

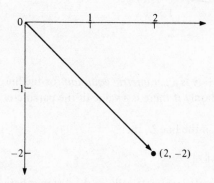

A *unit vector* $\mathbf{u}$ is a vector of length 1, i.e. which has $\|\mathbf{u}\| = 1$. When using a vector $\mathbf{v}$ to specify a direction, it is often convenient to replace $\mathbf{v}$ by the unit vector which points in the same direction. For example, the vector $\mathbf{v} = (1, 1, 2)$ is *not* a unit vector because

$$\|\mathbf{v}\| = \{1^2 + 1^2 + 2^2\}^{1/2} = \sqrt{6}.$$

But the vector

$$\mathbf{u} = \|\mathbf{v}\|^{-1}\mathbf{v} = \frac{1}{\sqrt{6}}\mathbf{v} = \left(\frac{1}{\sqrt{6}}, \frac{1}{\sqrt{6}}, \frac{2}{\sqrt{6}} \right)$$

is a unit vector which points in the same directions as $\mathbf{v}$.

18.13 Lines

If $\boldsymbol{\xi}$ and $\mathbf{v}$ are in $\mathbb{R}^n$ and $\mathbf{v} \neq 0$, then the set

$$L = \{\boldsymbol{\xi} + t\mathbf{v}: t \in \mathbb{R}\}$$

may be interpreted geometrically as the straight *line* which passes through the point $\boldsymbol{\xi}$ in the direction $\mathbf{v}$. The diagram indicates three points on this line corresponding to the values $t = t_1$, $t = t_2$ and $t = t_3$. The point $\boldsymbol{\xi}$ itself, of course, corresponds to the value $t = 0$.

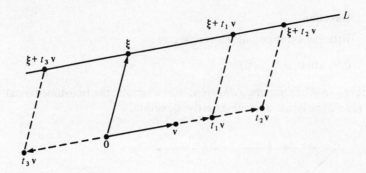

We sometimes say that $\mathbf{x} = \boldsymbol{\xi} + t\mathbf{v}$ is a *parametric equation* for the line L because $\mathbf{x}$ lies on the line L if and only if there is a value of the parameter t such that $\mathbf{x} = \boldsymbol{\xi} + t\mathbf{v}$.

If $\mathbf{v}$ is a **unit** vector and $\mathbf{x}$ lies on the line L, then

$$\|\mathbf{x} - \boldsymbol{\xi}\| = \|t\mathbf{v}\| = |t|.\|\mathbf{v}\| = |t|.$$

Thus, in this case, t may be interpreted geometrically as the distance between $\mathbf{x}$ and $\boldsymbol{\xi}$. Note that t is positive for values $\mathbf{x}$ on L which lie to one side of $\boldsymbol{\xi}$ and negative for values of $\mathbf{x}$ to the other side of $\boldsymbol{\xi}$.

18.14 *Example* The line in $\mathbb{R}^3$ passing through $(1, 2, 3)$ in the direction $(2, 0, 5)$ has the parametric equation

$$(x_1, x_2, x_3) = (1, 2, 3) + t(2, 0, 5)$$

$$(x_1, x_2, x_3) = (1 + 2t, 2, 3 + 5t).$$

This vector equation may be written more painfully as three scalar equations:

$$\left.\begin{array}{l} x_1 = 1 + 2t \\ x_2 = 2 \\ x_3 = 3 + 5t. \end{array}\right\}$$

Eliminating t, we obtain that the line is determined by the two equations

$$\left.\begin{array}{l} 5x_1 - 2x_3 = -1 \\ x_2 = 2. \end{array}\right\}$$

18.15 Hyperplanes

If $\boldsymbol{\xi}$ and $\mathbf{v}$ are in $\mathbb{R}^n$ and $\mathbf{v} \neq \mathbf{0}$, the set

$$H = \{\mathbf{x} : \mathbf{x} \in \mathbb{R}^n \text{ and } \langle \mathbf{x} - \boldsymbol{\xi}, \mathbf{v} \rangle = 0\}$$

may be interpreted geometrically as the *hyperplane* which passes through $\boldsymbol{\xi}$ and is orthogonal to the vector $\mathbf{v}$. We say that $\mathbf{v}$ is a *normal* to this hyperplane.

In $\mathbb{R}^3$, a hyperplane is the same thing as a plane. The diagram below shows a typical $\mathbf{x}$ in $\mathbb{R}^3$ for which $\mathbf{x} - \boldsymbol{\xi}$ is orthogonal to $\mathbf{v}$. The set H of all such $\mathbf{x}$ is a plane in $\mathbb{R}^3$ with normal $\mathbf{v}$.

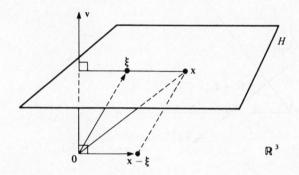

In $\mathbb{R}^2$, a hyperplane is the same thing as a line. The diagram below shows a typical $\mathbf{x}$ in $\mathbb{R}^2$ for which $\mathbf{x} - \boldsymbol{\xi}$ is orthogonal to $\mathbf{v}$. The set H of all such $\mathbf{x}$ is a line in $\mathbb{R}^2$ with normal $\mathbf{v}$.

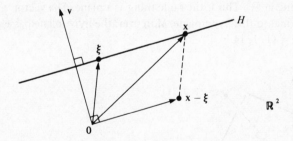

If we put $c = \langle \boldsymbol{\xi}, \mathbf{v} \rangle$, then the equation $\langle \mathbf{x} - \boldsymbol{\xi}, \mathbf{v} \rangle = 0$ becomes

$$\langle \mathbf{x}, \mathbf{v} \rangle = c$$

which we may rewrite more clumsily in the form

$$x_1 v_1 + x_2 v_2 + \ldots + x_n v_n = c.$$

Thus a hyperplane is determined by one linear equation.

18.16   Example The equation

$$3x_1 + 4x_2 = 5$$

describes a hyperplane in $\mathbb{R}^2$. This is the same thing as a line. The vector **v** = (3, 4) is a normal to this hyperplane. Moreover, the hyperplane passes through the point **ξ** = (3, −1) because $3 \times 3 + 4 \times (-1) = 5$.

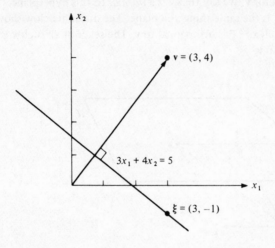

18.17   Example The equation

$$x_1 + 2x_2 + 2x_3 = 5$$

describes a hyperplane in $\mathbb{R}^3$. This is the same thing as a plane. The vector **v** = (1, 2, 2) is a normal to this hyperplane. Moreover, the hyperplane passes through the point **ξ** = (1, 1, 1).

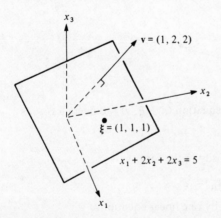

18.18 Flats

A set of hyperplanes in $\mathbb{R}^n$ determines a *flat* (or *affine set*) which consists of the set of all points which belong to each of the given hyperplanes, i.e. it is the set in which the hyperplanes meet.

By considering sets of hyperplanes containing one, two or three elements, we see that a flat in $\mathbb{R}^3$ may be a plane, a line or a point.

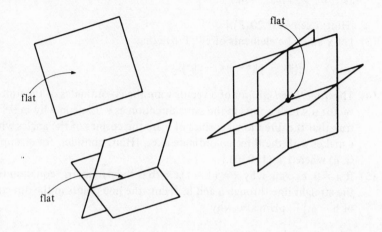

We also classify the empty set $\emptyset$ and the whole space $\mathbb{R}^n$ as (degenerate) flats. In $\mathbb{R}^2$ the only non-degenerate flats are points and lines.

18.19 Example In example 18.14 we saw that the parametric equation

$$(x_1, x_2, x_3) = (1, 2, 3) + t(2, 0, 5)$$

for the line through $(1, 2, 3)$ in the direction $(2, 0, 5)$ can be re-expressed as the two scalar equations

$$\left. \begin{array}{r} 5x_1 - 2x_3 = -1 \\ x_2 = 2. \end{array} \right\}$$

Each of these two equations define a hyperplane in $\mathbb{R}^3$ and the line with which we began is the flat consisting of all points which are common to both hyperplanes.

18.20 Exercise

(1) Let $\mathbf{x} = (0, 1, 0)$ and $\mathbf{y} = (1, 1, 0)$ be vectors in $\mathbb{R}^3$. Calculate the quantities

(i) $\mathbf{x} + \mathbf{y}$ (ii) $\mathbf{x} - \mathbf{y}$ (iii) $2\mathbf{x}$
(iv) $\|\mathbf{x}\|$ (v) $\|\mathbf{x} - \mathbf{y}\|$ (vi) $\langle \mathbf{x}, \mathbf{y} \rangle$.

What is the length of the vector $\mathbf{x}$? What are the distance and the angle between the vectors $\mathbf{x}$ and $\mathbf{y}$?

(2) Let $\mathbf{c}$ and $\mathbf{d}$ be elements of $\mathbb{R}^n$. Prove that

$$\|\mathbf{c} - \mathbf{d}\| \geqslant |\|\mathbf{c}\| - \|\mathbf{d}\||.$$

[Hint: exercise 1.20(2).]

(3) Let $\mathbf{x}$ and $\mathbf{y}$ be elements of $\mathbb{R}^n$. Prove that

$$\langle \mathbf{x}, \mathbf{y} \rangle = \tfrac{1}{4}\{\|\mathbf{x} + \mathbf{y}\|^2 - \|\mathbf{x} - \mathbf{y}\|^2\}.$$

(4) The *directional cosines* of a vector $\mathbf{v}$ are the co-ordinates of the unit vector $\mathbf{u}$ which points in the same direction as $\mathbf{v}$. Explain why in $\mathbb{R}^3$ it is true that the directional cosines of $\mathbf{v}$ are the cosines of the angles which $\mathbf{v}$ makes with the three co-ordinate axes. [Hint: consider, for example, $\langle \mathbf{i}, \mathbf{u} \rangle$ where $\mathbf{i} = (1, 0, 0)$.]

(5) If $\mathbf{a} \neq \mathbf{b}$, explain why $\mathbf{x} = (1 - t)\mathbf{a} + t\mathbf{b}$ is the parametric equation for the straight line through $\mathbf{a}$ and $\mathbf{b}$. [Hint: the line points in the direction of $\mathbf{b} - \mathbf{a}$.] Explain also why

$$\{(1 - t)\mathbf{a} + t\mathbf{b} : 0 \leqslant t \leqslant 1\}$$

is the set of all points which lie on the straight line segment joining $\mathbf{a}$ and $\mathbf{b}$.

(6) Offer a geometric argument to justify the assertion that the shortest distance from $\mathbf{0}$ to the hyperplane described by $\langle \mathbf{x}, \mathbf{u} \rangle = c$ is equal to $|c|$ provided that $\mathbf{u}$ is a unit vector.

18.21   Vector functions

Previously we have considered only functions $f: A \to B$ in the case when A and B are sets of real numbers. In this chapter, however, we shall be concerned with functions $f: A \to B$ for which A is a subset of $\mathbb{R}^n$ and B is a subset of $\mathbb{R}^m$. Such a function assigns to each vector $\mathbf{x} \in A$ a unique vector $\mathbf{y} \in B$. We call $\mathbf{y}$ the *image* of $\mathbf{x}$ and write $\mathbf{y} = f(\mathbf{x})$.

The graph of such a function is the set G in $\mathbb{R}^{n+m}$ defined by

$$G = \{(\mathbf{x}, f(\mathbf{x})): \mathbf{x} \in A\}.$$

If $m + n = 3$, the graph of the function f can be illustrated by a three-dimensional diagram. If $n = 2$ and $m = 1$, the graph is a surface and if $n = 1$ and $m = 2$ the graph is a curve.

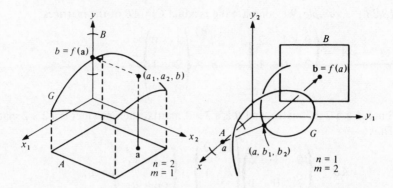

If $m + n > 3$, it is not possible to draw a useful representation of the graph of the function f. Occationally, however, some insight can be obtained by studying a picture like that drawn below for the case $n = 2$ and $m = 2$.

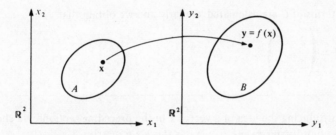

In the next section we shall illustrate the idea of a vector function by giving some examples of affine functions.

18.22 Linear and affine functions

We begin by recalling the definition of the product $C = AB$ of two *matrices* A and B. For this to make sense, the number of columns of A must equal the number of rows of B. In this case

$$\begin{pmatrix} c_{11} & c_{12} & \ldots c_{1p} \\ c_{21} & c_{22} & \ldots c_{2p} \\ \vdots \\ c_{m1} & c_{m2} \ldots c_{mp} \end{pmatrix} = \begin{pmatrix} a_{11} & a_{12} & \ldots a_{1n} \\ a_{21} & a_{22} & \ldots a_{2n} \\ \vdots \\ a_{m1} & a_{m2} & \ldots a_{2mn} \end{pmatrix} \begin{pmatrix} b_{11} & b_{12} \ldots b_{1p} \\ b_{21} & b_{22} \ldots b_{2p} \\ \vdots \\ b_{n1} & b_{n2} \ldots b_{np} \end{pmatrix}$$

where

$$c_{ij} = \sum_{k=1}^{n} a_{ik}b_{kj},$$

18.23   Example We compute the product $C = AB$ of the matrices

$$A = \begin{pmatrix} 0 & 1 & 2 \\ 2 & 0 & 1 \end{pmatrix} \qquad B = \begin{pmatrix} 1 & 0 \\ 2 & 1 \\ 0 & 2 \end{pmatrix}.$$

Since A is a 2×3 matrix and B is a 3×2 matrix, their product is a 2×2 matrix. Thus

$$AB = \begin{pmatrix} 0 & 1 & 2 \\ 2 & 0 & 1 \end{pmatrix} \begin{pmatrix} 1 & 0 \\ 2 & 1 \\ 0 & 2 \end{pmatrix} = \begin{pmatrix} c_{11} & c_{12} \\ c_{21} & c_{22} \end{pmatrix} = C.$$

The term in the ith row and jth column of C is calculated using the ith row of A and the jth column of B. Thus

$$c_{21} = 2 \times 1 + 0 \times 2 + 1 \times 0 = 2.$$

The other terms of C are calculated similarly and we obtain that

$$C = \begin{pmatrix} 2 & 5 \\ 2 & 2 \end{pmatrix}.$$

When introducing the idea of a vector $\mathbf{x}$ in the preceding sections, we choose to write its co-ordinates $x_1, x_2, \ldots, x_n$ as the $1 \times n$ matrix

$$(x_1, x_2, \ldots, x_n).$$

Such a matrix has only a single row and is therefore called a *row vector*. An alternative would have been to write the co-ordinates in the form of the *column vector*

$$\begin{pmatrix} x_1 \\ x_2 \\ \vdots \\ x_n \end{pmatrix}$$

At first sight, this second alternative seems a rather clumsy piece of notation but, when working with matrices, it turns out to be the more convenient of the two possibilities. We shall therefore always interpret a vector which appears in a matrix equation as a *column* and to emphasise this fact we shall use the notation x for a vector in a matrix equation rather than the boldface $\mathbf{x}$ used hitherto.

A function $f: \mathbb{R}^n \to \mathbb{R}^m$ is said to be *linear* if there exists an $m \times n$ matrix L such that $y = f(\mathbf{x})$ if and only if

$$y = Lx.$$

This last equation is an abbreviation for

$$
\begin{pmatrix} y_1 \\ y_2 \\ \vdots \\ y_m \end{pmatrix} = \begin{pmatrix} l_{11} & l_{12} & \ldots & l_{1n} \\ l_{21} & l_{22} & \ldots & l_{2n} \\ \vdots & & & \\ l_{m1} & l_{m2} & \ldots & l_{mn} \end{pmatrix} \begin{pmatrix} x_1 \\ x_2 \\ \vdots \\ x_n \end{pmatrix}
$$

which, in turn, is an abbreviation for the system

$$
\left.\begin{aligned}
y_1 &= l_{11}x_1 + l_{12}x_2 + \ldots + l_{1n}x_n \\
y_2 &= l_{21}x_1 + l_{22}x_2 + \ldots + l_{2n}x_n \\
&\ldots\ldots\ldots\ldots\ldots\ldots\ldots\ldots\ldots \\
y_m &= l_{m1}x_1 + l_{m2}x_2 + \ldots + l_{mn}x_n
\end{aligned}\right\} \tag{1}
$$

of linear equations. A linear function $f: \mathbb{R}^n \to \mathbb{R}^m$ is therefore determined by a system of m linear equations of the form (1).

(The introduction of the notion of a linear function makes it possible to justify the seemingly arbitrary manner in which matrix multiplication is defined. The definition has been specifically chosen so that, if L represents the linear function $f: \mathbb{R}^m \to \mathbb{R}^p$ and M represents the linear function $g: \mathbb{R}^n \to \mathbb{R}^m$, then the function $f \circ g: \mathbb{R}^n \to \mathbb{R}^p$ will be represented by LM (see §7.9). This explains, among other things, why matrix multiplication is associative, i.e. $A(BC) = (AB)C$ but not necessarily commutative, i.e. $AB \neq BA$.)

A function $g: \mathbb{R}^n \to \mathbb{R}^m$ is said to be *affine* if there exists an $m \times n$ matrix L and an $m \times 1$ column vector c such that $\mathbf{y} = g(\mathbf{x})$ if and only if

$$
y = Lx + c
$$

which we may write less compactly as

$$
\begin{aligned}
y_1 &= l_{11}x_1 + l_{12}x_2 + \ldots + l_{1n}x_n + c_1 \\
y_2 &= l_{21}x_1 + l_{22}x_2 + \ldots + l_{2n}x_n + c_2 \\
&\ldots\ldots\ldots\ldots\ldots\ldots\ldots\ldots\ldots\ldots\ldots\ldots \\
y_m &= l_{m1}x_1 + l_{m2}x_2 + \ldots + l_{mn}x_n + c_m.
\end{aligned} \tag{2}
$$

18.24 Examples

(i) An affine function $g: \mathbb{R}^1 \to \mathbb{R}^1$ is defined by an equation of the form

$$
y = lx + c.
$$

Its graph is therefore a straight line in $\mathbb{R}^2$.

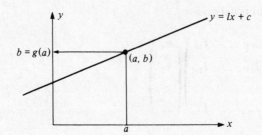

The function g is linear if and only if $c = 0$.

(ii) An affine function $g\colon \mathbb{R}^2 \to \mathbb{R}^1$ is defined by an equation of the form

$$y = (l_1, l_2)\begin{pmatrix} x_1 \\ x_2 \end{pmatrix} + c$$

i.e.

$$y = l_1 x_1 + l_2 x_2 + c.$$

As we known from §18.15, the graph of g therefore is a plane in $\mathbb{R}^3$ (with normal $(l_1, l_2, -1)$).

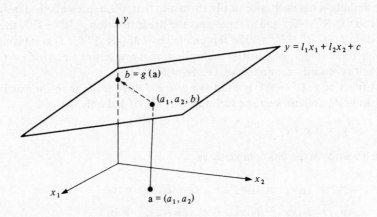

The function g is linear if and only if $c = 0$.

(iii) An affine function $g\colon \mathbb{R}^1 \to \mathbb{R}^2$ is defined by an equation of the form

$$\begin{pmatrix} y_1 \\ y_2 \end{pmatrix} = \begin{pmatrix} l_1 \\ l_2 \end{pmatrix} x + \begin{pmatrix} c_1 \\ c_2 \end{pmatrix}$$

i.e.

$$\left.\begin{aligned} y_1 &= l_1 x + c_1 \\ y_2 &= l_2 x + c_2. \end{aligned}\right\}$$

These two equations define a line in $\mathbb{R}^3$ (see example 18.19) which is the graph of the function g.

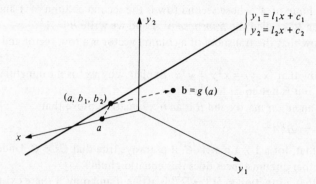

The function f is linear if and only if $c_1 = c_2 = 0$.

(iv) An affine function $g: \mathbb{R}^2 \rightarrow \mathbb{R}^2$ is defined by an equation of the form

$$\begin{pmatrix} y_1 \\ y_2 \end{pmatrix} = \begin{pmatrix} l_{11} & l_{12} \\ l_{21} & l_{22} \end{pmatrix} \begin{pmatrix} x_1 \\ x_2 \end{pmatrix} + \begin{pmatrix} c_1 \\ c_2 \end{pmatrix}$$

i.e.

$$y_1 = l_{11}x_1 + l_{12}x_2 + c_1$$

$$y_2 = l_{21}x_1 + l_{22}x_2 + c_2.$$

It is not possible to draw a graph of g but it is often useful to note that g maps any system of parallel lines in $\mathbb{R}^2$ onto another system of parallel lines in $\mathbb{R}^2$.

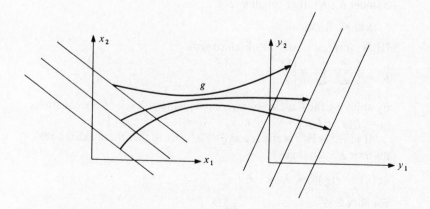

18.25   Exercise

(1) If A is an $m \times n$ matrix and B is the $n \times m$ matrix whose first row is the first column of A, whose second row is the second column of A and so on, then B is called the *transpose* of A and we write $B = A^T$.

(i) Show that the transpose of a column vector is a row vector and vice versa.

(ii) Show that $\langle x, y \rangle = x^T y = y^T x$. Explain why yx^T is meaningful as a matrix but is not equal to $x^T y$.

(2) If A is an $m \times n$ matrix and B is an $n \times p$ matrix, prove that

$$(AB)^T = B^T A^T.$$

Show that, for a 1×1 matrix C, it is always true that $C^T = C$. Under what other circumstances does this equation hold?

(3) Prove that a function $g: \mathbb{R}^n \to \mathbb{R}^m$ is affine if and only if there exists a linear function $f: \mathbb{R}^n \to \mathbb{R}^m$ and constant vectors $\xi \in \mathbb{R}^n$, $\eta \in \mathbb{R}^m$ such that

$$g(x) = f(x - \xi) + \eta$$

for all $x \in \mathbb{R}^n$.

(4) Explain why the graph of an affine function $g: \mathbb{R}^n \to \mathbb{R}^m$ is a flat in $\mathbb{R}^{n+m}$. [Hint: rewrite equations (2) of §18.22 in terms of inner products involving the vector $(x_1, x_2, \ldots, x_n, y_1, y_2, \ldots, y_m)$.]

(5) A function $f: \mathbb{R}^n \to \mathbb{R}^m$ is usually defined to be linear if and only if

$$f(\alpha x + \beta y) = \alpha f(x) + \beta f(y)$$

for all x and y in $\mathbb{R}^n$ and for all scalars α and β. Show that this definition is equivalent to that of §18.22. [Hint: take L to be the matrix whose jth column is the vector $f(e_j)$ where the co-ordinates of e_j are all zero except for the jth which is equal to one.]

(6) If $f: \mathbb{R}^n \to \mathbb{R}^m$ is a linear function, prove that there exists a real number K such that for all $x \in \mathbb{R}^n$,

$$\|f(x)\| \leqslant K\|x\|.$$

[Hint: this result may be obtained with

$$K = \left\{ \sum_{i=1}^{m} \sum_{j=1}^{n} l_{ij}^2 \right\}^{1/2}$$

by applying the Cauchy–Schwarz inequality to each of the equations $y_i = l_{i1}x_1 + l_{i2}x_2 + \ldots + l_{in}x_n \quad (i = 1, 2, \ldots, m)$.]

If $g: \mathbb{R}^n \to \mathbb{R}^m$ is affine and $\xi \in \mathbb{R}^n$, prove that there exists a real number K such that

$$\|g(x) - g(\xi)\| \leqslant K\|x - \xi\|$$

for all $x \in \mathbb{R}^n$.

18.26 Convergence of sequences in $\mathbb{R}^n$

The convergence of a sequence $\langle x_k \rangle$ of vectors in $\mathbb{R}^n$ is defined exactly as for sequences of real numbers. We say that

$$x_k \to 1 \text{ as } k \to \infty$$

if and only if, given any $\epsilon > 0$, we can find a K such that for any $k > K$

$$\| x_k - 1 \| < \epsilon.$$

Recall that the real number $\| x_k - 1 \|$ is the *distance* between the vector x_k and the vector 1. The definition simply requires that we can make this distance as small as we choose by taking k large enough, i.e.

$$\| x_k - 1 \| \to 0 \text{ as } k \to \infty.$$

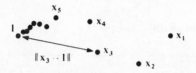

18.27 *Example* Consider the sequence $\langle x_k \rangle$ in $\mathbb{R}^2$ defined by

$$x_k = \left(\frac{k-1}{k}, \frac{k+1}{k} \right) \quad (k = 1, 2, \ldots).$$

We shall show that $x_k \to e$ as $k \to \infty$, where $e = (1, 1)$. Observe that

$$x_k - e = \left\{ \left(\frac{k-1}{k} - 1 \right)^2 + \left(\frac{k+1}{k} - 1 \right)^2 \right\}^{1/2}$$

$$= \sqrt{(2/k)} \to 0 \text{ as } k \to \infty$$

and the result follows.

Although the proof given in the preceding example is very easy, the following theorem shows that an even easier proof is available.

18.28 *Theorem* A sequence $\langle x_k \rangle$ of vectors in $\mathbb{R}^n$ converges to $1 \in \mathbb{R}^n$ if and only if the co-ordinates of x_k converge to the corresponding co-ordinates of 1.

Proof To avoid cumbersome notation, we work in $\mathbb{R}^2$ although the proof for $\mathbb{R}^n$ is just the same. Let $x_k = (a_k, b_k)$ and let $1 = (\lambda, \mu)$. We need to show that $x_k \to 1$ as $k \to \infty$ if and only if $a_k \to \lambda$ as $k \to \infty$ and $b_k \to \mu$ as $k \to \infty$.

(i) Suppose that $x_k \to 1$ as $k \to \infty$. Then $\| x_k - 1 \| \to 0$ as $k \to \infty$. Hence

$$|a_k - \lambda| \leqslant \{(a_k - \lambda)^2 + (b_k - \mu)^2\}^{1/2}$$
$$= \|\mathbf{x}_k - \mathbf{l}\| \to 0 \text{ as } k \to \infty$$

and so $a_k \to \lambda$ as $k \to \infty$ by Corollary 4.11. Similarly $b_k \to \mu$ as $k \to \infty$.

(ii) Suppose that $a_k \to \lambda$ as $k \to \infty$ and $b_k \to \mu$ as $k \to \infty$. Then

$$\|\mathbf{x}_k - \mathbf{l}\| = \{(a_k - \lambda)^2 + (b_k - \mu)^2\}^{1/2}$$
$$\to \{0^2 + 0^2\}^{1/2} = 0 \quad \text{as } k \to \infty$$

and so $\mathbf{x}_k \to \mathbf{l}$ as $k \to \infty$.

18.29 *Example* Returning to example 18.27, we observe that

$$\frac{k-1}{k} \to 1 \text{ as } k \to \infty; \qquad \frac{k+1}{k} \to 1 \text{ as } k \to \infty$$

and hence $\mathbf{x}_k \to \mathbf{e}$ as $k \to \infty$ by theorem 18.28.

Since the definition of convergence in $\mathbb{R}^n$ is the same as that given for real numbers, it is not surprising that some theorems which are true for sequences of real numbers should also be true for sequences of vectors. In fact, all of the results proved in chapters 4, 5 and 6 which make sense in $\mathbb{R}^n$ are true in $\mathbb{R}^n$. Usually exactly the same proof suffices to prove the theorems in $\mathbb{R}^n$ except that the norm sign $\|$ replaces the modulus sign $|$ where appropriate. Often, however, it is possible to construct an essentially trivial proof by appealing to the corresponding result for real numbers. As an example, we give the generalization of proposition 4.8(i) to $\mathbb{R}^n$.

18.30 *Theorem* Let $\langle \mathbf{x}_k \rangle$ and $\langle \mathbf{y}_k \rangle$ be sequences in $\mathbb{R}^n$ for which $\mathbf{x}_k \to \mathbf{l}$ as $k \to \infty$ and $\mathbf{y}_k \to \mathbf{m}$ as $k \to \infty$. Then, for any scalars λ and μ,

$$\lambda \mathbf{x}_k + \mu \mathbf{y}_k \to \lambda \mathbf{l} + \mu \mathbf{m} \text{ as } k \to \infty.$$

Proof The theorem follows immediately from proposition 4.8(i) and theorem 18.28.

Note that proposition 4.8(ii) and 4.8(iii) do not make any immediate sense when translated into results about $\mathbb{R}^n$. (See exercise 18.45(2).) This is because multiplication and division in $\mathbb{R}^n$ have no natural analogues. Usually, however, the reason that a theorem which is valid for real numbers fails in $\mathbb{R}^n$ is that its statement depends (sometimes implicitly) on the fact that the real numbers can be ordered in terms of increasing magnitude whereas vectors cannot.

Finally, we observe in passing that the very important Bolzano–Weierstrass theorem generalises to $\mathbb{R}^n$ but that the proof given in chapter 5 does not since it depends on theorem 5.9 which makes no sense in $\mathbb{R}^n$. (See exercise 18.45(3).)

18.31 Convergence of functions

As with sequences, the definition of convergence for vector functions is the same as that given in the case of real numbers. We say that

$$f(\mathbf{x}) \to \mathbf{l} \text{ as } \mathbf{x} \to \boldsymbol{\xi}$$

if and only if, given any $\epsilon > 0$ we can find a $\delta > 0$ such that

$$\|f(\mathbf{x}) - \mathbf{l}\| < \epsilon$$

provided that $0 < \|\mathbf{x} - \boldsymbol{\xi}\| < \delta$. This definition makes sense for any function $f: A \to B$ for which $A \subset \mathbb{R}^n$ and $B \subset \mathbb{R}^m$ provided that there exists a real number $\Delta > 0$ such that A contains all vectors $\mathbf{x}$ satisfying $0 < \|\mathbf{x} - \boldsymbol{\xi}\| < \Delta$.

The definition requires that we can make the distance between $f(\mathbf{x})$ and $\mathbf{l}$ as small as we choose by taking $\mathbf{x}$ sufficiently close to $\boldsymbol{\xi}$. (As explained in §8.6, we exclude consideration of what happens when $\mathbf{x} = \boldsymbol{\xi}$.) The diagram illustrates the case $m = n = 2$. However small the disc E centred at $\mathbf{l}$ we must be able to find a disc centred at $\boldsymbol{\xi}$ such that $\mathbf{x} \in D$ implies $f(\mathbf{x}) \in E$ (unless $\mathbf{x} = \boldsymbol{\xi}$).

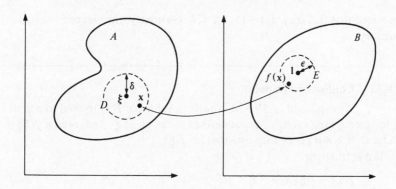

It is often useful to note that $f(\mathbf{x}) \to f(\boldsymbol{\xi})$ as $\mathbf{x} \to \boldsymbol{\xi}$ if and only if

$$\|f(\mathbf{x}) - f(\boldsymbol{\xi})\| \to 0 \text{ as } \mathbf{x} \to \boldsymbol{\xi}.$$

18.32   *Example* Consider the linear function $f: \mathbb{R}^2 \to \mathbb{R}^2$ defined by $\mathbf{y} = f(\mathbf{x})$ where

$$\begin{pmatrix} y_1 \\ y_2 \end{pmatrix} = \begin{pmatrix} 1 & 1 \\ 1 & -1 \end{pmatrix} \begin{pmatrix} x_1 \\ x_2 \end{pmatrix}$$

i.e.

$$\left. \begin{aligned} y_1 &= x_1 + x_2 \\ y_2 &= x_1 - x_2. \end{aligned} \right\}$$

We prove that $f(x_1, x_2) \to (2, 0)$ as $(x_1, x_2) \to (1, 1)$. It is necessary to show that, given any $\epsilon > 0$, we can find a $\delta > 0$ such that

$$\|(x_1 + x_2, x_1 - x_2) - (2, 0)\| < \epsilon$$

provided that $0 < \|(x_1, x_2) - (1, 1)\| < \delta$.

Observe to begin with that

$$\|(x_1, x_2) - (1, 1)\| = \{(x_1 - 1)^2 + (x_2 - 1)^2\}^{1/2}.$$

Also

$$
\begin{aligned}
\|(x_1 + x_2, x_1 - x_2) - (2, 0)\| &= \{(x_1 + x_2 - 2)^2 + (x_1 - x_2)^2\}^{1/2} \\
&= \{((x_1 - 1) + (x_2 - 1))^2 + ((x_1 - 1) - (x_2 - 1))^2\}^{1/2} \\
&= \{2(x_1 - 1)^2 + 2(x_2 - 1)^2\}^{1/2} \\
&= \sqrt{2}\|(x_1, x_2) - (1, 1)\|.
\end{aligned}
$$

Given any $\epsilon > 0$, we therefore have to choose $\delta > 0$ so that

$$\sqrt{2}\|(x_1, x_2) - (1, 1)\| < \epsilon$$

provided that $0 < \|(x_1, x_2) - (1, 1)\| < \delta$. Obviously, the choice $\delta = \epsilon/\sqrt{2}$ suffices.

18.33   Continuity at a point

As explained in §8.6, it is quite possible for $f(\mathbf{x})$ to tend to a limit l as $\mathbf{x}$ tends to $\boldsymbol{\xi}$ even though f is not defined at the point $\boldsymbol{\xi}$. And, even if $f(\boldsymbol{\xi})$ *is* defined, it is *not* necessarily true that $l = f(\boldsymbol{\xi})$.

If it *is* true that

$$f(\mathbf{x}) \to f(\boldsymbol{\xi}) \text{ as } \mathbf{x} \to \boldsymbol{\xi}$$

we say that f is *continuous at the point* $\boldsymbol{\xi}$.

18.34   Example Suppose that $g \colon \mathbb{R}^n \to \mathbb{R}^m$ is affine. Then f is continuous at every point $\boldsymbol{\xi}$.

Proof By exercise 18.25(6), there exist a real number K such that, for all $\mathbf{x} \in \mathbb{R}^n$,

$$\|g(\mathbf{x}) - g(\boldsymbol{\xi})\| \leqslant K\|\mathbf{x} - \boldsymbol{\xi}\|.$$

Given any $\epsilon > 0$, we have to find a $\delta > 0$ such that $\|g(\mathbf{x}) - g(\boldsymbol{\xi})\| < \epsilon$ provided that $0 < \|\mathbf{x} - \boldsymbol{\xi}\| < \delta$. Obviously the choice $\delta = \epsilon/K$ suffices (unless $K = 0$ in which case any value of $\delta > 0$ will do).

Note that this example generalises both example 18.32 and example 8.7.

18.35   Properties of limits

All of the results of chapter 8 which make sense for functions
$f: \mathbb{R}^n \to \mathbb{R}^m$ remain true and with the same proofs. Note in particular theorems
8.9 and 8.17. The generalisation of proposition 8.12 is particularly important
and so we quote it separately here.

18.36   *Proposition*   Suppose that $f: A \to \mathbb{R}^p$ and $g: A \to \mathbb{R}^q$, where A is a set in
$\mathbb{R}^n$ containing all points $\mathbf{x} \in \mathbb{R}^n$ satisfying $0 < \|\mathbf{x} - \boldsymbol{\xi}\| < \Delta$. Let $f(\mathbf{x}) \to \mathbf{l}$ as $\mathbf{x} \to \boldsymbol{\xi}$
and $g(\mathbf{x}) \to \mathbf{m}$ as $\mathbf{x} \to \boldsymbol{\xi}$ and suppose that λ and μ are any real numbers. Then

> (i) $\lambda f(\mathbf{x}) + \mu g(\mathbf{x}) \to \lambda \mathbf{l} + \mu \mathbf{m}$ (provided $p = q$)
>
> (ii) $f(\mathbf{x}) g(\mathbf{x}) \to \mathbf{lm}$ (provided $p = 1$ or $q = 1$)
>
> (iii) $f(\mathbf{x})/g(\mathbf{x}) \to \mathbf{l}/\mathbf{m}$ (provided $q = 1$ and $\mathbf{m} \neq 0$)

as $\mathbf{x} \to \boldsymbol{\xi}$.

Theorem 18.28, of course, has an analogue for the convergence of functions
which we quote below. This proposition may be deduced from theorem 18.28
with the help of the vector generalisation of theorem 8.9.

18.37   *Proposition*   Suppose that $f: A \to \mathbb{R}^m$, where A is a set in $\mathbb{R}^n$ containing
all points x satisfying $0 < \|\mathbf{x} - \boldsymbol{\xi}\| < \Delta$. Then $f(\mathbf{x}) \to \mathbf{l}$ as $\mathbf{x} \to \boldsymbol{\xi}$ if and only if the
co-ordinates of $f(\mathbf{x})$ converge to the corresponding co-ordinates of l as $\mathbf{x} \to \boldsymbol{\xi}$.

A function $P: \mathbb{R}^n \to \mathbb{R}$ is called a *polynomial* if the value of $P(x_1, x_2, \ldots, x_n)$
is obtained from $x_1, x_2, \ldots, x_n$ and a finite number of constants using a finite
number of additions and multiplications. Thus the function $P: \mathbb{R}^2 \to \mathbb{R}$ defined
by

$$P(x_1, x_2) = x_1^2 + 3x_1 x_2 - x_1^2 x_2^3 + 1$$

is a polynomial. Suppose that $P: \mathbb{R}^n \to \mathbb{R}$ and $Q: \mathbb{R}^n \to \mathbb{R}$ are polynomials and
that $A = \{\mathbf{x}: Q(\mathbf{x}) \neq 0\}$. Then the function $R: A \to \mathbb{R}$ defined by
$R(\mathbf{x}) = P(\mathbf{x})/Q(\mathbf{x})$ is called a *rational function*. Thus the function R defined
everywhere on $\mathbb{R}^2$ except at $(0, 0)$ by

$$R(x_1, x_2) = \frac{x_1^2 + 3x_1 x_2 - x_1^2 x_2^3 + 1}{x_1^2 + x_1 x_2 + x_2^2}$$

is a rational function.

These definitions generalise those of §7.6 and allow us to quote a
generalisation of theorem 8.13. The proof is much the same as that of theorem
8.13 but relies on propositions 18.36 and 18.37 rathre than proposition 8.12.

18.38   Proposition A function whose co-ordinates are all polynomials is continuous at every point. A function whose co-ordinates are all rational functions is continuous at every point at which it is defined.

18.39   Examples

(i) $\displaystyle\lim_{(x_1, x_2) \to (1, 1)} (x_1^2 - 1, x_1 x_2) = (1^2 - 1, 1.1) = (0, 1)$

because both co-ordinates are polynomials.

(ii) $\displaystyle\lim_{(x_1, x_2) \to (1, 1)} \left(\frac{x_1^2 + 3x_1 x_2 - x_1^2 x_2^3 + 1}{x_1^2 + x_1 x_2 + x_2^2} \right) = \frac{1 + 3 - 1 + 1}{1 + 1 + 1} = \frac{4}{3}$

because the single co-ordinate is a rational function defined at the point $(1, 1)$.

18.40   Limits along a path

Left and right hand limits do not make any sense when discussing functions $f \colon \mathbb{R}^n \to \mathbb{R}^m$ unless $n = 1$. When $n \geqslant 2$, the words 'left' and 'right' become meaningless as they depend on the notion of an ordering on $\mathbb{R}$. However, proposition 8.4 has a natural analogue when $n \geqslant 2$ in terms of the limits of functions along a path. To discuss this we first need to say something about the notion of a path in $\mathbb{R}^n$.

We define a *path* in $\mathbb{R}^n$ to be a function $g \colon [0, 1] \to \mathbb{R}^n$ which is continuous on $[0, 1]$. (See §9.1.) The idea is that as the real number t increases from 0 to 1, the vector $\mathbf{x} = g(t)$ describes a continous curve in $\mathbb{R}^n$ joining the points $\boldsymbol{\xi} = g(0)$ and $\boldsymbol{\eta} = g(1)$. The diagram illustrates the case $n = 2$.

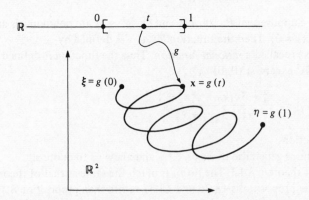

18.41 *Example* The path $g: [0, 1] \to \mathbb{R}^n$ defined by

$$g(t) = \boldsymbol{\xi} + t\mathbf{v}$$

corresponds geometrically to the straight line joining the points $\boldsymbol{\xi}$ and $\boldsymbol{\xi} + \mathbf{v}$. (See §18.13.)

Let A be a set in $\mathbb{R}^n$ which contains all points $\mathbf{x} \in \mathbb{R}^n$ which satisfy $0 < \|\mathbf{x} - \boldsymbol{\xi}\| < \Delta$ and let $g: [0, 1] \to \mathbb{R}^n$ be a path with $g(0) = \boldsymbol{\xi}$.

If $f: A \to \mathbb{R}^m$, we say that

$$f(\mathbf{x}) \to l \text{ as } \mathbf{x} \to \boldsymbol{\xi}$$

along the path defined by g
if and only if

$$f(g(t)) \to l \text{ as } t \to 0+.$$

The diagram illustrates the case of a function $f: \mathbb{R}^2 \to \mathbb{R}^1$.

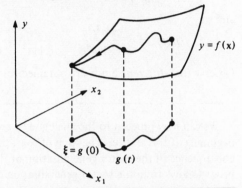

18.42 *Example* Consider the function f defined everywhere in $\mathbb{R}^2$ *except* at $(0, 0)$ by the formula

$$f(x_1, x_2) = \frac{x_2^2 - x_1^2}{x_2^2 + x_1^2}.$$

We shall consider what happens as (x_1, x_2) approaches $(0, 0)$ along the line $x_2 = \alpha x_1$. The appropriate path g is given by the equations

$$\left.\begin{array}{l} x_1 = t \\ x_2 = \alpha t \end{array}\right\}$$

and so

$$f(g(t)) = \frac{\alpha^2 t^2 - t^2}{\alpha^2 t^2 + t^2} = \frac{\alpha^2 - 1}{\alpha^2 + 1} \to \frac{\alpha^2 - 1}{\alpha^2 + 1} \text{ as } t \to 0+.$$

Note that f tends to *different* limits as (x_1, x_2) tends to $(0, 0)$ along different straight line paths.

18.43 *Example* Consider the function defined everywhere in $\mathbb{R}^2$ *except* at $(0, 0)$ by the formula

$$f(x_1, x_2) = \frac{x_1 x_2^2}{x_1^2 + x_2^4}.$$

For the straight line path of example 18.42 we have that

$$f(g(t)) = \frac{\alpha^2 t^3}{t^2 + \alpha^4 t^4} = \frac{\alpha^2 t}{1 + \alpha^4 t^2} \to 0 \text{ as } t \to 0+.$$

Note that f tends to the *same* limit as (x_1, x_2) approaches $(0, 0)$ along all straight line paths.

Next consider what happens as (x_1, x_2) approaches $(0, 0)$ along the parabola $x_2^2 = ax_1$. In this case the appropriate path h is given by the equations

$$\left. \begin{array}{l} x_1 = at^2 \\ x_2 = at \end{array} \right\}$$

and so

$$f(h(t)) = \frac{at^2 a^2 t^2}{a^2 t^4 + a^4 t^4} = \frac{a}{1 + a^2} \to \frac{a}{1 + a^2} \text{ as } t \to 0+.$$

Observe that *different* limits are obtained for different parabolas.

We can now proceed to the analogue of proposition 8.4 mentioned at the beginning of this section. The 'only if' part of the proof is an immediate consequence of the vector generalisation of theorem 8.17. The 'if' part of the proof follows from the vector generalisation of theorem 8.9.

18.44   Proposition Suppose that $f: A \to \mathbb{R}$ where A is a set in $\mathbb{R}^n$ containing all points $x \in \mathbb{R}^n$ which satisfy $0 < \|x - \xi\| < \Delta$. Then

$$f(x) \to l \text{ as } x \to \xi$$

if and only if $f(x) \to l$ as $x \to \xi$ *along all paths* g *with* $g(0) = \xi$.

Proposition 18.44 is often useful in detecting when limits do *not* exist. Thus in examples 18.42 and 18.43 we have that

$$\lim_{x \to 0} f(x)$$

does *not* exist. If the limit did exist the function would tend to this along *all* paths but the function tends to different values depending on which path is considered. Note that example 18.43 shows that it is *not* enough to consider only straight line paths.

18.45   Exercise

(1) Find the following limits:

(i) $\displaystyle \lim_{k \to \infty} \left(\frac{1}{k}, \frac{k-1}{k}, \frac{k^2-1}{k^2} \right)$

(ii) $\lim_{(x_1, x_2) \to (1, 1)} \left(\dfrac{x_1^2 - x_2^2}{x_1^2 + x_2^2}, \dfrac{x_1 - x_2}{x_1 + x_2} \right)$

(iii) $\lim_{(x_1, x_2) \to (0, 0)} (\sin(x_1 x_2), \cos(x_1 + x_2))$.

(2) Let $\langle x_k \rangle$ and $\langle y_k \rangle$ be sequences in $\mathbb{R}^n$ with the property that $x_k \to l$ as $k \to \infty$ and $y_k \to m$. Prove that

$\langle x_k, y_k \rangle \to \langle l, m \rangle$ as $k \to \infty$,

(3) Show that every sequence $\langle x_k \rangle$ in $\mathbb{R}^n$ which satisfies $\| x_k \| \leqslant H$ $(k = 1, 2, \ldots)$ has a convergent subsequence. This is the Bolzano–Weierstrass theorem for $\mathbb{R}^n$. (See theorem 5.10.) [Hint: apply the Bolzano–Weierstrass theorem for real numbers to each co-ordinate successively.]

(4) Prove directly from the definition of convergence (i.e. using ϵ and δ) that

$$\frac{x_1^3}{x_1^2 + x_2^2} \to 0 \text{ as } (x_1, x_2) \to (0, 0).$$

The rational function R given by

$$R(x_1, x_2) = \frac{x_1^3}{x_1^2 + x_2^2}$$

is defined everywhere in $\mathbb{R}^2$ except at $(0, 0)$. What value should be assigned to $R(0, 0)$ to make the function continuous at the point $(0, 0)$?

(5) A function $f \colon \mathbb{R}^2 \to \mathbb{R}$ is defined by

$$f(x_1, x_2) = \begin{cases} 1, & x_1 + x_2 \geqslant 2 \\ -1, & x_1 + x_2 < 2. \end{cases}$$

Determine the limits f approaches as (x_1, x_2) approaches $(1, 1)$ along all straight 'ine paths. Deduce that

$$\lim_{(x_1, x_2) \to (1, 1)} f(x_1, x_2)$$

does not exist. At what points in $\mathbb{R}^2$ is f continuous?

(6) Construct a function f defined everywhere in $\mathbb{R}^2$ except at $(0, 0)$ which approaches 0 as (x_1, x_2) approaches $(0, 0)$ along all paths considered in example 18.43 but for which

$$\lim_{(x_1, x_2) \to (0, 0)} f(x_1, x_2)$$

does not exist.

19 VECTOR DERIVATIVES

19.1 Directional derivatives

Suppose that $\boldsymbol{\xi}$ and $\mathbf{u}$ are vectors in $\mathbb{R}^n$ and that $\mathbf{u}$ is of unit length (i.e. $\|\mathbf{u}\| = 1$). We shall think of $\boldsymbol{\xi}$ as determining a point in $\mathbb{R}^n$ and of $\mathbf{u}$ as determining a direction. The set of all vectors of the form

$$\mathbf{x} = \boldsymbol{\xi} + t\mathbf{u} \quad (t \in \mathbb{R})$$

can then be thought of as the straight line through the point $\boldsymbol{\xi}$ in the direction of $\mathbf{u}$. The parameter t measures the distance from $\boldsymbol{\xi}$ to $\mathbf{x}$ along this line. (See §18.13.)

If $f: \mathbb{R}^n \to \mathbb{R}$, we can define a function $F: \mathbb{R} \to \mathbb{R}$ by

$$F(t) = f(\boldsymbol{\xi} + t\mathbf{u}).$$

In the case when $n = 2$, the geometry of the situation is very simple. One may regard $y = F(t)$ as the equation of the curve obtained by slicing the surface $y = f(\mathbf{x})$ with a vertical plane through the point $\boldsymbol{\xi}$ which is parallel to $\mathbf{u}$. The diagram below shows the surface $y = f(\mathbf{x})$ sliced by the appropriate vertical plane. The following diagram suppresses everything outside this vertical plane and relabels the point $\mathbf{x} = \boldsymbol{\xi} + t\mathbf{u}$ with the real number t. Correspondingly, $\boldsymbol{\xi}$ is relabelled with the real number 0.

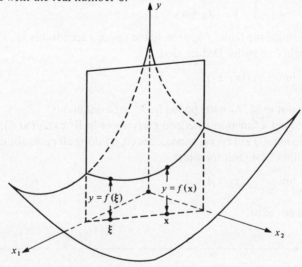

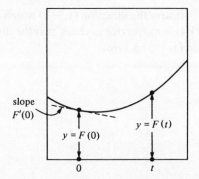

It is apparent from these diagrams that $F'(0)$ is the slope (or rate of increase) of the surface $y = f(\mathbf{x})$ at the point $\boldsymbol{\xi}$ in the direction $\mathbf{u}$. We call this important quantity the *directional derivative* of f at the point $\boldsymbol{\xi}$ in the direction $\mathbf{u}$ and denote it by $D_{\mathbf{u}}f(\boldsymbol{\xi})$. Recalling the definition of a derivative given in §10.1, we have that

$$D_{\mathbf{u}}f(\boldsymbol{\xi}) = \lim_{t \to 0} \frac{f(\boldsymbol{\xi} + t\mathbf{u}) - f(\boldsymbol{\xi})}{t}.$$

The same notation is also used when f is a vector-valued function and $\mathbf{u}$ is not necessarily a unit vector. But then of course the preceding geometric interpretation does not apply.

19.2 **Example** Let $f: \mathbb{R}^2 \to \mathbb{R}$ be defined by $f(x_1, x_2) = x_1 x_2$. Find the slope of the surface $y = f(x_1, x_2)$ at the point $(1, 2)$ in the direction $(3, 4)$.

The vector $(3, 4)$ is not a unit vector and so we begin by replacing it by the unit vector $(3/5, 4/5)$ which points in the same direction. We have to calculate

$$D_{\mathbf{u}}f(\boldsymbol{\xi}) = \lim_{t \to 0} \frac{f(\boldsymbol{\xi} + t\mathbf{u}) - f(\boldsymbol{\xi})}{t}$$

in the case when $\boldsymbol{\xi} = (1, 2)$ and $\mathbf{u} = (3/5, 4/5)$. We obtain that

$$D_{\mathbf{u}}f(\boldsymbol{\xi}) = \lim_{t \to 0} \frac{(1 + 3t/5)(2 + 4t/5) - (1)(2)}{t}$$

$$= \lim_{t \to 0} \left(2 + \frac{12t}{25}\right) = 2.$$

It follows that the slope of the surface $y = x_1 x_2$ at the point $(1, 2)$ in the direction $(3, 4)$ is equal to 2. The diagram illustrates the curve $x_1 x_2 = 2$ which is called a *contour* of the function f. Points on the surface $y = f(x_1, x_2)$ above this contour are all at the same height. The unbroken arrow drawn at $(1, 2)$ indicates the direction $(3, 4)$. The broken arrow indicates the direction $(1, -2)$ which is tangent to the contour $x_1 x_2 = 2$ (why?). It is instructive to check that the directional derivative at $(1, 2)$ in the direction $(1, -2)$ is zero.

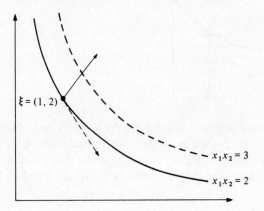

19.3 Partial derivatives

Suppose that $f: \mathbb{R}^n \to \mathbb{R}$. To avoid cumbersome notation we shall assume in what follows that $n = 2$ but the general case is entirely analogous.

If $\boldsymbol{\xi} \in \mathbb{R}^2$ the *partial derivatives* of f at $\boldsymbol{\xi}$ are defined by

$$f_1(\xi_1, \xi_2) = \lim_{h \to 0} \frac{f(\xi_1 + h, \xi_2) - f(\xi_1, \xi_2)}{h}$$

$$f_2(\xi_1, \xi_2) = \lim_{k \to 0} \frac{f(\xi_1, \xi_2 + k) - f(\xi_1, \xi_2)}{k}$$

provided that these limits exist.

It is no more difficult to calculate the partial derivatives of a function of a vector variable than it is to calculate the derivative of a function of a real variable . To calculate f_1, for example, one simply differentiates with respect to x_1 treating x_2 as a constant.

19.4   *Examples*

(i) Let $f: \mathbb{R}^2 \to \mathbb{R}$ be defined by $f(x_1, x_2) = x_1{}^2 x_2 + x_2{}^3 x_1$. Then

$$\left. \begin{aligned} f_1(x_1, x_2) &= 2x_1 x_2 + x_2{}^3 \\ f_2(x_1, x_2) &= x_1{}^2 + 3x_2{}^2 x_1. \end{aligned} \right\}$$

(ii) Let $g: \mathbb{R}^2 \to \mathbb{R}$ be defined by $g(x_1, x_2) = \sin(x_1 x_2 + x_2{}^2)$. Then

$$\left.\begin{aligned}
g_1(x_1, x_2) &= x_2 \cos(x_1 x_2 + x_2{}^2) \\
g_2(x_1, x_2) &= (x_1 + 2x_2) \cos(x_1 x_2 + x_2{}^2).
\end{aligned}\right\}$$

(iii) Let $h: \mathbb{R}^3 \to \mathbb{R}$ be defined by $h(x_1, x_2, x_3) = x_1 + x_2 x_3$. Then

$$\left.\begin{aligned}
h_1(x_1, x_2, x_3) &= 1 \\
h_2(x_1, x_2, x_3) &= x_3 \\
h_3(x_1, x_2, x_3) &= x_2.
\end{aligned}\right\}$$

The geometric interpretation of a partial derivative is important. Let $\mathbf{i} = (1, 0)$ and $\mathbf{j} = (0, 1)$. Observe that $\mathbf{i}$ is a unit vector which points in the direction of the x_1-axis while $\mathbf{j}$ is a unit vector which points in the direction of the x_2-axis. We have that

$$\begin{aligned}
D_{\mathbf{i}} f(\boldsymbol{\xi}) &= \lim_{t \to 0} \frac{f(\boldsymbol{\xi} + t\mathbf{i}) - f(\boldsymbol{\xi})}{t} \\
&= \lim_{t \to 0} \frac{f(\xi_1 + t, \xi_2) - f(\xi_1, \xi_2)}{t} \\
&= f_1(\xi_1, \xi_2).
\end{aligned}$$

Similarly,

$$\begin{aligned}
D_{\mathbf{j}} f(\boldsymbol{\xi}) &= \lim_{t \to 0} \frac{f(\boldsymbol{\xi} + t\mathbf{j}) - f(\boldsymbol{\xi})}{t} \\
&= \lim_{t \to 0} \frac{f(\xi_1, \xi_2 + t) - f(\xi_1, \xi_2)}{t} \\
&= f_2(\xi_1, \xi_2).
\end{aligned}$$

Thus the partial derivative $f_1(\xi_1, \xi_2)$ is the slope of the surface $y = f(x_1, x_2)$ at the point $\boldsymbol{\xi}$ in the direction of the x_1-axis while $f_2(\xi_1, \xi_2)$ is the slope in the direction of the x_2-axis.

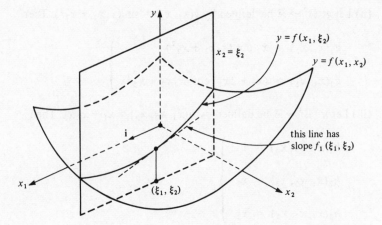

The diagram above illustrates the geometric significance of the partial derivative $f_1(\xi_1, \xi_2)$. It shows the surface $y = f(x_1, x_2)$ sliced by the vertical plane $x_2 = \xi_2$ (which passes through (ξ_1, ξ_2) and is parallel to $\mathbf{i} = (1, 0)$). On the plane $x_2 = \xi_2$, the graph $y = f(x_1, \xi_2)$ can be seen. The partial derivative $f_1(\xi_1, \xi_2)$ is simply the slope of this graph at the point $x_1 = \xi_1$.

The diagrams below illustrate both partial derivatives simultaneously.

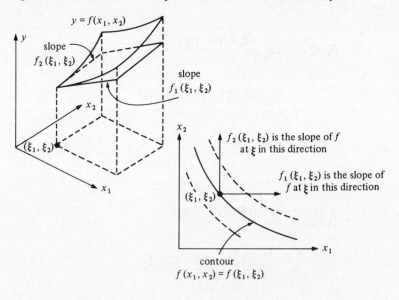

19.5   Notation

As in the case of ordinary derivatives, there are numerous alternative notations which have to be kept in mind when working with partial derivatives. Which is used will depend on the context.

We note first that it is common to denote the partial derivative $f_1(\xi_1, \xi_2)$ by

$$f_{x_1}(\xi_1, \xi_2)$$

with a similar convention for $f_2(\xi_1, \xi_2)$. This may seem a clumsy piece of notation but it is quite useful when, as often happens, it is convenient to write x rather than x_1 and y rather than x_2.

However, the most important alternative notation is the use of

$$\frac{\partial y}{\partial x_1}$$

for the partial derivative $f_1(x_1, x_2)$ in the case when $y = f(x_1, x_2)$. Similarly

$$\frac{\partial y}{\partial x_2}$$

is used for the partial derivative $f_2(x_1, x_2)$. As noted under similar circumstances in §10.4, this notation suffers from the disadvantage that x_1 and x_2 are used ambiguously both as the co-ordinates of the point at which the partial derivative is to be evaluated and as the variables with respect to which one is to differentiate. This ambiguity is often irrelevant in practical applications but can be very confusing when theorems are to be proved.

When a number of variables are in play, it is not always obvious which are to be held constant when differentiating partially. But it is of the *greatest* importance to be clear on this point. Where necessary, the notation

$$\left(\frac{\partial y}{\partial x_1}\right)_{x_2}$$

is available. This means 'differentiate partially with respect to x_1 treating x_2 as constant'.

19.6   *Example*   Consider the equations

$$\left.\begin{array}{l} u = x + y \\ v = x - y. \end{array}\right\}$$

From the first equation

$$\frac{\partial u}{\partial x} = 1$$

If the equations are solved for x and y in terms of u and v, we obtain that

$$\left.\begin{array}{l} x = \tfrac{1}{2}(u + v) \\ y = \tfrac{1}{2}(u - v) \end{array}\right\}$$

and therefore that

$$\frac{\partial x}{\partial u} = \tfrac{1}{2}.$$

Note that

$$\frac{\partial u}{\partial x} \neq \left(\frac{\partial x}{\partial u}\right)^{-1}.$$

This is not in the least surprising since *different* variables were held constant during the two differentiations. What has been shown is that

$$\left(\frac{\partial u}{\partial x}\right)_y = 1 \qquad \left(\frac{\partial x}{\partial u}\right)_v = \tfrac{1}{2}.$$

It is instructive to calculate

$$\left(\frac{\partial u}{\partial x}\right)_v.$$

Eliminating y from the first set of equations, we obtain that $u = 2x - v$ and hence

$$\left(\frac{\partial u}{\partial x}\right)_v = 2 = \left(\frac{\partial x}{\partial u}\right)_v^{-1}.$$

(Note: the symbol ∂ used in the notation for a partial derivative discussed above must be carefully distinguished from the Greek letter δ and the Roman letter d.)

19.7   Local maxima and minima

An immediate application of the ideas introduced above is to the location of local maxima and minima. These are defined as in §11.1. Suppose that $f: A \to \mathbb{R}$ where A contains every $\mathbf{x} \in \mathbb{R}^n$ satisfying $\|\mathbf{x} - \boldsymbol{\xi}\| < \Delta$. Then we say that f has a *local maximum* at the point $\boldsymbol{\xi}$ if and only if there exists a $\delta > 0$ such that

$$f(\mathbf{x}) \leqslant f(\boldsymbol{\xi})$$

for each $\mathbf{x}$ satisfying $\|\mathbf{x} - \boldsymbol{\xi}\| < \delta$. Similarly, f has a *local minimum* at $\boldsymbol{\xi}$ if and only if there exists a $\delta > 0$ such that

$$f(\mathbf{x}) \geqslant f(\boldsymbol{\xi})$$

for each $\mathbf{x}$ satisfying $\|\mathbf{x} - \boldsymbol{\xi}\| < \delta$.

Very roughly, if f has a local maximum at $\boldsymbol{\xi}$, then its graph has a little hill above $\boldsymbol{\xi}$ and, if f has a local minimum at $\boldsymbol{\xi}$, then its graph has a little valley above $\boldsymbol{\xi}$.

19.8   *Theorem*   Suppose that $f: A \to \mathbb{R}$ where A contains every $\mathbf{x} \in \mathbb{R}^n$ satisfying $\|\mathbf{x} - \boldsymbol{\xi}\| < \Delta$ and that the directional derivative $D_{\mathbf{u}}f(\boldsymbol{\xi})$ exists. If f has a local maximum or minimum at $\boldsymbol{\xi}$, then

$$D_{\mathbf{u}}f(\boldsymbol{\xi}) = 0.$$

Proof   A function F is defined on some open interval containing 0 by the formula

$$F(t) = f(\boldsymbol{\xi} + t\mathbf{u}).$$

If f has a local maximum or minimum at $\boldsymbol{\xi}$, then F has a local maximum or minimum at 0. From theorem 11.2, it follows that $F'(0) = 0$, i.e.

$$D_{\mathbf{u}}f(\boldsymbol{\xi}) = 0.$$

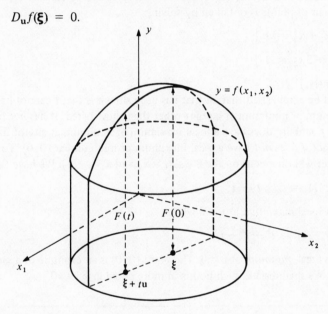

19.9   *Corollary*   Suppose that $f: A \to \mathbb{R}^n$ where A contains every $\mathbf{x} \in \mathbb{R}^n$ satisfying $\|\mathbf{x} - \boldsymbol{\xi}\| < \Delta$. If f has a local maximum or minimum at $\boldsymbol{\xi}$, then all the partial derivatives of f which exist at $\boldsymbol{\xi}$ are zero

Proof   As we know from § 19.3, a partial derivative is simply a directional derivative taken parallel to one of the co-ordinate axes.

19.10   *Example*   Consider the function $f: \mathbb{R}^2 \to \mathbb{R}$ defined by

$$f(x_1, x_2) = x_1^2 x_2 + x_2^3 x_1 - x_1 x_2.$$

By the preceding corollary, a necessary condition that f have a local maximum or minimum at (x_1, x_2) is that the equations

$$2x_1x_2 + x_2^3 - x_2 = 0$$
$$x_1^2 + 3x_2^2x_1 - x_1 = 0$$

hold simultaneously. This means that

$$(x_2 = 0 \text{ or } 2x_1 + x_2^2 = 1) \text{ and } (x_1 = 0 \text{ or } x_1 + 3x_2^2 = 1)$$

i.e. $(x_2 = 0 \text{ and } x_1 = 0)$ or $(x_2 = 0 \text{ and } x_1 + 3x_2^2 = 1)$ or

$(2x_1 + x_2^2 = 1 \text{ and } x_1 = 0)$ or $(2x_1 + x_2^2 = 1 \text{ and } x_1 + 3x_2^2 = 1)$.

The only points at which it is possible for f to have a local maximum or minimum are therefore $(0, 0)$, $(1, 0)$, $(0, 1)$, $(0, -1)$, $(2/5, 1/\sqrt{5})$, $(2/5, -1/\sqrt{5})$. (The final pair of points is obtained by solving

$$2x_1 + x_2^2 = 1$$
$$x_1 + 3x_2^2 = 1$$

simultaneously.)

It should be emphasised that all that has been shown is that f cannot have a local maximum or minimum at a point other than those listed. It has *not* been shown that f actually does have a local maximum or minimum at any of these points. In fact, f does *not* have a local maximum or minimum at $(0, 0)$. To see this, consider what happens on the lines $x_1 = x_2$ and $x_1 = -x_2$. We have that

$$f(t, t) = t^3 + t^4 - t^2$$

which has a local *maximum* at $t = 0$ while

$$f(t, -t) = t^3 - t^4 + t^2$$

which has a local *minimum* at $t = 0$. The point $(0, 0)$ is an example of a *saddle point* for f. We shall discuss such points in more detail in §19.40.

19.11   Exercise

(1) Let $f: \mathbb{R}^2 \to \mathbb{R}$ be defined by $f(x_1, x_2) = x_1^2x_2 + x_2^3x_1$. Compute the directional derivative of f at the point $(1, 2)$ in the direction $(3, 4)$.

(2) Compute the partial derivatives of the functions $f: \mathbb{R}^2 \to \mathbb{R}$ defined by

(i) $f(x_1, x_2) = x_1 + 2x_2$ (ii) $f(x_1, x_2) = x_1^2x_2^5$

(iii) $f(x_1, x_2) = \log(x_2 + e^{x_1x_2})^2$ (iv) $f(x_1, x_2) = \cos(x_2 \sin x_1)$.

(3) Find all the points in $\mathbb{R}^2$ at which all partial derivatives of the function $f: \mathbb{R}^2 \to \mathbb{R}$ defined by $f(x_1, x_2) = x_1^4x_2 - x_1^2x_2^3$ are zero.

(4) Find all points in $\mathbb{R}^3$ at which all partial derivatives of the function $f: \mathbb{R}^3 \to \mathbb{R}$ defined by $f(x_1, x_2, x_3) = x_1x_2x_3$ are zero. Show that none of these points corresponds to a local maximum or minimum.

(5) The *component functions* $f^i: \mathbb{R}^n \to \mathbb{R}$ of a function $f: \mathbb{R}^n \to \mathbb{R}^m$ are defined by

$$f(\mathbf{x}) = (f^1(\mathbf{x}), f^2(\mathbf{x}), \ldots, f^m(\mathbf{x})).$$

Explain why

$$D_{\mathbf{u}}f(\boldsymbol{\xi}) = (D_{\mathbf{u}}f^1(\boldsymbol{\xi}), D_{\mathbf{u}}f^2(\boldsymbol{\xi}), \ldots, D_{\mathbf{u}}f^m(\boldsymbol{\xi})).$$

(6) Prove that the function $f: \mathbb{R}^2 \to \mathbb{R}$ defined by

$$f(x_1, x_2) = \begin{cases} \dfrac{x_1^2 x_2}{x_1^4 + x_2^2} & (x_1, x_2) \neq (0, 0) \\[2mm] 0 & (x_1, x_2) = (0, 0) \end{cases}$$

has directional derivatives in every direction at $(0, 0)$ but is *not* continuous at $(0, 0)$. [Hint: example 18.43.] Does f have a local maximum or minimum at $(0, 0)$?

19.12 Differentiable functions

Consider a function $f: A \to \mathbb{R}^m$ where A is a set which contains each $\mathbf{x} \in \mathbb{R}^n$ satisfying $\|\mathbf{x} - \boldsymbol{\xi}\| < \Delta$. What should it mean to say that f is differentiable at $\boldsymbol{\xi}$? We dealt with the case $m = n = 1$ in §10.1. Our definition required the existence of the limit

$$l = \lim_{x \to \xi} \frac{f(x) - f(\xi)}{x - \xi} = \lim_{h \to 0} \frac{f(\xi + h) - f(\xi)}{h}.$$

This may be rewritten as

$$f(x) - f(\xi) - l(x - \xi) = o(x - \xi) \quad (x \to \xi)$$

or as

$$f(\xi + h) - f(\xi) - lh = o(h) \quad (h \to 0)$$

where $o(h)$ denotes a quantity which tends to zero when divided by h. The number l, of course, is called the derivative of f at ξ and denoted by $f'(\xi)$.

An affine function $g: \mathbb{R} \to \mathbb{R}$ is defined by an equation of the form $g(x) = lx + c$. (See example 18.24(i).) We can therefore restate the definition yet again as being the requirement that there exist an affine function $g: \mathbb{R} \to \mathbb{R}$ with $g(\xi) = f(\xi)$ such that

$$f(x) - g(x) = o(x - \xi) \quad (x \to \xi)$$

The geometric interpretation of this condition is that the graphs $y = f(x)$ and $y = g(x)$ are tangent where $x = \xi$.

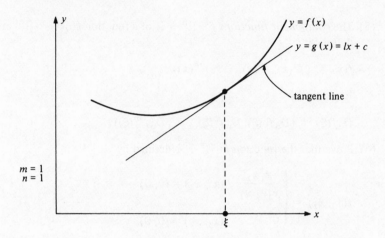

The graph of an affine function is a straight line and so our analytic definition of differentiability corresponds to the geometric idea that it be possible to draw a tangent line to the graph of f where $x = \xi$.

We are now in a position to define differentiability in the general case when m and n are not necessarily equal to one. We say that $f: A \to \mathbb{R}^m$ is *differentiable* at ξ if and only if there exists an affine function $g: A \to \mathbb{R}^m$ with $g(\xi) = f(\xi)$ such that

$$f(\mathbf{x}) - g(\mathbf{x}) = o(\mathbf{x} - \xi) \qquad (\mathbf{x} \to \xi).$$

One cannot, of course, divide by a vector and $o(\mathbf{x} - \xi)$ in this equation denotes a quantity which tends to zero when divided by the real number $\|\mathbf{x} - \xi\|$. The criterion can therefore be rewritten as

$$\lim_{\mathbf{x} \to \xi} \frac{\|f(\mathbf{x}) - g(\mathbf{x})\|}{\|\mathbf{x} - \xi\|} = 0.$$

As in the $m = n = 1$ case, the geometric interpretation is that the graphs $\mathbf{y} = f(\mathbf{x})$ and $\mathbf{y} = g(\mathbf{x})$ are tangent at $\mathbf{x} = \xi$. The graph of an affine function is a *flat*. (See §18.18 and exercise 18.25(4).) The definition therefore corresponds to the geometric idea that it be possible to draw a *tangent flat* to the graph of f where $\mathbf{x} = \xi$.

In the case $m = n = 1$, a tangent flat at $x = \xi$ is simply a straight line as illustrated above. In example 18.24(ii) we saw that an affine function $g: \mathbb{R}^2 \to \mathbb{R}^1$ is defined by an equation of the form

$$y = l_1 x_1 + l_2 x_2 + c$$

and its graph is therefore a plane in $\mathbb{R}^3$. In the case $m = 1$ and $n = 2$ it therefore follows that a tangent flat at $(x_1, x_2) = (\xi_1, \xi_2)$ is a plane as in the diagram below.

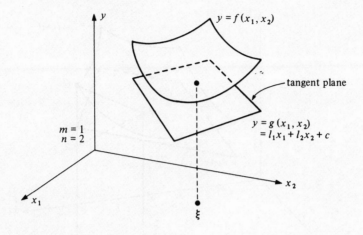

In example 18.24(iii) we saw that an affine function $g: \mathbb{R}^1 \to \mathbb{R}^2$ is defined by two equations of the form

$$y_1 = l_1 x_1 + c_1$$

$$y_2 = l_2 x_2 + c_2$$

and its graph is therefore a line in $\mathbb{R}^3$. In the case $m = 2$ and $n = 1$, it therefore follows that a tangent flat at $x = \xi$ is a line as indicated below.

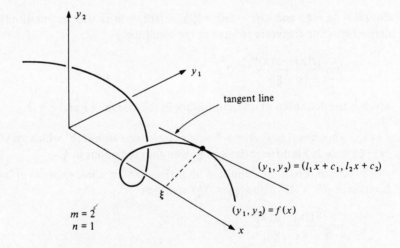

In §10.1 we gave a geometric justification for our identification of the idea of differentiability with the existence of a tangent line in the case $m = n = 1$. We now give a similar justification for the case $m = 1$ and $n = 2$.

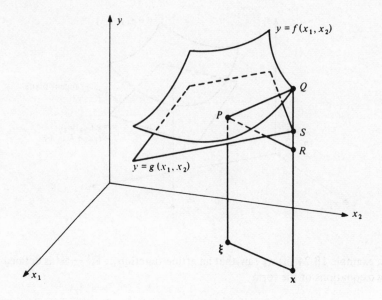

In the diagram, Q lies on the surface $y = f(x_1, x_2)$ and S on the surface $y = g(x_1, x_2) = l_1 x_1 + l_2 x_2 + c$. The point R is at the same height as P. The condition for $y = g(x_1, x_2)$ to be tangent to $y = f(x_1, x_2)$ at P is that

$$\lim_{\mathbf{x} \to \boldsymbol{\xi}} \left\{ \frac{QR}{PR} - \frac{SR}{PR} \right\} = 0.$$

But $PR = \|\mathbf{x} - \boldsymbol{\xi}\|$ and $|QR - SR| = |QS| = |f(\mathbf{x}) - g(\mathbf{x})|$. Our geometrically derived criterion therefore reduces to the condition

$$\lim_{\mathbf{x} \to \boldsymbol{\xi}} \frac{|f(\mathbf{x}) - g(\mathbf{x})|}{\|\mathbf{x} - \boldsymbol{\xi}\|} = 0$$

which is the definition of differentiability in the case $m = 1$ and $n = 2$.

19.13   Theorem Let $f: A \to \mathbb{R}^m$ where A contains all $\mathbf{x} \in \mathbb{R}^n$ which satisfy $\|\mathbf{x} - \boldsymbol{\xi}\| < \Delta$. If f is differentiable at $\boldsymbol{\xi}$, then f is continuous at $\boldsymbol{\xi}$.

Proof From the definition of differentiability there exists an affine function $g: \mathbb{R}^n \to \mathbb{R}^m$ with $g(\boldsymbol{\xi}) = f(\boldsymbol{\xi})$ such that

$$\lim_{\mathbf{x} \to \boldsymbol{\xi}} \frac{\|f(\mathbf{x}) - g(\mathbf{x})\|}{\|\mathbf{x} - \boldsymbol{\xi}\|} = 0$$

It follows (as in theorem 10.6) that

$$\|f(\mathbf{x}) - g(\mathbf{x})\| \to 0 \text{ as } \mathbf{x} \to \boldsymbol{\xi}.$$

But g is necessarily continuous at $\boldsymbol{\xi}$ (example 18.34). Hence

$$\|f(x) - f(\xi)\| = \|f(x) - g(\xi)\|$$
$$= \|f(x) - g(x) + g(x) - g(\xi)\|$$
$$\leqslant \|f(x) - g(x)\| + \|g(x) - g(\xi)\|$$
$$\to 0 + 0 \text{ as } x \to \xi.$$

Thus $f(x) \to f(\xi)$ as $x \to \xi$ and so f is continuous at ξ.

19.14 Derivatives

In this section we shall need the ideas introduced in §18.22. In particular, we shall find it convenient to replace $\mathbf{x}$ and $\mathbf{y}$ by *column* vectors. To emphasise this substitution we shall write x and y where previously we have written $\mathbf{x}$ and $\mathbf{y}$.

An affine function $g\colon \mathbb{R}^n \to \mathbb{R}^m$ is defined by an equation of the form

$$y = g(x) = Lx + c$$

where L is an $m \times n$ matrix and c is an $m \times 1$ column vector. If $g(\xi) = f(\xi)$, we may rewrite this equation in the form

$$y = g(x) = L(x - \xi) + f(\xi).$$

It follows that f is differentiable at ξ if and only if there exists an $m \times n$ matrix L such that

$$f(x) - f(\xi) - L(x - \xi) = o(x - \xi) \qquad (x \to \xi)$$

or, what is the same thing,

$$f(\xi + h) - f(\xi) - Lh = o(h) \qquad (h \to 0).$$

These equations are identical in form to the second and third equations of §19.12 except that L replaces l. This correspondence is very significant but it must not be forgotten that, in the current equations, the small letters all represent column *vectors* and L represents a *matrix* whereas in §19.12 all the letters in the equations represent real numbers. Note in particular that the real number l in §19.12 is the derivative $f'(\xi)$ of f at ξ. This leads us to a natural definition for the derivative of a differentiable function in the general case. If f is differentiable at ξ, we define its *derivative* at ξ by

$$f'(\xi) = L.$$

Sometimes the notation $Df(\xi)$ is used instead.

The derivative of f at ξ is therefore an $m \times n$ *matrix* in the general case. A number of advantages follow from the use of this definition. The tangent flat at ξ, for example, is given by the equation

$$y = f(\xi) + f'(\xi)(x - \xi)$$

exactly as in the case $m = n = 1$.

How is the derivative of a vector function calculated? We begin to answer this question with the following theorem which connects the idea of a derivative introduced above with that of a directional derivative introduced earlier.

19.15   Theorem Suppose that $f: A \to \mathbb{R}^m$ where A contains all $x \in \mathbb{R}^n$ satisfying $\|x - \xi\| < \Delta$. If f is differentiable at ξ, then all directional derivatives of f at ξ exist and

$$D_u f(\xi) = f'(\xi)u.$$

Proof We have that

$$\frac{\|f(x) - f(\xi) - f'(\xi)(x - \xi)\|}{\|x - \xi\|} \to 0$$

as x approaches ξ and, in particular, as x approaches ξ along the straight line through ξ in the direction u. Writing $x = \xi + tu$, we obtain that

$$\lim_{t \to 0} \left\| \frac{f(\xi + tu) - f(\xi)}{t} - f'(\xi)u \right\| = 0$$

and hence $D_u f(\xi) = f'(\xi)u$.

With the assumptions of the preceding theorem, $f(x)$ denotes an $m \times 1$ column vector. The formula

$$f(x) = \begin{pmatrix} f^1(x) \\ f^2(x) \\ \vdots \\ f^m(x) \end{pmatrix}$$

therefore determines m real-valued functions $f^1, f^2, \ldots, f^m$ which we call the *component* functions of f. In the next theorem we explain how the entries in the matrix

$$f'(\xi) = \begin{pmatrix} l_{11} & l_{12} \ldots l_{1n} \\ l_{21} & l_{22} \ldots l_{2n} \\ \vdots \\ l_{m1} & l_{m2} \ldots l_{mn} \end{pmatrix}$$

can be calculated from the partial derivatives of these component functions. (An incidental corollary is that the derivative of a differentiable function is uniquely defined.)

19.16   Theorem With the assumptions of the preceding theorem,

$$f'(\xi) = \begin{pmatrix} f_1^1(\xi) & f_2^1(\xi) \cdots f_n^1(\xi) \\ f_1^2(\xi) & f_2^2(\xi) \cdots f_n^2(\xi) \\ \vdots \\ f_1^m(\xi) & f_2^m(\xi) \cdots f_n^m(\xi) \end{pmatrix}$$

Proof We shall only prove that $l_{12} = f_2^1(\xi)$. We take $u = (0, 1, 0, \ldots, 0)$. Then, by exercise 19.11(5),

$$D_u f(\xi) = \begin{pmatrix} D_u f^1(\xi) \\ D_u f^2(\xi) \\ \vdots \\ D_u f^m(\xi) \end{pmatrix} = \begin{pmatrix} f_2^1(\xi) \\ f_2^2(\xi) \\ \vdots \\ f_2^m(\xi) \end{pmatrix}$$

Also

$$f'(\xi)u = \begin{pmatrix} l_{11} & l_{12} \cdots l_{1n} \\ l_{21} & l_{22} \cdots l_{2n} \\ \vdots \\ l_{m1} & l_{m2} \cdots l_{mn} \end{pmatrix} \begin{pmatrix} 0 \\ 1 \\ \vdots \\ 0 \end{pmatrix} = \begin{pmatrix} l_{12} \\ l_{22} \\ \vdots \\ l_{m2} \end{pmatrix}$$

Comparing the first co-ordinates of each column and appealing to theorem 19.15, we obtain that $l_{12} = f_2^1(\xi)$.

If $y = f(x)$, it is natural to use the notation

$$\frac{\mathrm{d}y}{\mathrm{d}x}$$

for the derivative of a vector function (with our usual reservations about the ambiguous use of x). The formula of theorem 19.16 then takes the pleasing and easily remembered form

$$\frac{\mathrm{d}y}{\mathrm{d}x} = \begin{pmatrix} \dfrac{\partial y_1}{\partial x_1} & \dfrac{\partial y_1}{\partial x_2} \cdots \dfrac{\partial y_1}{\partial x_n} \\ \dfrac{\partial y_2}{\partial x_1} & \dfrac{\partial y_2}{\partial x_2} \cdots \dfrac{\partial y_2}{\partial x_n} \\ \vdots \\ \dfrac{\partial y_m}{\partial x_1} & \dfrac{\partial y_m}{\partial x_2} \cdots \dfrac{\partial y_m}{\partial x_n} \end{pmatrix}$$

We have seen that the derivative of a differentiable vector function f is a matrix whose entries consist of all partial derivatives of the co-ordinates of f. It is important, however, to bear in mind that the mere existence of this matrix does *not* guarantee that f is differentiable. The function of exercise 19.11(6) provides a suitable example. All its partial derivatives exist at **O** but the function is not even continuous at **O** and hence is certainly not differentiable at **O** (theorem 19.13). Thus, although the matrix of partial derivatives can be written down, it does not help very much since the function simply does not have a tangent flat at **O**.

How can one tell whether a vector function is differentiable? The following theorem provides a criterion which is adequate for most practical purposes.

19.17   *Theorem* Suppose that all partial derivatives of f are *continuous* at ξ. Then f is differentiable at ξ.

Proof To avoid cumbersome notation, we consider only the case $m = 1$ and $n = 2$. From the mean value theorem (11.6) we have that

$$f(x_1, x_2) - f(\xi_1, \xi_2) = \{f(x_1, x_2) - f(\xi_1, x_2)\} + \{f(\xi_1, x_2) - f(\xi_1, \xi_2)\}$$
$$= f_1(u_1, x_2)(x_1 - \xi_1) + f_2(\xi_1, u_2)(x_2 - \xi_2)$$

where u_1 lies between x_1 and ξ_1 while u_2 lies between x_2 and ξ_2. Thus $|u_1 - \xi_1| < |x_1 - \xi_1| \leqslant \|x - \xi\|$ and so $u_1 \to \xi_1$ as $x \to \xi$. It follows that $(u_1, x_2) \to (\xi_1, \xi_2)$ as $x \to \xi$. Similarly $(\xi_1, u_2) \to (\xi_1, \xi_2)$ as $x \to \xi$. Since f_1 and f_2 are continuous at ξ, we deduce that

$$f_1(u_1, x_2) \to f_1(\xi_1, \xi_2) \text{ as } x \to \xi,$$
$$f_2(\xi_1, u_2) \to f_2(\xi_1, \xi_2) \text{ as } x \to \xi$$

(a conclusion which requires the vector analogue of theorem 8.17).

We now show that

$$f(x) - f(\xi) - L(x - \xi) = o(x - \xi) \quad (x \to \xi)$$

with $L = (f_1(\xi), f_2(\xi))$. This follows from the fact that

$$|f(x) - f(\xi) - f_1(\xi)(x_1 - \xi_1) - f_2(\xi)(x_2 - \xi_2)|$$
$$= |(f_1(u_1, x_2) - f_1(\xi_1, \xi_2))(x_1 - \xi_1) + (f_2(\xi_1, u_2)$$
$$- f_2(\xi_1, \xi_2))(x_2 - \xi_2)| \leqslant E\|x - \xi\|$$

by the Cauchy–Schwarz inequality, where

$$E = \{(f_1(u_1, x_2) - f_1(\xi_1, \xi_2))^2 + (f_2(\xi_1, u_2) - f_2(\xi_1, \xi_2))^2\}^{1/2}$$
$$\to \{0^2 + 0^2\}^{1/2} = 0 \text{ as } x \to \xi.$$

19.18   *Examples*

(i) Let $f: \mathbb{R}^2 \to \mathbb{R}$ be defined by $f(x_1, x_2) = x_1^2 x_2 + x_2^3 x_1$. Then

$$\left.\begin{array}{l} f_1(x_1, x_2) = 2x_1 x_2 + x_2^3 \\ f_2(x_1, x_2) = x_1^2 + 3x_2^2 x_1. \end{array}\right\}$$

These partial derivatives are polynomials and hence are continuous everywhere (proposition 18.38). Thus f is differentiable at every point. The derivative at the point $(2, 1)$ is the 1×2 matrix

$$f'(2, 1) = (f_1(2, 1), f_2(2, 1))$$

$$= (5, 10).$$

Various facts about the behaviour of f at $(2, 1)$ may be deduced from this information. For example, the slope (or rate of increase) of f at the point $(2, 1)$ in the direction $(4, 3)$ is given by

$$D_u f(1, 2)$$

where u is the unit vector which points in the same direction as $(4, 3)$. From theorem 19.15 we have that

$$D_u f(1, 2) = f'(1, 2)u$$

$$= (5, 10)\begin{pmatrix} 4/5 \\ 3/5 \end{pmatrix} = 4 + 6 = 10.$$

We can also obtain the equation of the tangent plane to the surface $y = x_1^2 x_2 + x_2^3 x_1$ where $(x_1, x_2) = (1, 2)$. This is given by writing $(\xi_1, \xi_2) = (1, 2)$ in the formula

$$y = f(\xi) + f'(\xi)(x - \xi).$$

This yields

$$y = 6 + (5, 10)\begin{pmatrix} x_1 - 1 \\ x_2 - 2 \end{pmatrix}$$

i.e.

$$y - 6 = 5(x_1 - 1) + 10(x_2 - 2).$$

(ii) Let $f: \mathbb{R}^3 \to \mathbb{R}$ be defined by $f(x_1, x_2, x_3) = x_1 x_2^2 + x_2 x_3^3 + x_3 x_1^4$. Then

$$\left.\begin{aligned} f_1(x_1, x_2, x_3) &= x_2^2 + 4x_3 x_1^3 \\ f_2(x_1, x_2, x_3) &= 2x_1 x_2 + x_3^3 \\ f_3(x_1, x_2, x_3) &= 3x_2 x_3^2 + x_1^4. \end{aligned}\right\}$$

These partial derivatives are continuous at every point and hence f is differentiable everywhere. The derivative at $(1, 1, 1)$ is the 1×3 matrix

$$f'(1, 1, 1) = (5, 3, 4).$$

The slope (or rate of increase) of f at $(1, 1, 1)$ in the direction $(2, 1, 2)$ is equal to

$$f'(1, 1, 1)u = (5, 3, 4)\begin{pmatrix} 2/3 \\ 1/3 \\ 2/3 \end{pmatrix} = 7$$

(where u is the unit vector which points in the same direction as $(2, 1, 2)$).

The tangent flat to $y = x_1 x_2^2 + x_2 x_3^3 + x_3 x_1^4$ where $(x_1, x_2, x_3) = (1, 1, 1)$ is the hyperplane

$$y = 3 + (5, 3, 4)\begin{pmatrix} x_1 - 1 \\ x_2 - 1 \\ x_3 - 1 \end{pmatrix}$$

i.e.

$$y - 3 = 5(x_1 - 1) + 3(x_2 - 1) + 4(x_3 - 1).$$

(iii) Let $f: \mathbb{R} \to \mathbb{R}^3$ be defined by

$$f(x) = \begin{pmatrix} x \\ x^2 \\ x^3 \end{pmatrix}.$$

The component functions $f^1: \mathbb{R} \to \mathbb{R}, f^2: \mathbb{R} \to \mathbb{R}$ and $f^3: \mathbb{R} \to \mathbb{R}$ are given by

$$\left.\begin{aligned} f^1(x) &= x \\ f^2(x) &= x^2 \\ f^3(x) &= x^3 \end{aligned}\right\}$$

and hence

$$\left.\begin{aligned} f_1^1(x) &= 1 \\ f_1^2(x) &= 2x \\ f_1^3(x) &= 3x^2. \end{aligned}\right\}$$

These functions are all continuous everywhere and hence f is differentiable at every point. Its derivative at -1 is the 3×1 matrix

$$f'(-1) = \begin{pmatrix} 1 \\ -2 \\ 3 \end{pmatrix}.$$

The graph of f is determined by the system of three equations

$$\left.\begin{aligned} y_1 &= x \\ y_2 &= x^2 \\ y_3 &= x^3. \end{aligned}\right\}$$

The tangent flat where $x = -1$ is given by

$$\begin{pmatrix} y_1 \\ y_2 \\ y_3 \end{pmatrix} = \begin{pmatrix} -1 \\ 1 \\ -1 \end{pmatrix} + \begin{pmatrix} 1 \\ -2 \\ 3 \end{pmatrix} (x + 1)$$

and hence is the straight line determined by the equations

$$\left. \begin{array}{l} y_1 = x \\ y_2 = -2x - 1 \\ y_3 = 3x + 2. \end{array} \right\}$$

(iv) Let $f \colon \mathbb{R}^2 \to \mathbb{R}^2$ be defined by

$$f(x_1, x_2) = \begin{pmatrix} x_1^2 + x_2^2 \\ x_1^2 - x_2^2 \end{pmatrix}.$$

The component functions $f^1 \colon \mathbb{R}^2 \to \mathbb{R}$ and $f^2 \colon \mathbb{R}^2 \to \mathbb{R}$ are then given by

$$f^1(x_1, x_2) = x_1^2 + x_2^2$$
$$f^2(x_1, x_2) = x_1^2 - x_2^2$$

and hence

$$f_1^1(x_1, x_2) = 2x_1 \qquad f_2^1(x_1, x_2) = 2x_2$$
$$f_1^2(x_1, x_2) = 2x_1 \qquad f_2^2(x_1, x_2) = -2x_2.$$

These functions are continuous everywhere and hence f is differentiable at every point. Its derivative at $(1, 1)$ is the 2×2 matrix

$$f'(1, 1) = \begin{pmatrix} 2 & 2 \\ 2 & -2 \end{pmatrix}.$$

The graph of f is determined by the two equations

$$\left. \begin{array}{l} y_1 = x_1^2 + x_2^2 \\ y_2 = x_1^2 - x_2^2. \end{array} \right\}$$

The tangent flat where $(x_1, x_2) = (1, 1)$ is given by

$$\begin{pmatrix} y_1 \\ y_2 \end{pmatrix} = \begin{pmatrix} 2 \\ 0 \end{pmatrix} + \begin{pmatrix} 2 & 2 \\ 2 & -2 \end{pmatrix} \begin{pmatrix} x_1 - 1 \\ x_2 - 1 \end{pmatrix}$$

i.e.

$$\left. \begin{array}{l} y_1 - 2 = 2(x_1 - 1) + 2(x_2 - 1) \\ y_2 = 2(x_1 - 1) - 2(x_2 - 1). \end{array} \right\}$$

19.19   Gradient

As though there were not enough notation associated with vector functions, there is a further notation which is used for the derivative of a real-valued function. A derivative $f'(\xi)$ in this case is a $1 \times n$ matrix, i.e. a row vector. When matrix notation is not being systematically used especially in applied mathematics, one sometimes writes

$$\nabla f(\boldsymbol{\xi}) = f'(\boldsymbol{\xi})$$

and refers to $\nabla f(\boldsymbol{\xi})$ as the *gradient* of f at $\boldsymbol{\xi}$.

The geometric interpretation of the gradient (i.e. the derivative) of a real-valued function of a vector variable is worth noting. Recall that

$$y - f(\xi) = f'(\xi)(x - \xi)$$

is the equation of the hyperplane in $\mathbb{R}^{n+1}$ which is the tangent to $y = f(x)$ where $x = \xi$. It follows that

$$f'(\xi)(x - \xi) = 0$$

is the equation of the hyperplane in $\mathbb{R}^n$ which is the tangent to the contour $f(x) = f(\xi)$ where $x = \xi$. Rewriting the equation $f'(\xi)(x - \xi) = 0$ in the form

$$\langle \nabla f(\boldsymbol{\xi}), x - \boldsymbol{\xi} \rangle = 0$$

and referring to §18.15, we observe that $\nabla f(\boldsymbol{\xi})$ is a *normal* to the contour $f(\mathbf{x}) = f(\boldsymbol{\xi})$ at $\mathbf{x} = \boldsymbol{\xi}$.

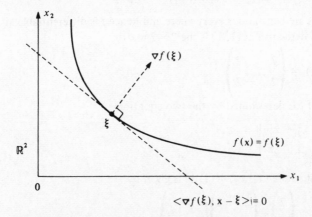

$\langle \triangledown f(\xi), x - \xi \rangle = 0$

This provides a geometric interpretation for the direction in which $\nabla f(\boldsymbol{\xi})$ points. Its length can also be interpreted geometrically. The slope (or rate of increase) of f at $\boldsymbol{\xi}$ in the direction of the unit vector $\mathbf{u}$ is given by

$$D_{\mathbf{u}}f(\boldsymbol{\xi}) = f'(\xi)u = \langle \nabla f(\boldsymbol{\xi}), \mathbf{u} \rangle.$$

By the Cauchy–Schwarz inequality,

$$|D_{\mathbf{u}} f(\boldsymbol{\xi})| = |\langle \nabla f(\boldsymbol{\xi}), \mathbf{u} \rangle| \leqslant \|\nabla f(\boldsymbol{\xi})\| . \|\mathbf{u}\| = \|\nabla f(\boldsymbol{\xi})\|.$$

But, taking $\mathbf{u} = \|\nabla f(\boldsymbol{\xi})\|^{-1} \nabla f(\boldsymbol{\xi})$, we obtain that

$$D_{\mathbf{u}} f(\boldsymbol{\xi}) = \|\nabla f(\boldsymbol{\xi})\|.$$

Thus $\nabla f(\boldsymbol{\xi})$ points in the direction in which the slope of f at $\boldsymbol{\xi}$ is *greatest* and the magnitude of the slope in this direction is equal to $\|\nabla f(\boldsymbol{\xi})\|$.

19.20 *Example* Consider the function $f \colon \mathbb{R}^2 \to \mathbb{R}$ of example 19.18(i). We have that

$$\nabla f(2, 1) = f'(2, 1) = (5, 10).$$

It follows that $(5, 10)$ is a normal to the contour $x_1^2 x_2 + x_2^3 x_1 = 6$ where $(x_1, x_2) = (2, 1)$. Also $(5, 10)$ is the direction in which f increases fastest at $(2,1)$ and the rate of increase in this direction is $\{5^2 + 10^2\}^{1/2} = 5\sqrt{5}$.

19.21 Manipulation of derivatives

For real-valued functions of a real variable, the following formulae are familiar.

(I) $\quad \dfrac{d}{dx}(Lx + c) = L$

(II) $\quad \dfrac{d}{dx}(Lx^2) = 2Lx$

(III) $\quad \dfrac{d}{dx}(Ly + Mz) = L\dfrac{dy}{dx} + M\dfrac{dz}{dx}$

(IV) $\quad \dfrac{d}{dx}(yz) = z\dfrac{dy}{dx} + y\dfrac{dz}{dx}$

(V) $\quad \dfrac{dz}{dx} = \dfrac{dz}{dy} \cdot \dfrac{dy}{dx}$

(VI) $\quad \dfrac{dy}{dx} = \left(\dfrac{dx}{dy}\right)^{-1}.$

The range of validity of these formulae was discussed in chapter 10. It is a very convenient fact that versions of all these formulae remain valid when the small letters are re-interpreted as column vectors and the large letters as matrices, *provided* that the formula obtained after this re-interpretation makes sense. With this re-interpretation, for example, formula (V) asserts that one matrix is the product of two other matrices and formula (VI) asserts that one matrix is the inverse of another. (The latter assertion, of course, makes no sense unless

$$\frac{dy}{dx}$$

is a square, non-singular matrix which requires in particular that x and y have the same number of co-ordinates.)

We leave further consideration of (V) and (VI) until the next section and focus attention for the moment on the other formulae. As far as these are concerned, (I) and (III) are straightforward to interpret and easy to prove (exercise 19.26(4)). Formulae (II) and (IV), however, make no immediate sense when x, y and z are interpreted as column vectors. We therefore re-arrange the formulae slightly so that they do make sense and obtain

$$\text{(II)}' \quad \frac{d}{dx}(x^{\mathrm{T}}Ax) = x^{\mathrm{T}}(A^{\mathrm{T}} + A)$$

$$\text{(IV)}' \quad \frac{d}{dx}(y^{\mathrm{T}}z) = z^{\mathrm{T}}\frac{dy}{dx} + y^{\mathrm{T}}\frac{dz}{dx}.$$

As an example, we shall deduce (II)$'$ from (IV)$'$. The proof of the latter formula is left as an exercise (19.26(4)).

19.22   Example Observe to begin with that (II)$'$ makes no sense unless A is a *square* matrix. From formula (IV)$'$ and (I),

$$\frac{d}{dx}x^{\mathrm{T}}Ax = \frac{d}{dx}x^{\mathrm{T}}(Ax) = (Ax)^{\mathrm{T}}\frac{dx}{dx} + x^{\mathrm{T}}\frac{d}{dx}(Ax)$$

$$= x^{\mathrm{T}}A^{\mathrm{T}}I + x^{\mathrm{T}}A = x^{\mathrm{T}}(A^{\mathrm{T}} + A).$$

Here I denotes the *identity matrix*, i.e.

$$I = \begin{pmatrix} 1 & 0 \ldots 0 \\ 0 & 1 \ldots 0 \\ \vdots & \\ 0 & 0 \ldots 1 \end{pmatrix}.$$

Note that in the special case when A is *symmetric*, i.e. $A^{\mathrm{T}} = A$, we have that

$$\frac{d}{dx}(x^{\mathrm{T}}Ax) = 2Ax.$$

19.23 Chain rule

As we know from §10.12 that formula (V), i.e.

$$\frac{dz}{dx} = \frac{dz}{dy} \cdot \frac{dy}{dx}$$

is the formula for differentiating a *composite function*. To justify this formula in the vector case, we therefore need to establish the following result. (See theorem 10.13.)

19.24 Proposition Suppose that g is differentiable at ξ and f is differentiable at $\eta = g(\xi)$. Then $f \circ g$ is differentiable at ξ and

$$(f \circ g)'(\xi) = f'(\eta)g'(\xi).$$

Proof We are given that there exist matrices $f'(\eta)$ and $g'(\xi)$ for which

$$f(y) - f(\eta) - f'(\eta)(y - \eta) = o(y - \eta) \quad (y \to \eta)$$

$$g(x) - g(\xi) - g'(\xi)(x - \xi) = o(x - \xi) \quad (x \to \xi).$$

Writing $y = g(x)$ and $\eta = g(\xi)$ in the first equation, we obtain that

$$f(g(x)) - f(g(\xi)) - f'(\eta)(g(x) - g(\xi)) = o(g(x) - g(\xi))$$

and hence

$$f(g(x)) - f(g(\xi)) - f'(\eta)\{g'(\xi)(x - \xi) + o(x - \xi)\}$$
$$= o\{g'(\xi)(x - \xi) + o(x - \xi)\}.$$

It follows that

$$f(g(x)) - f(g(\xi)) - f'(\eta)g'(\xi)(x - \xi) = o(x - \xi) \quad (x \to \xi)$$

and so $f \circ g$ is differentiable at ξ with derivative $f'(\eta)g'(\xi)$. As in theorem 10.13, the final step is not entirely straightforward but we prefer not to hold up the discussion by presenting a detailed proof at this stage.

Written in full, formula (V) takes the form

$$\begin{pmatrix} \dfrac{\partial z_1}{\partial x_1} & \dfrac{\partial z_1}{\partial x_2} & \dots & \dfrac{\partial z_1}{\partial x_n} \\[2mm] \dfrac{\partial z_2}{\partial x_1} & \dfrac{\partial z_2}{\partial x_2} & \dots & \dfrac{\partial z_2}{\partial x_n} \\[2mm] \vdots & \vdots & & \\[2mm] \dfrac{\partial z_m}{\partial x_1} & \dfrac{\partial z_m}{\partial x_2} & \dots & \dfrac{\partial z_m}{\partial x_n} \end{pmatrix} = \begin{pmatrix} \dfrac{\partial z_1}{\partial y_1} & \dots & \dfrac{\partial z_1}{\partial y_p} \\[2mm] \dfrac{\partial z_2}{\partial y_1} & \dots & \dfrac{\partial z_2}{\partial y_p} \\[2mm] \vdots & & \\[2mm] \dfrac{\partial z_m}{\partial y_1} & \dots & \dfrac{\partial z_m}{\partial y_p} \end{pmatrix} \begin{pmatrix} \dfrac{\partial y_1}{\partial x_1} & \dfrac{\partial y_1}{\partial x_2} & \dots & \dfrac{\partial y_1}{\partial x_n} \\[2mm] \vdots & & & \\[2mm] \dfrac{\partial y_p}{\partial x_1} & \dfrac{\partial y_p}{\partial x_2} & \dots & \dfrac{\partial y_p}{\partial x_n} \end{pmatrix}$$

from which we deduce that

$$\frac{\partial z_i}{\partial x_j} = \frac{\partial z_i}{\partial y_1}\frac{\partial y_1}{\partial x_j} + \frac{\partial z_i}{\partial y_2}\frac{\partial y_2}{\partial x_j} + \ldots + \frac{\partial z_i}{\partial y_p}\frac{\partial y_p}{\partial x_j}.$$

This formula is called the *chain rule* and is often very useful in calculating partial derivatives.

19.25   Examples

(i) Let $f: \mathbb{R}^2 \to \mathbb{R}^1$ and $g: \mathbb{R}^1 \to \mathbb{R}^2$ be defined by

$$z = f(y_1, y_2) = y_1^2 y_2 + y_2^3 y_1 \, ; \, y = g(x) = \begin{pmatrix} x^2 \\ x^3 \end{pmatrix}.$$

From formula (V)

$$\frac{dz}{dx} = \frac{dz}{dy} \cdot \frac{dy}{dx} = (2y_1 y_2 + y_2^3, y_1^2 + 3y_2^2 y_1) \begin{pmatrix} 2x \\ 3x^2 \end{pmatrix}.$$

This has to be calculated when $y = g(x)$, i.e. when $y_1 = x^2$ and $y_2 = x^3$. Thus

$$\frac{dz}{dx} = (2x^5 + x^9, x^4 + 3x^8) \begin{pmatrix} x^2 \\ x^3 \end{pmatrix} = 7x^6 + 11x^{10}.$$

The same result can be easily obtained directly in this case because $f \circ g: \mathbb{R}^1 \to \mathbb{R}^1$ is given by the simple formula $f \circ g(x) = (x^2)^2 x^3 + (x^3)^3 x = x^7 + x^{11}$.

(ii) Suppose that $w = xy^2 z$ where

$$x = r + 2s + 3t$$

$$y = 2r + 3s + t$$

$$z = 3r + s + t.$$

We can calculate the partial derivative of w with respect to r, s or t using the chain rule. For example

$$\frac{\partial w}{\partial s} = \frac{\partial w}{\partial x}\frac{\partial x}{\partial s} + \frac{\partial w}{\partial y}\frac{\partial y}{\partial s} + \frac{\partial w}{\partial z}\frac{\partial z}{\partial s}.$$

It is as well to appreciate that every symbol in this formula is used ambiguously. This can naturally lead to confusion unless one is quite clear about what everything means. It is sometimes helpful to rewrite the equation in the form

$$\left(\frac{\partial w}{\partial s}\right)_{r,t} = \left(\frac{\partial w}{\partial x}\right)_{y,z}\left(\frac{\partial x}{\partial s}\right)_{r,t} + \left(\frac{\partial w}{\partial y}\right)_{x,z}\left(\frac{\partial y}{\partial s}\right)_{r,t} + \left(\frac{\partial w}{\partial z}\right)_{x,y}\left(\frac{\partial z}{\partial s}\right)_{r,t}$$

to make it clear what is to be held constant in each differentiation. (See §19.5.) Thus

$$\frac{\partial w}{\partial s} = y^2 z^3 2 + 2xyz^3 3 + 3xy^2 z^2 = 2(2r + 3s + t)^2 (3r + s + t)^3$$
$$+ 6(r + 2s + 3t)(2r + 3s + t)(3r + s + t)^3$$
$$+ 3(r + 2s + 3t)(2r + 3s + t)^2 (3r + s + t)^2.$$

We now turn to formula (VI), i.e.

$$\frac{dy}{dx} = \left(\frac{dx}{dy}\right)^{-1}.$$

As we know from §12.9, this formula concerns the derivative of an *inverse function*. Assuming that f has an inverse function f^{-1}, then

$$f \circ f^{-1}(x) = x$$

(exercise 7.16(5)). If f is differentiable at ξ and f^{-1} is differentiable at $\eta = f(\xi)$, then the formula (V) yields that

$$f'(\xi)(f^{-1})'(\eta) = I$$

from which formula (VI) follows.

When considering the real case we also gave conditions under which a suitable differentiable inverse exists. It follows from theorem 12.10 that it is sufficient in the real case if f' exists and is positive on some open interval containing ξ. The same is, of course, true if the word positive is replaced by the word negative. In the particular case when f' is continuous at ξ, it is therefore enough if $f'(\xi) \neq 0$.

A similar result holds in the vector case. Note that $f'(\xi) \neq 0$ in the real case corresponds to the requirement that the affine function tangent to f at ξ has an inverse function. The condition in the vector case which corresponds to this requirement is that the $n \times n$ matrix $f'(\xi)$ be *non-singular*, i.e.

$$\det f'(\xi) \neq 0.$$

We call the determinant $\det f'(\xi)$ the *Jacobian* of f at ξ. It is an important result that a suitable differentiable inverse of f exists provided that all the first order partial derivatives of f are continuous at ξ and $\det f'(\xi) \neq 0$. However, we shall neither explain the meaning of the word 'suitable' in this context nor offer a proof of the result. A proof would require apparatus that has not been developed in this book.

19.26 Exercise

(1) Explain why all functions $f: \mathbb{R}^n \to \mathbb{R}^m$ whose co-ordinates are all polynomials are differentiable at every point. What is the appropriate result when the co-ordinates are rational functions?

(2) Explain why each of the following functions is differentiable at the given point and compute its derivative there.

(i) $f: \mathbb{R}^2 \to \mathbb{R}$ at $(1, 2)$ where $f(x_1, x_2) = x_1^2 + 2x_1x_2 + x_2^3$.

(ii) $f: \mathbb{R}^2 \to \mathbb{R}^2$ at $(\pi, 1)$ where

$$f(x_1, x_2) = \begin{pmatrix} \cos(x_1 x_2) \\ \sin(x_1 x_2) \end{pmatrix}.$$

(iii) $f: \mathbb{R}^2 \to \mathbb{R}^3$ at $(1, 1)$ where

$$f(x_1, x_2) = \begin{pmatrix} x_1 \\ x_1 + x_2 \\ x_1 x_2 \end{pmatrix}.$$

(iv) $f: \mathbb{R}^3 \to \mathbb{R}^2$ at $(1, 2, 3)$ where

$$f(x_1, x_2, x_3) = \begin{pmatrix} x_1 x_2^2 \\ x_2^2 x_3^3 \end{pmatrix}.$$

(v) $f: \mathbb{R} \to \mathbb{R}^3$ at 0 where

$$f(x) = \begin{pmatrix} 1 \\ x \\ x^2 \end{pmatrix}.$$

For each of these functions, write down the equations for the appropriate tangent flat.

(3) What is the slope (or rate of increase) of the function f of question 2(i) at the point $(1, 2)$ in the direction $(1, -1)$? What is the direction at $(1, 2)$ in which the slope is greatest? What is the value of the slope in this direction? Write down a vector which is normal to the contour $x_1^2 + 2x_1 x_2 + x_2^3 = 13$ at $(1, 2)$. Also write down the equation of the tangent line at this point.

(4) Explain the circumstances under which formulae (I), (III) and (IV)′ of §19.21 make sense when L and M are matrices and x, y, z and c are column vectors. Prove that the formulae are valid when they make sense.

(5) Suppose that $f: \mathbb{R} \to \mathbb{R}$ is differentiable everywhere and that the function $\phi: \mathbb{R}^2 \to \mathbb{R}$ is defined by $\phi(x_1, x_2) = f(x_1 x_2)$. If $y = \phi(x_1, x_2)$, prove that

$$x_1 \frac{\partial y}{\partial x_1} = x_2 \frac{\partial y}{\partial x_2}$$

for all $(x_1, x_2) \in \mathbb{R}^2$ for which $f'(x_1 x_2) \neq 0$.

(6) Suppose that $f: \mathbb{R}^2 \to \mathbb{R}$ is differentiable everywhere and that $y = f(x_1, x_2)$ where

$$x_1 = r \cos \theta$$

$$x_2 = r \sin \theta.$$

Express $\partial y / \partial x_2$ in terms of $\partial y / \partial r$ and $\partial y / \partial \theta$.

___

19.27 Stationary points

In §11.3 we defined a point ξ to be a *stationary point* of f if $f'(\xi) = 0$. We shall use precisely the same definition in the case of a real-valued function of a vector variable. In this case, of course, $f'(\xi)$ is a $1 \times n$ row vector and $f'(\xi) = 0$ means that all the co-ordinates of this row vector (i.e. the partial derivatives of f at ξ) are zero. In geometric terms, the fact that ξ is a stationary point means that the tangent hyperplane

$$y = f(\xi) + f'(\xi)(x - \xi)$$

reduces to $y = f(\xi)$, i.e. the tangent hyperplane is horizontal at a stationary point. The diagram shows some points on a surface $y = f(x_1, x_2)$ at which the tangent plane is horizontal.

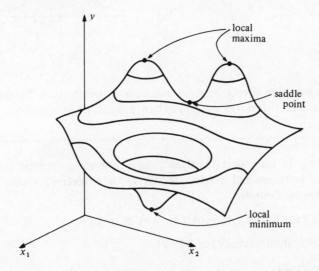

Any local maximum or minimum at which f is differentiable is, of course, a stationary point. This follows immediately from corollary 19.9 and we quote the result as a theorem below. It is not at all uncommon, however, for stationary points in the vector case to correspond to *saddle points* as indicated in the diagram. Less common, but not to be ignored, are other more exotic stationary points similar to those mentioned in §11.3.

19.28   *Theorem* Suppose that $f: A \to \mathbb{R}$ where A is a set in $\mathbb{R}^n$. If f is differentiable at $\xi \in A$ and f has a local maximum or minimum at ξ, then

$$f'(\xi) = 0.$$

Proof This follows from corollary 19.9 and theorem 19.16.

19.29   *Example* Consider the function $f: \mathbb{R}^2 \to \mathbb{R}$ defined by $f(x_1, x_2) = x_1 x_2$. Observe that

$$f'(0, 0) = (f_1(0, 0), f_2(0, 0)) = (0, 0)$$

but f does *not* have a local maximum or minimum at $(0, 0)$. On the line $x_1 = x_2$ we have that $f(x_1, x_2) = x_1^2 > 0$ unless $x_1 = 0$ while, on the line $x_1 = -x_2$, $f(x_1, x_2) = -x_1^2 < 0$ unless $x_1 = 0$.

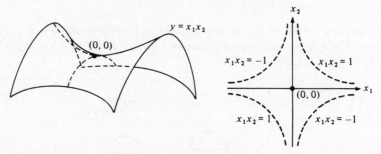

The two diagrams are alternative representations of the situation. The left hand diagram makes it clear why f is said to have a saddle point at $(0, 0)$.

19.30   *Example* In 'least squares analysis', one is given an *inconsistent* system $Ax = b$ of linear equations and asked to find the vector or vectors x which minimises $\|Ax - b\|$. Consider

$$y = \|Ax - b\|^2 = (Ax - b)^{\mathrm{T}}(Ax - b) = u^{\mathrm{T}}u$$

where $u = Ax - b$. By formula (V) of §19.21,

$$\frac{dy}{dx} = \frac{dy}{du}\frac{du}{dx} = (2u^{\mathrm{T}}I)A = 2(Ax - b)^{\mathrm{T}}A.$$

At a local minimum, the derivative must be the zero vector and so, if x minimises $\|Ax - b\|$, then

$$2(Ax - b)^{\mathrm{T}}A = 0.$$

Recalling exercise 18.25(2), we rewrite this as

$$A^{\mathrm{T}}(Ax - b) = 0.$$

i.e.

$$A^{\mathrm{T}}Ax = A^{\mathrm{T}}b.$$

If $A^{\mathrm{T}}A$ is non-singular, it follows that there is only one candidate for a minimising x, namely

$$x = (A^{\mathrm{T}}A)^{-1}A^{\mathrm{T}}b.$$

In the preceding example, the geometry of the situation makes it clear that a minimising x must exist. However, a geometrical argument is no substitute for a proof and, in any case, the geometry of the situation is seldom as transparent as in example 19.30. Some method is therefore necessary for classifying stationary points into local maxima, local minima, saddle points or other more exotic categories. Our experience with the real case would indicate an examination of the *second* derivative at a stationary point. But how is this defined?

19.31 Second derivatives

We consider only the second derivative of *real-valued* functions. The reason for this is that the derivative of a real-valued function is a vector and we know how to differentiate vector-valued functions. In order to discuss second derivatives in general, we should need first to discuss the derivative of matrix-valued functions and for this purpose the notation of matrix algebra is not adequate. (See §19.48.)

Even restricting ourselves to real-valued functions, we have a small problem since §19.14 assumes that we have a column vector to differentiate whereas $f'(x)$ is a row vector. However, this problem is easily overcome. If $\phi(x)$ is a row vector we define $\phi'(x)$ by

$$\phi'(x)^{\mathrm{T}} = \frac{d}{dx}\{\phi(x)^{\mathrm{T}}\}$$

provided the latter derivative exists. If ϕ is differentiable at ξ, we then have that

$$\phi(x) - \phi(\xi) - (x - \xi)^{\mathrm{T}}\phi'(\xi) = o(x - \xi)$$

i.e. the theory for row vectors is the same as that for column vectors except that it is necessary to write $(x - \xi)^{\mathrm{T}}\phi'(\xi)$ rather than $f'(\xi)(x - \xi)$ wherever the latter expression occurs.

With this understanding, we can now define the second derivative of a real-valued function f at the point x by

$$f''(x)^{\mathrm{T}} = \frac{d}{dx} f'(x)^{\mathrm{T}}$$

provided the latter derivative exists. Since

$$f'(x)^{\mathrm{T}} = \begin{pmatrix} f_1(x) \\ f_2(x) \\ \vdots \\ f_n(x) \end{pmatrix}$$

it follows immediately from theorem 19.16 that $f''(x)$ is the $n \times n$ matrix

$$f''(x) = \begin{pmatrix} f_{11}(x) & f_{12}(x) \ldots f_{1n}(x) \\ f_{21}(x) & f_{22}(x) \ldots f_{2n}(x) \\ \cdots\cdots\cdots\cdots\cdots\cdots \\ f_{n1}(x) & f_{n2}(x) \ldots f_{nn}(x) \end{pmatrix} \qquad (1)$$

where

$$f_{ij}(x) = \frac{\partial}{\partial x_i} \{f_j(x)\} = \frac{\partial}{\partial x_i} \frac{\partial}{\partial x_j} f(x)\}.$$

It is customary to abbreviate the latter notation and to write

$$f_{ij}(x) = \frac{\partial^2}{\partial x_i \partial x_j} f(x).$$

Thus, if $y = f(x)$,

$$\frac{d^2 y}{dx^2} = \begin{pmatrix} \dfrac{\partial^2 y}{\partial x_1^2} & \dfrac{\partial^2 y}{\partial x_1 \partial x_2} & \cdots & \dfrac{\partial^2 y}{\partial x_1 \partial x_n} \\ \dfrac{\partial^2 y}{\partial x_2 \partial x_1} & \dfrac{\partial^2 y}{\partial x_2^2} & \cdots & \dfrac{\partial^2 y}{\partial x_2 \partial x_n} \\ \vdots & & & \\ \dfrac{\partial^2 y}{\partial x_n \partial x_1} & \dfrac{\partial^2 y}{\partial x_n \partial x_2} & \cdots & \dfrac{\partial^2 y}{\partial x_n^2} \end{pmatrix}$$

and so the second derivative of a real-valued function consists of an $n \times n$ matrix whose entries consist of all second order partial derivatives of the function. From theorem 19.17 we know that a sufficient condition that f be twice differentiable at a point ξ is that all these second order partial derivatives exist and are *continuous* at ξ.

19.32   *Example* Consider the function $f \colon \mathbb{R}^2 \to \mathbb{R}$ defined by $f(x_1, x_2) = x_1^2 x_2 + x_2^3 x_1$. We have that

$$\frac{\partial y}{\partial x_1} = 2x_1 x_2 + x_2^3 \qquad \frac{\partial y}{\partial x_2} = x_1^2 + 3x_2^2 x_1$$

and so

$$\frac{\partial^2 y}{\partial x_1^2} = 2x_2 \qquad \frac{\partial^2 y}{\partial x_1 \partial x_2} = 2x_1 + 3x_2^2$$

$$\frac{\partial^2 y}{\partial x_2 \partial x_1} = 2x_1 + 3x_2^2 \qquad \frac{\partial^2 y}{\partial x_2^2} = 6x_1 x_2.$$

All these are continuous everywhere and hence f is twice differentiable at every point with

$$f''(x_1, x_2) = \begin{pmatrix} 2x_2 & 2x_1 + 3x_2^2 \\ 2x_1 + 3x_2^2 & 6x_1 x_2 \end{pmatrix}.$$

It is no accident that the matrix $f''(x_1, x_2)$ in the preceding example is *symmetric*, i.e.

$$\frac{\partial^2 y}{\partial x_1 \partial x_2} = \frac{\partial^2 y}{\partial x_2 \partial x_1}.$$

19.33 Proposition If all the second order partial derivatives in the matrix (1) of §19.31 are continuous at ξ, then the $n \times n$ matrix $f''(\xi)$ is symmetric, i.e. for all i and j

$$f_{ij}(\xi) = f_{ji}(\xi).$$

19.34 Mean value theorems

The mean value theorem of §11.5 generalises immediately to the case of *real-valued* functions of a vector variable. Note, however, that it does *not* generalise to vector-valued functions. (See exercise 19.47(5).)

19.35 Theorem (mean value theorem) Let L denote the straight line segment joining $a \in \mathbb{R}^n$ and $b \in \mathbb{R}^n$ and let L^0 be the set obtained by deleting a and b from L. If f is a real-valued function which is continuous at each point of L and differentiable at each point of L^0, then

$$f(b) - f(a) = f'(\xi)(b - a)$$

for some $\xi \in L^0$.

Proof Define $F: [0, 1] \to \mathbb{R}$ by

$$F(t) = f((1 - t)a + tb) \quad \mathbb{R} \ (0 \le t \le 1).$$

Then F is continuous on $[0, 1]$ and differentiable on $(0, 1)$. (See exercise 18.20(5).) By theorem 11.6

$$\frac{F(1) - F(0)}{1 - 0} = F'(\tau)$$

for some $\tau \in (0, 1)$. Thus $f(b) - f(a) = F(1) - F(0) = F'(\tau)$. It remains to express $F'(\tau)$ in terms of the function f. Write $y = f(x)$ and $x = (1 - t)a + tb$. Then

$$F'(t) = \frac{dy}{dt} = \frac{dy}{dx} \cdot \frac{dx}{dt} = f'(x)(b - a).$$

Writing $\xi = (1 - \tau)a + \tau b$, it follows that

$$f(b) - f(a) = F'(\tau) = f'(\xi)(b - a).$$

In chapter 11, we went on to prove Taylor's theorem which, as we pointed out, may be regarded as the nth order mean value theorem. This also generalises to the case of *real-valued* functions but we consider only the case $n = 2$ since we have not discussed derivatives of higher order.

19.36   Theorem (Taylor's theorem of order 2) Suppose that f is a real-valued function which is twice differentiable at each point $x \in \mathbb{R}^n$ which satisfies $\|x - \xi\| < \Delta$. Then for any such x,

$$f(x) = f(\xi) + \frac{1}{1!}f'(\xi)(x - \xi) + \frac{1}{2!}(x - \xi)^{\mathrm{T}}f''(\eta)(x - \xi)$$

for some η on the line segment L^0 between x and ξ.

Proof We apply theorem 11.10 to the function $F: [0, 1] \to \mathbb{R}$ defined by $F(t) = f((1 - t)\xi + tx) \, (0 \leqslant t \leqslant 1)$ and obtain

$$F(1) = F(0) + \frac{1}{1!}F'(0) + \frac{1}{2!}F''(\tau)$$

for some τ satisfying $0 < \tau < 1$. We have that $F(1) = f(x)$, $F(0) = f(\xi)$ and, as in the previous theorem,

$$F'(t) = f'((1 - t)\xi + tx)(x - \xi)$$

so that $F'(0) = f'(\xi)(x - \xi)$. It remains to express $F''(\tau)$ in terms of f. This requires a little care with transposition signs. (See exercise 18.25(2).) Observe that

$$F''(t)^{\mathrm{T}} = \frac{d}{dt}\{F'(t)\}^{\mathrm{T}}$$

$$= \frac{d}{dt}\{f'((1 - t)\xi + tx)(x - \xi)\}^{\mathrm{T}}$$

$$= \frac{d}{dt}(x - \xi)^{\mathrm{T}} f'((1 - t)\xi + tx)^{\mathrm{T}}$$

$$= (x - \xi)^{\mathrm{T}} f''((1 - t)\xi + tx)^{\mathrm{T}}(x - \xi).$$

Writing $\eta = (1 - \tau)\xi + tx$, we obtain that $F''(\tau)^{\mathrm{T}} = (x - \xi)^{\mathrm{T}} f''(\eta)^{\mathrm{T}}(x - \xi)$. Transposing both sides then yields $F''(\tau) = (x - \xi)^{\mathrm{T}} f''(\eta)(x - \xi)$ as required.

If ξ is a stationary point, then $f'(\xi) = 0$ and so the preceding theorem reduces to

$$f(\xi + h) - f(\xi) = \tfrac{1}{2} h^{\mathrm{T}} f''(\eta) h$$

where η lies on the line segment between ξ and $\xi + h$. In order that ξ correspond to a local maximum it is therefore clearly sufficient that $h^{\mathrm{T}} f''(\eta) h$ be *negative* for all h of sufficiently small length. The analogous condition for a local minimum is that $h^{\mathrm{T}} f''(\eta) h$ be *positive* for all h of sufficiently small length. What properties of the matrix $f''(\eta)$ guarantee one or other of these criteria? To answer this question reasonably elegantly requires the use of rather more linear algebra than we have been assuming so far.

19.37   Eigenvalues

If M is an $n \times n$ *symmetric* matrix, there exists an orthogonal matrix H (i.e. a matrix H satisfying $H^{\mathrm{T}} H = I$) such that

$$H^{\mathrm{T}} M H = D$$

where D is a diagonal matrix, i.e.

$$D = \begin{pmatrix} \lambda_1 & 0 \ldots 0 \\ 0 & \lambda_2 \ldots 0 \\ \vdots & \\ 0 & 0 \ldots \lambda_n \end{pmatrix}$$

The entries $\lambda_1, \lambda_2, \ldots, \lambda_n$ on the diagonal of D are called the *eigenvalues* of M and can be found by solving the polynomial equation

$$\det (M - \lambda I) = 0$$

obtained by setting the determinant of the square matrix $M - \lambda I$ equal to zero. It is a useful fact that the eigenvalues of a symmetric matrix with real entries are always real numbers.

We shall take this theory for granted. Its interest in the current context is that it allows us to discuss quadratic forms in a coherent fashion. If M is an $n \times n$ symmetric matrix and x is an $n \times 1$ column vector, we call

$$x^{\mathrm{T}} M x$$

a *quadratic form.* If $x^T M x > 0$ for all $x \neq 0$ we say that M is *positive definite.* If $x^T M x < 0$ for all $x \neq 0$ we say that M is *negative definite.* In view of the remarks at the end of §19.34, these notions are clearly relevant to the classification of stationary points.

If we write $x = Hy$, then $x^T = y^T H^T$ and so

$$x^T M x = y^T H^T M H y = y^T D y$$
$$= \lambda_1 y_1^2 + \lambda_2 y_2^2 + \ldots + \lambda_n y_n^2.$$

Since $\|x\|^2 = x^T x = y^T H^T H y = y^T I y = y^T y = \|y\|^2$, we have that $x = 0$ if and only if $y = 0$. It follows that M is positive definite if and only if all its eigenvalues are positive. Similarly M is negative definite if and only if all its eigenvalues are negative. We may also deduce the useful inequality

$$\lambda_1 \|x\|^2 \leqslant x^T M x \leqslant \lambda_n \|x\|^2 \tag{1}$$

where λ_1 is the smallest and λ_n is the largest of the eigenvalues of M.

If all the eigenvalues of M are non-negative, then (1) shows that $x^T M x$ has a minimum when $x = 0$. Similarly, if all the eigenvalues of M are non-positive, then $x^T M x$ has a maximum when $x = 0$. If one or more of the eigenvalues are zero, however, there are straight lines through the point 0 on which $x^T M x$ is identically zero. To see this we need the idea of an eigenvector.

The unit vector $u^j = He^j$, where e^j is the column vector whose co-ordinates are all zero except for the jth which is equal to one, is an *eigenvector* corresponding to λ_j. If $x = tu^j$, then

$$x^T M x = \lambda_j t^2. \tag{2}$$

Thus, if $\lambda_j = 0$, then $x^T M x = 0$ for all x on the line through 0 which points in the direction of the eigenvector u^j.

It is also instructive to see what happens when some eigenvalues are positive and others are negative. Suppose in particular that $\lambda_1 < 0$ while $\lambda_n > 0$. From (2) it follows that $x^T M x < 0$ for $x \neq 0$ on the straight line through 0 in the direction of u^1. Similarly, $x^T M x > 0$ for $x \neq 0$ on the straight line through 0 in the direction of u^n. Thus, if we restrict our attention to values of x along one line through 0, we find that $x^T M x$ is strictly smallest when $x = 0$. But along a second line, $x^T M x$ is strictly largest when $x = 0$. Under these circumstances, we call 0 a *saddle point.*

The diagrams below illustrate the various cases we have considered.

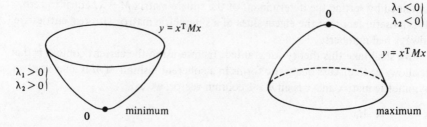

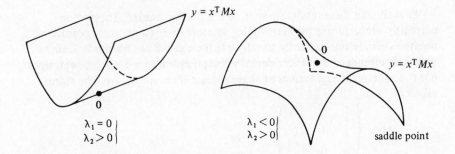

$y = x^{\mathrm{T}} M x$

$y = x^{\mathrm{T}} M x$

$$\left.\begin{array}{l} \lambda_1 = 0 \\ \lambda_2 > 0 \end{array}\right\}$$ $$\left.\begin{array}{l} \lambda_1 < 0 \\ \lambda_2 > 0 \end{array}\right\}$$ saddle point

19.38 Principal minors

If A and B are square $n \times n$ matrices, then det $(AB) = (\det A)(\det B)$. It follows that

$$\det M = \det (H^{\mathrm{T}} D H) = (\det H^{\mathrm{T}})(\det D)(\det H)$$

$$= \det (H^{\mathrm{T}} H) \det D = \det I \det D.$$

But the determinant of a diagonal matrix is simply the product of its diagonal entries. Hence

$$\det M = \lambda_1 \lambda_2 \ldots \lambda_n.$$

Various useful consequences follow. For example, M has a zero eigenvalue if and only if det $M = 0$. Also det M is positive when M is positive definite and det M has the sign of $(-1)^n$ when M is negative definite. This last sentence is part of a more general result concerning the principal minors of M.

If

$$M = \begin{pmatrix} m_{11} & m_{12} & \ldots & m_{1n} \\ m_{21} & m_{22} & \ldots & m_{2n} \\ \vdots & & & \\ m_{n1} & m_{n2} & \ldots & m_{nn} \end{pmatrix}$$

then its *principal minors* are the determinants

$$P_1 = m_{11}, \quad P_2 = \begin{vmatrix} m_{11} & m_{12} \\ m_{21} & m_{22} \end{vmatrix}, \quad P_3 = \begin{vmatrix} m_{11} & m_{12} & m_{13} \\ m_{21} & m_{22} & m_{23} \\ m_{31} & m_{32} & m_{33} \end{vmatrix}$$

and so on, ending up with $P_n = \det M$. We then have the following results:

(I) The matrix M is positive definite if and only if

$$P_j > 0 \quad (j = 1, 2, \ldots, n).$$

(II) The matrix M is negative definite if and only if

$$(-1)^j P_j > 0 \quad (j = 1, 2, \ldots, n).$$

We shall take these results from linear algebra for granted. They are an extremely useful means for determining whether or not a matrix is positive or negative definite since actually to calculate the eigenvalues when $n \geqslant 3$ can be very difficult indeed. (Note incidentally that, if $P_j \geqslant 0$ $(j = 1, 2, \ldots, n)$ it does *not* follow that the eigenvalues of M are necessarily non-negative. The eigenvalues of

$$M = \begin{pmatrix} 1 & 1 & 0 \\ 1 & 1 & 0 \\ 0 & 0 & -1 \end{pmatrix}$$

are $-1, 0$ and 2 but all the principal minors are non-negative. Nor is it true that $(-1)^j P_j \geqslant 0$ implies that the eigenvalues are necessarily non-positive.)

19.39   Examples

(i) The matrix

$$A = \begin{pmatrix} 5 & -3 \\ -3 & 5 \end{pmatrix}$$

is positive definite because its principal minors are 5 and

$$\begin{vmatrix} 5 & -3 \\ -3 & 5 \end{vmatrix} = 25 - 9 = 16.$$

(ii) The matrix

$$B = \begin{pmatrix} -2 & 1 \\ 1 & -2 \end{pmatrix}$$

is negative definite because its principal minors are -2 and

$$\begin{vmatrix} -2 & 1 \\ 1 & -2 \end{vmatrix} = 4 - 1 = 3.$$

(iii) The matrix

$$C = \begin{pmatrix} 1 & 6 \\ 6 & 4 \end{pmatrix}$$

is neither positive definite nor negative definite because

$$\det C = \begin{vmatrix} 1 & 6 \\ 6 & 4 \end{vmatrix} = 4 - 36 = -32 < 0.$$

Note that, since $\det C \neq 0$, neither eigenvalue is zero.

19.40 Classification of stationary points

Suppose that f is a real-valued function with a stationary point at ξ. If all the second order partial derivatives of f exist and are continuous at each x satisfying $\|x - \xi\| < \Delta$, it is often possible to determine whether ξ corresponds to a local maximum, a local minimum or a saddle point of f by examining the matrix $f''(\xi)$. Since f has a stationary point at ξ, $f'(\xi) = 0$ and so theorem 19.36 reduces to

$$f(\xi + h) - f(\xi) = \tfrac{1}{2}h^T f''(\eta)h$$

provided $\|h\|$ is sufficiently small. We need to relate the quadratic form $h^T f''(\eta)h$ to the quadratic form $h^T f''(\xi)h$. For this purpose we use the inequality

$$|h^T L h| \leqslant \left\{ \sum_{i=1}^{n} \sum_{j=1}^{n} l_{ij}^2 \right\}^{1/2} \|h\|^2$$

which holds for any $n \times n$ matrix L and any $n \times 1$ column vector h. The proof consists of two applications of the Cauchy–Schwarz inequality. We have that

$$
\begin{aligned}
|h^T L h| &= \left| \sum_i h_i \sum_j l_{ij} h_j \right| \\
&\leqslant \sum_i |h_i| \left\{ \sum_j l_{ij}^2 \right\}^{1/2} \|h\| \\
&\leqslant \left\{ \sum_i \sum_j l_{ij}^2 \right\}^{1/2} \|h\|^2.
\end{aligned}
$$

19.41 Lemma Suppose that f is a real-valued function with second order partial derivatives which exist and are continuous at each $x \in \mathbb{R}^n$ satisfying $\|x - \xi\| < \Delta$. Then, given any $\epsilon > 0$, we can find a $\delta > 0$ such that

$$|h^T(f''(\xi) - f''(\eta))h| < \epsilon\|h\|^2$$

provided that $0 < \|h\| < \delta$ and η is as in theorem 19.36.

Proof We simply observe that

$$|h^T(f''(\xi) - f''(\eta))h| \leqslant \left\{ \sum_i \sum_j (f_{ij}(\xi) - f_{ij}(\eta))^2 \right\}^{1/2} \|h\|^2$$

and then use the fact that each of the second order partial derivatives f_{ij} is continuous at ξ.

19.42 Theorem Suppose that f is a real-valued function with second order partial derivatives which exist and are continuous at each $x \in \mathbb{R}^n$ satisfying $\|x - \xi\| < \Delta$. Then
(i) If $f'(\xi) = 0$ and $f''(\xi)$ is negative definite then f has a local maximum at ξ.
(ii) If $f'(\xi) = 0$ and $f''(\xi)$ is positive definite, then f has a local minimum at ξ.

Proof We prove (ii). The proof of (i) is entirely analogous.

Let λ_1 be the smallest eigenvalue of $f''(\xi)$. It follows from (1) of §19.37 that

$$h^T f''(\xi)h \geqslant \lambda_1 \|h\|^2$$

for all h.

Since $f''(\xi)$ is positive definite, $\lambda_1 > 0$ and so we may take $\epsilon = \frac{1}{2}\lambda_1$ in lemma 19.41. From theorem 19.36, we then obtain that

$$\begin{aligned} f(\xi + h) - f(\xi) &= \tfrac{1}{2}h^T f''(\eta)h \\ &= \tfrac{1}{2}h^T f''(\xi)h + \tfrac{1}{2}h^T(f''(\eta) - f''(\xi))h \\ &> \tfrac{1}{2}\lambda_1 \|h\|^2 - \tfrac{1}{2}\epsilon\|h\|^2 \end{aligned}$$

provided that $0 < \|h\| < \delta$. Hence, for $0 < \|h\| < \delta$,

$$f(\xi + h) - f(\xi) > \tfrac{1}{4}\lambda_1 \|h\|^2$$

and so f has a local minimum at ξ.

19.43   Theorem Suppose that f is a real-valued function with second order partial derivatives which exist and are continuous at each $x \in \mathbb{R}^n$ satisfying $\|x - \xi\| < \Delta$.

If $f'(\xi) = 0$ but $f''(\xi)$ is neither positive definite nor negative definite, then f has a (local) saddle point at ξ provided that

$$\det f(\xi) \neq 0.$$

Proof The conditions guarantee that the smallest eigenvalue λ_1 of $f''(\xi)$ is negative while the largest eigenvalue λ_n is positive. Using (2) of §19.37 together with the argument of theorem 19.42 we obtain that

$$f(\xi + h) - f(\xi) < \tfrac{1}{4}\lambda_1 \|h\|^2$$

for values of h which satisfy $0 < \|h\| < \delta$ and lie on the line through 0 in the direction u^1. Similarly

$$f(\xi + h) - f(\xi) > \tfrac{1}{4}\lambda_n \|h\|^2$$

for values of h which satisfy $0 < \|h\| < \delta$ and lie on the line through 0 in the direction u^n.

19.44   *Note* The condition $\det f(\xi) \neq 0$ in theorem 19.43 guarantees that no eigenvalues of $f''(\xi)$ are zero. It is worth pointing out that nothing very useful can be deduced about the behaviour of f at a stationary point ξ from the fact that an eigenvalue λ_j of $f''(\xi)$ is zero. In this case

$$f(\xi + h) - f(\xi) = \tfrac{1}{2}h^T(f''(\eta) - f''(\xi))h$$

for sufficiently small values of h on the line through 0 in the direction u^j and we have no information about the sign of the right hand side. (When $\lambda_j \neq 0$ this is irrelevant since the term is then dominated by $\frac{1}{2}\lambda_j\|h\|^2$.)

In the case $n = 1$, $f''(\xi)$ is just a real number and theorem 19.42 is the familiar result which relates the existence of a local maximum or minimum at a stationary point with the sign of the real number $f''(\xi)$. Theorem 19.43 does not apply in the case $n = 1$ because, if $f''(\xi)$ is neither positive or negative it must be zero.

In the case $n = 2$, it is best to begin examining $f''(\xi)$ at a stationary point ξ by looking at

$$\det f''(\xi) = \begin{vmatrix} f_{11}(\xi) & f_{12}(\xi) \\ f_{21}(\xi) & f_{22}(\xi) \end{vmatrix}$$

$$= f_{11}(\xi)f_{22}(\xi) - f_{12}(\xi)^2.$$

We know from §19.38 that for $f''(\xi)$ to be positive definite or negative definite it is necessary that $\det f''(\xi) > 0$. If $\det f''(\xi) > 0$ and $f_{11}(\xi) > 0$, then $f''(\xi)$ is positive definite and we have a local minimum. If $\det f''(\xi) > 0$ and $f_{11}(\xi) < 0$, then $f''(\xi)$ is negative definite and we have a local maximum. If $\det f''(\xi) < 0$, theorem 19.43 assures the existence of a saddle point.

Do *not* be tempted to draw conclusions when $\det f''(\xi) = 0$. One of the two eigenvalues is then zero and, as explained in note 19.44, the method does not work in this case. It is not possible to be tempted to draw a wrong conclusion from $\det f''(\xi) > 0$ and $f_{11}(\xi) = 0$ because the latter equation implies that $\det f''(\xi) = -f_{12}(\xi)^2$.

19.45   *Example* Consider the function $f: \mathbb{R}^2 \to \mathbb{R}$ defined by $f(x_1, x_2) = x_1^3 - 3x_1x_2^2 + x_2^4$. The stationary points are found by solving

$$\left. \begin{aligned} f_1(x_1, x_2) &= 3x_1^2 - 3x_2^2 = 0 \\ f_2(x_1, x_2) &= -6x_1x_2 + 4x_2^3 = 0 \end{aligned} \right\}$$

simultaneously. The solutions are $(3/2, 3/2)$, $(3/2, -3/2)$ and $(0, 0)$. The second derivative is

$$f''(x_1, x_2) = \begin{pmatrix} f_{11}(x_1, x_2) & f_{12}(x_1, x_2) \\ f_{21}(x_1, x_2) & f_{22}(x_1, x_2) \end{pmatrix} = \begin{pmatrix} 6x_1 & -6x_2 \\ -6x_2 & -6x_1 + 12x_2^2 \end{pmatrix}$$

and so the principal minors are

$$\det f''(x_1, x_2) = \begin{vmatrix} 6x_1 & -6x_2 \\ -6x_2 & -6x_1 + 12x_2^2 \end{vmatrix} = 36(-x_1^2 + 2x_1x_2^2 - x_2^2)$$

and

$$f_{11}(x_1, x_2) = 6x_1.$$

(i) Since $\det f''(3/2, 3/2) = 81 > 0$ and $f_{11}(3/2, 3/2) = 9 > 0$, we have that $(3/2, 3/2)$ is a local minimum.

(ii) Also $\det f''(3/2, -3/2) = 81 > 0$ and $f_{11}(3/2, -3/2) = 9 > 0$ and so $(3/2, -3/2)$ is also a local minimum.

(iii) Since $\det f''(0,0) = 0$, the method is useless for classifying the stationary point $(0,0)$. Observe, however, that

$$y = f(0, x_2) = x_2^4$$
$$y = f(x_1, 0) = x_1^3$$

and so the function certainly does not have a local maximum or minimimum at $(0,0)$.

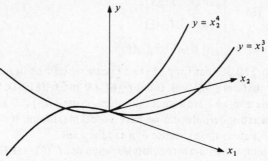

19.46   *Example* Consider the function $f \colon \mathbb{R}^3 \to \mathbb{R}$ defined by $f(x_1, x_2, x_3) = x_1 x_2 + x_2 x_3 + x_3 x_1$. The stationary points are found by solving the equations

$$\left. \begin{array}{l} x_2 + x_3 = 0 \\ x_1 + x_3 = 0 \\ x_2 + x_1 = 0 \end{array} \right\}$$

simultaneously. The only stationary point is therefore $(0,0,0)$. The second derivative at this point is

$$\begin{pmatrix} 0 & 1 & 1 \\ 1 & 0 & 1 \\ 1 & 1 & 0 \end{pmatrix}$$

which has principal minors $2, -1$ and 0. The matrix is neither positive definite nor negative definite because one of the principal minors is zero. But $f''(0,0,0) \neq 0$ and so we have a saddle point at $(0,0,0)$.

19.47   *Exercise*

(1) A function $f \colon \mathbb{R}^n \to \mathbb{R}$ is defined by

$$f(x) = (Ax + a)^{\mathrm{T}}(Bx + b)$$

where A and B are $m \times n$ matrices and x, a and b are $m \times 1$ column vectors. Find a system of linear equations satisfied by the local maxima and minima of f. [Hint: example 19.30.]

(2) Show that that the function $f: \mathbb{R}^2 \to \mathbb{R}$ defined by

$$f(x_1, x_2) = x_1 e^{-x_2}$$

is twice differentiable at every point and determine $f''(x_1, x_2)$.

(3) Show that the function $f: \mathbb{R}^2 \to \mathbb{R}$ defined by

$$f(x_1, x_2) = \begin{cases} \dfrac{x_1^3 x_2}{x_1^2 + x_2^2} & (x_1, x_2) \neq (0,0) \\ 0 & (x_1, x_2) = (0,0) \end{cases}$$

has a continuous first derivative $f': \mathbb{R}^2 \to \mathbb{R}^2$ at $(0,0)$ and that all its second order partial derivatives exist at $(0, 0)$. Prove that

$$f_{12}(0, 0) \neq f_{21}(0, 0)$$

and explain why this does not contradict proposition 19.33.

(4) Classify the stationary points of the function $f: \mathbb{R}^2 \to \mathbb{R}$ defined by

$$f(x_1, x_2) = x_1^2 x_2 + x_2^3 x_1 - x_1 x_2$$

using the second derivative. [Hint: example 19.10.]

(5) Show that there exists no value of $\xi \in (0, 1)$ such that

$$f(1) - f(0) = f'(\xi)(1 - 0)$$

in the case of the function $f: \mathbb{R}^1 \to \mathbb{R}^2$ defined by

$$f(t) = \begin{pmatrix} t^2 \\ t^3 \end{pmatrix}.$$

(See §19.34.)

(6) Suppose that λ is an eigenvalue of an $n \times n$ symmetric matrix M with real entries. Using the results quoted in this chapter, prove that

$$\lambda^2 \leqslant \sum_{i=1}^{n} \sum_{j=1}^{n} m_{ij}^2.$$

19.48[†] Differentials

The derivative $f'(x)$ or $Df(x)$ of a vector function f at a point x was defined to be a matrix. Recall that f is differentiable at x if and only if there exists an affine function g which is tangent to f at x. The affine function g is given by

$$g(x + h) = f(x) + f'(x)h.$$

The differential $df(x)$ of f at x is a distinct but closely related notion. It is the *linear function $df(x)$*: $\mathbb{R}^n \rightarrow \mathbb{R}^m$ defined by

$$df(x)(h) = f'(x)h.$$

Thus, as noted in §10.4, we require an equation involving two variables, x and h, in order to define the differential df of a function. Strictly speaking, however, we should first introduce the set X of all points $x \in \mathbb{R}^n$ at which f is differentiable and then the set $\mathcal{L}(\mathbb{R}^n, \mathbb{R}^m)$ of all linear functions $L: \mathbb{R}^n \rightarrow \mathbb{R}^m$. Then the *differential* of f is the function $df: X \rightarrow \mathcal{L}(\mathbb{R}^n, \mathbb{R}^m)$ whose value at $x \in X$ is the linear function $df(x)$.

As explained in §10.4, for a fixed value of x, the equation $k = f'(x)h$ represents the tangent flat at x provided that the h- and k-axes are drawn with the origin at the point $(x, f(x))$. It is usual to use the notation dx for the variable h and the notation dy for the variable k. The equation $k = f'(x)h$ then becomes

$$dy = f'(x)dx$$

which can be rewritten in the more appealing form

$$dy = \frac{dy}{dx}dx.$$

In §10.4 we were working with real variables but in the current context this equation is shorthand for the system

$$dy_1 = \frac{\partial y_1}{\partial x_1}dx_1 + \frac{\partial y_1}{\partial x_2}dx_2 + \ldots + \frac{\partial y_1}{\partial x_n}dx_n$$

$$dy_2 = \frac{\partial y_2}{\partial x_1}dx_1 + \frac{\partial y_2}{\partial x_2}dx_2 + \ldots + \frac{\partial y_2}{\partial x_n}dx_n$$

$$\cdots\cdots\cdots\cdots\cdots\cdots\cdots\cdots\cdots\cdots\cdots\cdots\cdots\cdots$$

$$dy_m = \frac{\partial y_m}{\partial x_1}dx_1 + \frac{\partial y_m}{\partial x_2}dx_2 + \ldots + \frac{\partial y_m}{\partial x_n}dx_n.$$

It is important to emphasise that this is a system of linear equation linking the *variables* $dx_1, dx_2, \ldots, dx_n$ with the *variables* $dy_1, dy_2, \ldots, dy_m$. It is sometimes more convenient to work with such a system than to work with the associated matrix.

19.49   *Example* Suppose that $f: \mathbb{R}^2 \rightarrow \mathbb{R}^2$ is defined by

$$\left.\begin{array}{l} y_1 = x_1 + x_2 \\ y_2 = x_1 - x_2. \end{array}\right\} \tag{1}$$

Then

$$f'(x) = \begin{pmatrix} \dfrac{\partial y_1}{\partial x_1} & \dfrac{\partial y_1}{\partial x_2} \\[2ex] \dfrac{\partial y_2}{\partial x_1} & \dfrac{\partial y_2}{\partial x_2} \end{pmatrix} = \begin{pmatrix} 1 & 1 \\ 1 & -1 \end{pmatrix}.$$

The function f has an inverse $f^{-1}: \mathbb{R}^2 \to \mathbb{R}^2$. As we know from formula (VI) of §19.21, its derivative at $y = f(x)$ is given by

$$(f^{-1})'(y) = \begin{pmatrix} \dfrac{\partial x_1}{\partial y_1} & \dfrac{\partial x_1}{\partial y_2} \\[2ex] \dfrac{\partial x_2}{\partial y_1} & \dfrac{\partial x_2}{\partial y_2} \end{pmatrix}$$

$$= \begin{pmatrix} \dfrac{\partial y_1}{\partial x_1} & \dfrac{\partial y_1}{\partial x_2} \\[2ex] \dfrac{\partial y_2}{\partial x_1} & \dfrac{\partial y_2}{\partial x_2} \end{pmatrix}^{-1} = \frac{1}{-2} \begin{pmatrix} -1 & -1 \\ -1 & 1 \end{pmatrix}.$$

In particular

$$\frac{\partial x_1}{\partial y_1} = \tfrac{1}{2}.$$

This result may also be obtained by writing

$$dy_1 = \frac{\partial y_1}{\partial x_1} dx_1 + \frac{\partial y_1}{\partial x_2} dx_2 = dx_1 + dx_2$$

$$dy_2 = \frac{\partial y_2}{\partial x_1} dx_1 + \frac{\partial y_2}{\partial x_2} dx_2 = dx_1 - dx_2.$$

Solving these equations for dx_1, we obtain that $dx_1 = \tfrac{1}{2} dy_1 + \tfrac{1}{2} dy_2$ and hence

$$\frac{\partial x_1}{\partial y_1} = \tfrac{1}{2}.$$

When n is large, the latter method can involve much less calculation than that required to invert the matrix. Of course, in this particularly simple example, it would have been easiest of all just to solve equations (1). Usually, however, this is impractical.

As in §10.4, we wish to emphasise that the symbols $dy_1, dy_2, \ldots, dy_m$ and $dx_1, dx_2, \ldots, dx_n$ in (1) do *not* stand for 'small changes' in $y_1, y_2, \ldots, y_m$ and $x_1, x_2, \ldots, x_n$. If $\delta y_1, \delta y_2, \ldots, \delta y_m$ represent the 'small changes' in $y_1, y_2, \ldots, y_m$ caused by the 'small changes' $\delta x_1, \delta x_2, \ldots, \delta x_n$ in $x_1, x_2, \ldots, x_n$, then it is certainly true that

$$\left. \begin{array}{l} \delta y_1 \doteqdot \dfrac{\partial y_1}{\partial x_1}\delta x_1 + \dfrac{\partial y_1}{\partial x_2}\delta x_2 + \ldots + \dfrac{\partial y_1}{\partial x_n}\delta x_n \\[2.5ex] \delta y_2 \doteqdot \dfrac{\partial y_2}{\partial x_1}\delta x_1 + \dfrac{\partial y_2}{\partial x_2}\delta x_2 + \ldots + \dfrac{\partial y_2}{\partial x_n}\delta x_n \\[1ex] \cdots\cdots\cdots\cdots\cdots\cdots\cdots\cdots\cdots\cdots\cdots\cdots \\[1ex] \delta y_m \doteqdot \dfrac{\partial y_m}{\partial x_1}\delta x_1 + \dfrac{\partial y_m}{\partial x_2}\delta x_2 + \ldots + \dfrac{\partial y_m}{\partial x_n}\delta x_n \end{array} \right\} \qquad (2)$$

because the fact that f is differentiable at x means that

$$\delta y - f'(x)\delta x = o(\delta x)$$

but *equality* will be obtained in (2) only when f is an affine function.

19.50   *Example* The cosine rule for a triangle asserts that
$a^2 = b^2 + c^2 - 2bc \cos A$.

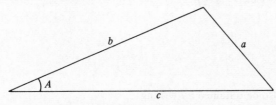

Find an approximation to the small change δa in a caused by small changes δA, δb and δc in A, b and c respectively. We have that

$$2a\,da = 2b\,db + 2c\,dc - 2c \cos A\,db - 2b \cos A\,dc + 2bc \sin A\,dA$$

$$a\,da = (b - c \cos A)db + (c - b \cos A)dc + bc \sin A\,dA$$

and so

$$\delta a \doteqdot \left(\frac{b - c \cos A}{a}\right)\delta b + \left(\frac{c - b \cos A}{a}\right)\delta c + \frac{bc \sin A}{a}\delta A.$$

In spite of the examples above, we have not introduced differentials just because they simplify certain calculations. Indeed, for those not fully at home with the background theory, the language of differentials is perhaps best avoided in elementary calculations since it allows much room for confusion and error. The reason for introducing differentials is so that something can be said about the general theory of differentiation.

In this chapter the derivative of a vector function was defined to be a matrix. This is entirely satisfactory when only first order derivatives and finite-dimensional vectors are to be considered and the exactness with which the theorems match those for the one-dimensional case is very pleasing. However, we were able to use the theory to define second derivatives only in the case of real-valued functions while third derivatives remained completely out of reach. What is more, the theory does not extend to the infinite-dimensional case. These problems all disappear if the ideas developed in this chapter are restated in terms of differentials. We shall not attempt a systematic account of how this is done. Such an account would require our working at a level of abstraction which would be inconsistent with the rest of the book. But there is no difficulty in explaining the general approach.

The space $\mathbb{R}^n$ is an example of a vector space which has real scalars and a norm. In what follows $\mathscr{X}$ and $\mathscr{Y}$ will denote any two vector spaces of this type. The notation $\mathscr{L}(\mathscr{X}, \mathscr{Y})$ will be used to the set of all continuous linear functions $L: \mathscr{X} \to \mathscr{Y}$. (Note that when $\mathscr{X}$ and $\mathscr{Y}$ are finite-dimensional *all* linear functions $L: \mathscr{X} \to \mathscr{Y}$ are continuous. See example 18.34. But this is not true in general.) Suppose that $f: A \to \mathscr{Y}$ where A is a set which contains every $\mathbf{x} \in \mathscr{X}$ which satisfies $\|\mathbf{x} - \boldsymbol{\xi}\| < \Delta$. Following §19.14, it is natural to say that f is differentiable if and only if there exists a continuous linear function $l: \mathscr{X} \to \mathscr{Y}$ such that

$$f(\boldsymbol{\xi} + \mathbf{h}) - f(\boldsymbol{\xi}) - l(\mathbf{h}) = o(\mathbf{h}) \qquad (\mathbf{h} \to \mathbf{0}).$$

We then define $df(\boldsymbol{\xi})$ to be the continuous linear function $l \in \mathscr{L}(\mathscr{X}, \mathscr{Y})$.

One reason for insisting that only *continuous* linear functions be considerd is that the set $\mathscr{L}(\mathscr{X}, \mathscr{Y})$ is then a vector space with real scalars and a norm just like $\mathscr{X}$ and $\mathscr{Y}$. The norm of an element l in the set $\mathscr{L}(\mathscr{X}, \mathscr{Y})$ is defined by

$$\|l\| = \sup_{\mathbf{x} \to \mathbf{0}} \frac{\|l(\mathbf{x})\|}{\|\mathbf{x}\|}.$$

(See exercise 18.25(6).) It follows that it is meaningful to ask whether df is differentiable at $\boldsymbol{\xi}$. If so, then its differential $ddf(\boldsymbol{\xi}) = d^2f(\boldsymbol{\xi})$ at $\boldsymbol{\xi}$ satisfies

$$df(\boldsymbol{\xi} + \mathbf{h}) - df(\boldsymbol{\xi}) - d^2f(\boldsymbol{\xi})(\mathbf{h}) = o(\mathbf{h}) \qquad (\mathbf{h} \to \mathbf{0}).$$

The object $d^2f(\boldsymbol{\xi})$ is an element of the set $\mathscr{L}(\mathscr{X}, \mathscr{L}(\mathscr{X}, \mathscr{Y}))$ which is again a vector space with real scalars and a norm. Again it is therefore meaningful to ask whether d^2f is differentiable at $\boldsymbol{\xi}$. If so, then its differential $dd^2f(\boldsymbol{\xi}) = d^3f(\boldsymbol{\xi})$ satisfies

$$d^2f(\boldsymbol{\xi} + \mathbf{h}) - d^2f(\boldsymbol{\xi}) - d^3f(\boldsymbol{\xi})(\mathbf{h}) = o(\mathbf{h}) \qquad (\mathbf{h} \to \mathbf{0})$$

Clearly we can continue in this way as often as we choose.

In the case when $\mathscr{X} = \mathbb{R}^n$ and $\mathscr{Y} = \mathbb{R}^1$, we have seen that

$$df(\boldsymbol{\xi})(\mathbf{h}) = f'(\xi)\mathbf{h} \tag{1}$$

where $f'(\xi)$ is an $m \times n$ matrix. Also we have that

$$d^2f(\boldsymbol{\xi})(\mathbf{h})(\mathbf{k}) = \mathbf{h}^{\mathsf{T}}f''(\xi)\mathbf{k}. \tag{2}$$

This last equation requires some explanation. Recall that $d^2f(\boldsymbol{\xi})$ is a function from $\mathscr{X}$ to $\mathscr{L}(\mathscr{X}, \mathscr{Y})$ and so $d^2f(\boldsymbol{\xi})(\mathbf{h})$ lies in $\mathscr{L}(\mathscr{X}, \mathscr{Y})$. It is therefore a function from $\mathscr{X}$ and $\mathscr{Y}$ and $d^2f(\xi)(h)(k)$ is the value of this function at k. Since $\mathscr{Y} = \mathbb{R}^1$ in this case, $d^2f(\xi)(h)(k)$ is a real number.

Unfortunately, matrix notation is not very helpful beyond this stage. However, there is no difficulty in writing.

$$df(\xi)h = \sum_{t=1}^{n} f_t(\xi)h_t$$

$$d^2f(\xi)(h)(k) = \sum_{s=1}^{n} \sum_{t=1}^{n} f_{st}(\xi)h_s k_t$$

$$d^3f(\xi)(h)(k)(l) = \sum_{r=1}^{n} \sum_{s=1}^{n} \sum_{t=1}^{n} f_{rst}(\xi)h_r k_s l_t$$

and so on. Taylor's theorem then takes the form

$$f(\xi + h) = f(\xi) + \frac{1}{1!}df(\xi)(h) + \frac{1}{2!}d^2f(\xi)(h)^2 + \frac{1}{3!}d^3f(\xi)(h)^3 + \ldots$$

where, for example,

$$d^3f(\xi)(h)^3 = d^3f(\xi)(h)(h)(h)$$

$$= \sum_{r=1}^{n} \sum_{s=1}^{n} \sum_{t=1}^{n} f_{rst}(\xi)h_r h_s h_t.$$

This brief outline is intended only to indicate that there is a theory beyond that given in the main body of the text. As far as the general theory is concerned, it is clearly the idea of a differential which is fundamental rather than that of a derivative. For this reason, the idea of a derivative is sometimes suppressed altogether in favour of that of a differential, and the two words used interchangeably. As commented in §10.4, however, this is not the practice in the real case and would be confusing in the context of the current chapter.

20 APPENDIX

20.1 Introduction

In the preceding chapters a number of results were stated without proof. These results were referred to as 'propositions'. The proofs were omitted from the main body of the text to avoid confusing the issue with too much detail. Instead we give the proofs here.

20.2 Bounded sets

Proposition 2.3 A set S of real numbers is bounded if and only if there exists a real number K such that $|x| \leqslant K$ for any $x \in S$.

Proof By exercise 1.20(1), $|x| \leqslant K$ if and only if $-K \leqslant x \leqslant K$. If $|x| \leqslant K$ for any $x \in S$, it follows that $-K$ is a lower bound for S and K is an upper bound for S. Thus S is bounded.

Suppose, on the other hand, that S is bounded. Let H be an upper bound and h a lower bound. Put $K = \max\{|H|, |h|\}$. Then

$$-K \leqslant h \leqslant x \leqslant H \leqslant K \quad (x \in S)$$

and hence $|x| \leqslant K$ for any $x \in S$.

20.3 Combination theorem

Proposition 4.8 Let $x_n \to l$ as $n \to \infty$ and $y_n \to m$ as $n \to \infty$. Let λ and μ be any real numbers. Then

(i) $\lambda x_n + \mu y_n \to \lambda l + \mu m$ as $n \to \infty$

(ii) $x_n y_n \to lm$ as $n \to \infty$

(iii) $\dfrac{x_n}{y_n} \to \dfrac{l}{m}$ as $n \to \infty$ (provided that $m \neq 0$).

Proofs of (ii) and (iii)

(ii) Since $\langle x_n \rangle$ converges, it is bounded (theorem 4.25). Suppose that $|x_n| \leqslant K$ ($n = 1, 2, 3, \ldots$). Consider

$$|x_n y_n - lm| = |x_n y_n - x_n m + x_n m - lm|$$

$$\leqslant |x_n| . |y_n - m| + |m| . |x_n - l| \text{ (triangle inequality)}$$

$$\leqslant K . |y_n - m| + |m| . |x_n - l|$$

$$= z_n.$$

But $x_n \to l$ as $n \to \infty$ and so $|x_n - l| \to 0$ as $n \to \infty$ (exercise 4.29(1i)). Similarly $|y_n - m| \to 0$ as $n \to \infty$. From proposition 4.8(i), $z_n \to 0$ as $n \to \infty$. The result then follows from corollary 4.11.

(iii) Since $y_n \to m$ as $n \to \infty$, $|y_n| \to |m|$ as $n \to \infty$ (exercise 4.29(1ii)). Since $m \neq 0$, $|m| > 0$. It follows from exercise 4.29(2) that we can find an N such that, for any $n > N$,

$$|y_n| > \tfrac{1}{2}|m|.$$

For $n > N$, consider

$$\left| \frac{x_n}{y_n} - \frac{l}{m} \right| = \left| \frac{mx_n - y_n l}{my_n} \right|$$

$$< \frac{2}{|m|^2} |mx_n - y_n l|.$$

By proposition 4.8(i), $mx_n - y_n l \to ml - ml = 0$ as $n \to \infty$. The result therefore follows from corollary 4.11.

20.4   Subsequences

Proposition 5.13 Let $\langle x_n \rangle$ be a bounded sequence and let L be the set of all real numbers which are the limit of some subsequence of $\langle x_n \rangle$. Then L has a maximum and a minimum.

Proof The set L is non-empty (Bolzano–Weierstrass theorem) and bounded (theorem 4.23). Hence L has a smallest upper bound $\bar{l}$. We seek to show that $\bar{l} \in L$, i.e. there exists a subsequence $\langle x_{n_r} \rangle$ such that $x_{n_r} \to \bar{l}$ as $r \to \infty$.

Let $\epsilon > 0$. Then $\tfrac{1}{2}\epsilon > 0$. Since $\bar{l}$ is the smallest upper bound of L, $\bar{l} - \tfrac{1}{2}\epsilon$ is not an upper bound of L. Hence there exists an $l \in L$ such that

$$\bar{l} \geqslant l > \bar{l} - \tfrac{1}{2}\epsilon.$$

and therefore

$$|l - \bar{l}| < \tfrac{1}{2}\epsilon. \tag{1}$$

Because $l \in L$, we can find a subsequence $\langle x_{m_r} \rangle$ which satisfies $x_{m_r} \to l$ as $r \to \infty$. Hence there exists an R such that, for any $r > R$,

$$|x_{m_r} - l| < \tfrac{1}{2}\epsilon. \tag{2}$$

Taking (1) and (2) together, we obtain, for any $r > R$,

$$|x_{m_r} - \bar{l}| = |x_{m_r} - l + l - \bar{l}|$$
$$\leqslant |x_{m_r} - l| + |l - \bar{l}|$$
$$< \tfrac{1}{2}\epsilon + \tfrac{1}{2}\epsilon$$

i.e. for any $r > R$,

$$|x_{m_r} - \bar{l}| < \epsilon.$$

This does *not* conclude the proof since the subsequence $\langle x_{m_r} \rangle$ will change as ϵ changes. All we have shown is that, given any $\epsilon > 0$, there exists an infinite collection of terms x_n of the sequence $\langle x_n \rangle$ which satisfy

$$|x_n - \bar{l}| < \epsilon. \tag{3}$$

With this information, we construct inductively a subsequence $\langle x_{n_r} \rangle$ which satisfies $x_{n_r} \to \bar{l}$ as $r \to \infty$.

Take $\epsilon = 1$ in (3). Then there exists an n_1 such that $|x_{n_1} - \bar{l}| < 1$. Take $\epsilon = \tfrac{1}{2}$ in (3). Then there exists an $n_2 > n_1$ such that $|x_{n_2} - \bar{l}| < \tfrac{1}{2}$. In this way we construct a subsequence $\langle x_{n_r} \rangle$ which satisfies

$$|x_{n_r} - \bar{l}| < \frac{1}{r}$$

and hence $x_{n_r} \to \bar{l}$ as $r \to \infty$ (sandwich theorem).

A similar argument shows that the infimum $\underline{l}$ of L also belongs to L.

Proposition 5.17 Any convergent sequence is a Cauchy sequence.

Proof Let $\epsilon > 0$ be given. Then $\tfrac{1}{2}\epsilon > 0$. If $x_n \to l$ as $n \to \infty$, then we can find an N such that, for any $n > N$,

$$|x_n - l| < \tfrac{1}{2}\epsilon.$$

Equally, for any $m > N$,

$$|x_m - l| < \tfrac{1}{2}\epsilon.$$

Thus, if $m > N$ and $n > N$,

$$|x_n - x_m| = |x_n - l + l - x_m|$$
$$\leqslant |x_n - l| + |x_m - l| < \tfrac{1}{2}\epsilon + \tfrac{1}{2}\epsilon = \epsilon$$

and thus $\langle x_n \rangle$ is a Cauchy sequence.

Proposition 5.18 Any Cauchy sequence is bounded.

Proof Let $\langle x_n \rangle$ be a Cauchy sequence. It is true that, for any $\epsilon > 0$, we can find an N such that, for any $n > N$ and any $m > N$, $|x_n - x_m| < \epsilon$. In particular, this is true when $\epsilon = 1$, i.e. there exists an N_1 such that, for any $n > N_1$

and any $m > N_1$.

$$|x_n - x_m| < 1.$$

Take $m = N_1 + 1$. Then, by theorem 1.18, for any $n > N_1$,

$$|x_n| - |x_{N_1+1}| \leqslant |x_n - x_{N_1+1}| < 1$$

$$|x_n| < |x_{N_1+1}| + 1.$$

Now take

$$K = \max \{|x_1|, |x_2|, \ldots, |x_{N_1}|, |x_{N_1+1}| + 1\}$$

and it follows that $|x_n| \leqslant K$ $(n = 1, 2, \ldots)$.

20.5   Tests for convergence of series

Proposition 6.17 (ratio test)

Let $\Sigma_{n=1}^{\infty} a_n$ be a series which satisfies

$$\lim_{n \to \infty} \left| \frac{a_{n+1}}{a_n} \right| = l.$$

If $l > 1$, the series diverges and, if $l < 1$, the series converges.

Proof The text indicates the proof for the case $l < 1$. If $l > 1$, we may take $\epsilon > 0$ so small that $l - \epsilon > 1$. Then, for a sufficiently large value of N,

$$|a_n| = \left| \frac{a_n}{a_{n-1}} \right| \cdot \left| \frac{a_{n-1}}{a_{n-2}} \right| \cdots \left| \frac{a_{N+2}}{a_{N+1}} \right| \cdot |a_{N+1}|$$

$$> (l-\epsilon)^{n-N+1} |a_{N+1}| \to +\infty \text{ as } n \to \infty.$$

Hence $a_n \not\to 0$ as $n \to \infty$ and so $\Sigma_{n=1}^{\infty} a_n$ diverges.

Proposition 6.18 (nth root test) Let $\Sigma_{n=1}^{\infty} a_n$ be a series which satisfies

$$\limsup_{n \to \infty} |a_n|^{1/n} = l.$$

If $l > 1$, the series diverges and, if $l < 1$, the series converges.

Proof Let $x_n = |a_n|^{1/n}$. If $l < 1$, we may choose $\epsilon > 0$ so small that $l + \epsilon < 1$. From exercise 5.15(4) it follows that we can find an N such that, for any $n > N$,

$$x_n < l + \epsilon$$

i.e.

$$|a_n| < (l+\epsilon)^n.$$

Since $\Sigma_{n=1}^{\infty} (l+\epsilon)^n$ converges, the convergence of $\Sigma_{n=1}^{\infty} a_n$ follows from the comparison test.

If $l > 1$, we may choose $\epsilon > 0$ so small that $l - \epsilon > 1$. Let $\langle x_{n_r} \rangle$ be a subsequence such that $x_{n_r} \to l$ as $r \to \infty$ (proposition 5.13). We can find an R such that, for any $r > R$,

i.e.
$$x_{n_r} > (l - \epsilon)$$
$$|a_{n_r}| > (l - \epsilon)^{n_r} \to + \infty \text{ as } r \to \infty.$$

Hence $a_n \nrightarrow 0$ as $n \to \infty$ (theorem 5.2) and thus $\sum_{n=1}^{\infty} a_n$ diverges.

20.6   Limits of functions

Proposition 8.4 Let f be defined on an interval (a, b) except possibly at a point $\xi \in (a, b)$. Then $f(x) \to l$ as $x \to \xi$ if and only if $f(x) \to l$ as $x \to \xi -$ and $f(x) \to l$ as $x \to \xi +$.

Proof (i) Suppose that $f(x) \to l$ as $x \to \xi$. Let $\epsilon > 0$ be given. Then we can find a $\delta > 0$ such that

$$|f(x) - l| < \epsilon$$

provided that $0 < |x - \xi| < \delta$. But $\xi - \delta < x < \xi$ implies that $0 < |x - \xi| < \delta$ (see §8.3). Hence $\xi - \delta < x < \xi$ implies that $|f(x) - l| < \epsilon$. Thus $f(x) \to l$ as $x \to \xi -$.

Similarly, $\xi < x < \xi + \delta$ implies that $0 < |x - \xi| < \delta$. Hence $\xi < x < \xi + \delta$ implies that $|f(x) - l| < \epsilon$. Thus $f(x) \to l$ as $x \to \xi +$.

(ii) Suppose that $f(x) \to l$ as $x \to \xi -$ and $f(x) \to l$ as $x \to \xi +$. Let $\epsilon > 0$ be given. We can find a $\delta_1 > 0$ such that $|f(x) - l| < \epsilon$ provided that

$$\xi - \delta_1 < x < \xi.$$

Also we can find a $\delta_2 > 0$ such that $|f(x) - l| < \epsilon$ provided that

$$\xi < x < \xi + \delta_2.$$

Let $\delta = \min \{\delta_1, \delta_2\}$. Then $|f(x) - l| < \epsilon$ provided that

i.e.
$$\xi - \delta < x < \xi \text{ and } \xi < x < \xi - \delta$$

i.e.
$$\xi - \delta < x < \xi + \delta \text{ and } x \neq \xi$$

$$0 < |x - \xi| < \delta.$$

Thus $f(x) \to l$ as $x \to \xi$.

Proposition 8.12 Let f and g be defined on an interval except possibly at $\xi \in (a, b)$. Suppose that $f(x) \to l$ as $x \to \xi$ and $g(x) \to m$ as $x \to \xi$ and suppose that λ and μ are any real numbers. Then

(i) $\lambda f(x) + \mu g(x) \to \lambda l + \mu m$ as $x \to \xi$

(ii) $f(x) g(x) \to lm$ as $x \to \xi$

(iii) $f(x)/g(x) \to l/m$ as $x \to \xi$ (provided $m \neq 0$).

Proof (i) Let $\langle x_n \rangle$ be any sequence of points of (a, b) such that $x_n \neq \xi$ $(n = 1, 2, \ldots)$ and $x_n \to \xi$ as $n \to \infty$. By theorem 8.9, $f(x_n) \to l$ as $n \to \infty$ and $g(x_n) \to m$ as $n \to \infty$. By proposition 4.8(i)

$$\lambda f(x_n) + \mu g(x_n) \to \lambda l + \mu m \text{ as } n \to \infty.$$

Appealing to theorem 8.9 again we obtain

$$\lambda f(x) + \mu g(x) \to \lambda l + \mu m \text{ as } x \to \xi.$$

Items (ii) and (iii) are proved in exactly the same way.

Proposition 8.14 Let f, g and h be defined on (a, b) except possibly at $\xi \in (a, b)$. Suppose that $g(x) \to l$ as $x \to \xi$, $h(x) \to l$ as $x \to \xi$ and that

$$g(x) \leqslant f(x) \leqslant h(x)$$

except possibly when $x = \xi$. Then $f(x) \to l$ as $x \to \xi$.

Proof Let $\langle x_n \rangle$ be a sequence of points of (a, b) such that $x_n \neq \xi$ $(n = 1, 2, \ldots)$ and $x_n \to \xi$ as $n \to \infty$. By theorem 8.9, $g(x_n) \to l$ as $n \to \infty$ and $h(x_n) \to l$ as $n \to \infty$. Since $g(x_n) \leqslant f(x_n) \leqslant h(x_n)$ it follows from theorem 4.10 that $f(x_n) \to l$ as $n \to \infty$. Hence $f(x) \to l$ as $x \to \xi$ by theorem 8.9.

20.7   Continuity

Proposition 9.3 Let f be defined on an interval. Then f is continuous on I if and only if, given any $x \in I$ and any $\epsilon > 0$, we can find a $\delta > 0$ such that

$$|f(x) - f(y)| < \epsilon$$

provided that $y \in I$ and $|x - y| < \delta$.

Proof If $x \in I$ but is not an endpoint of I, then the condition $y \in I$ and $|x - y| < \delta$ is just the same as $|x - y| < \delta$ provided that δ is sufficiently small. The criterion given in the proposition therefore reduces to the assertion that $f(y) \to f(x)$ as $y \to x$, i.e. f is continuous at x.

Suppose that $x \in I$ and is a left hand endpoint of I. The the condition $y \in I$ and $|x - y| < \delta$ reduces to $x \leqslant y < x + \delta$, provided that δ is sufficiently small. The criterion of the proposition therefore reduces to $f(y) \to f(x)$ as $y \to x +$, i.e. f is continuous on the right at x. Similarly if $x \in I$ and is a right hand endpoint.

Propositions 9.4, 9.5 and 9.6 are entirely trivial consequences of the results indicated in the text.

20.8   Integration

Proposition 13.4 is proved in §13.2.

Proposition 13.7 Let f be continuous on $[a, b]$ and let c be constant. Then

$$\int_a^b \{f(t) + c\}\, dt = \int_a^b f(t)\, dt + c(b - a).$$

Proof Let $P = \{y_0, y_1, y_2, \ldots, y_n\}$ denote a partition of $[a, b]$. We have

$$m_1^{(f+c)} = \inf_{y_0 \leqslant x \leqslant y_1} \{f(x) + c\} = c + \inf_{y_0 \leqslant x \leqslant y_1} \{f(x)\} = c + m_1^{(f)}.$$

Similarly $m_k^{(f+c)} = c + m_k^{(f)}$ $(k = 1, 2, \ldots, n)$. It follows that

$$S^{(f+c)}(P) = \sum_{k=1}^n m_k^{(f+c)}(y_k - y_{k-1})$$

$$= \sum_{k=1}^n m_k^{(f)}(y_k - y_{k-1}) + c \sum_{k=1}^n (y_k - y_{k-1})$$

$$= S^{(f)}(P) + c(b - a).$$

It follows that

$$\int_a^b \{f(t) + c\}\, dt = \sup_P \{S^{(f+c)}(P)\} = \sup_P \{S^{(f)}(P) + c(b - a)\}$$

$$= \sup_P \{S^{(f)}(P)\} + c(b - a) = \int_a^b f(t)\, dt + c(b - a).$$

Proposition 13.29 Suppose that ϕ is a function which is continuous and non-negative on the open interval I. Suppose also that f is continuous on I and satisfies

$$|f(x)| \leqslant \phi(x) \quad (x \in I).$$

If the improper integral of ϕ over the interval I exists, then so does that of f.

Proof We consider only the case when $I = (0, \infty)$ and the integral

$$l = \int_0^{\to \infty} \phi(x)\, dx$$

exists. Let

$$a_n = \int_{n-1}^n f(x)\, dx; \qquad b_n = \int_{n-1}^n \phi(x)\, dx \quad (n = 1, 2, \ldots).$$

Then the series $\Sigma_{n=1}^\infty b_n$ is a convergent series of non-negative terms (with sum l). Also

$$|a_n| \leqslant b_n \quad (n = 1, 2, \ldots)$$

and hence $\sum_{n=1}^{\infty} a_n$ converges by the comparison test, i.e.

$$m = \lim_{N \to \infty} \int_0^N f(x) \, dx$$

exists. Given any $X > 0$, we may take N to be the smallest natural number satisfying $N > X$. Then

$$\left| \int_1^X f(x) \, dx - \int_1^{N-1} f(x) \, dx \right| \leqslant \int_{N-1}^X |f(x)| \, dx$$

$$\leqslant b_N \to 0 \text{ as } N \to \infty.$$

It follows that

$$\lim_{X \to \infty} \int_1^X f(x) \, dx = m$$

as required.

20.9   Continuity and differentiation of power series

Proposition 15.8 Suppose that the power series

$$f(x) = \sum_{n=0}^{\infty} a_n (x - \xi)^n$$

has interval of convergence I. Then its sum is continuous on I and differentiable on I (except at the endpoints). Moreover

$$f'(x) = \sum_{n=1}^{\infty} n a_n (x - \xi)^{n-1}.$$

Proof Let the radius of convergence of the power series be R.

Suppose that x is any point of I other than an endpoint. We can then find another point x_0 of I so that x lies between ξ and x_0 and thus

$$|x - \xi| < |x_0 - \xi| < R.$$

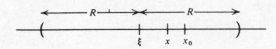

From the formula of §15.1

$$\frac{1}{R} = \limsup_{n \to \infty} |a_n|^{1/n} = \limsup_{n \to \infty} \left\{ \left| \frac{n(n-1)}{2} \right| |a_n| \right\}^{1/n}$$

we may deduce the convergence of the series

$$\sum_{n=2}^{\infty} \left| \frac{n(n-1)}{2} a_n X_0^{n-2} \right|$$

where $X_0 = x_0 - \xi$.

Put $\delta = |x - x_0|$. For values of y satisfying $0 < |x - y| < \delta$, consider

$$\Delta = \frac{f(y) - f(x)}{y - x} - \sum_{n=1}^{\infty} na_n (x - \xi)^{n-1} = \sum_{n=1}^{\infty} a_n \left(\frac{Y^n - X^n}{Y - X} - nX^{n-1} \right)$$

where $X = x - \xi$ and $Y = y - \xi$. We seek to show that $\Delta \to 0$ as $Y \to X$. From exercise 3.11(2),

$$\frac{Y^n - X^n}{Y - X} - nX^{n-1} = (Y^{n-1} + Y^{n-2}X + \ldots + YX^{n-2} + X^{n-1}) - nX^{n-1}$$

$$= (Y^{n-1} - X^{n-1}) + X(Y^{n-2} - X^{n-2}) + \ldots + X^{n-2}(Y - X).$$

But from each of these terms we can extract a factor of $(Y - X)$. The right hand side then reduces to the product of $(Y - X)$ with the sum of $\frac{1}{2}n(n-1)$ terms of the form $X^r Y^s$ where $r + s = n - 2$. The number $\frac{1}{2}n(n-1)$ arises from the use of the formula

$$1 + 2 + 3 + \ldots + (n-1) = \tfrac{1}{2}n(n-1) \quad \text{(example 3.9)}.$$

Since

$$|X| = |x - \xi| < |x_0 - \xi| = |X_0| \text{ and } |Y| = |y - \xi| < |x - \xi| + \delta = |X_0|$$

we have

$$\left| \frac{Y^n - X^n}{Y - X} - nX^{n-1} \right| < \frac{1}{2} n(n-1) |X_0|^{n-2} |Y - X|.$$

Hence

$$|\Delta| \leqslant |Y - X| \sum_{n=2}^{\infty} \frac{n(n-1)}{2} |a_n X_0^{n-2}|$$

$$\to 0 \text{ as } Y \to X.$$

It follows that f is differentiable at each x in I which is not an endpoint and

$$f'(x) = \sum_{n-1}^{\infty} na_n x^{n-1}.$$

There remains the question of left or right hand continuity at the endpoints of the interval of convergence I, if these happen to belong to I. A somewhat more subtle argument is required to deal with this question which is the subject matter of Abel's theorem quoted below.

Abel's theorem If the series $\Sigma_{k=0}^{\infty} a_k$ converges, then

$$\lim_{x \to 1-} \left\{ \sum_{k=0}^{\infty} a_k x^k \right\} = \sum_{k=0}^{\infty} a_k.$$

Proof Let $\epsilon > 0$ be given. Since the series $\Sigma \, a_k$ converges, its sequence of partial sums is a Cauchy sequence. We can therefore find an N such that, whenever $k \geqslant m \geqslant N$,

$$\left| \sum_{l=m}^{k} a_l \right| < \frac{1}{3} \epsilon. \tag{4}$$

Abel's lemma (which is easily proved by induction) asserts that

$$\sum_{k=m}^{n} u_k v_k = \sum_{k=m}^{n-1} \left\{ \left(\sum_{l=m}^{k} u_l \right) (v_k - v_{k+1}) \right\} + v_n \cdot \sum_{k=m}^{n} u_k.$$

We apply Abel's lemma with $u_k = a_k$ and $v_k = x^k$. Then

$$\sum_{k=m}^{n} a_k x^k = \sum_{k=m}^{n-1} \left\{ \left(\sum_{l=m}^{k} a_l \right) (x^k - x^{k+1}) \right\} + x^n \cdot \sum_{k=m}^{n} a_k.$$

It follows that, for $n \geqslant m \geqslant N$ and $0 < x < 1$,

$$\left| \sum_{k=m}^{n} a_k x^k \right| < (1-x) \sum_{k=m}^{n-1} \frac{1}{3} \epsilon x^k + \frac{1}{3} \epsilon x^n$$

$$< \frac{1}{3} \epsilon (1-x) \cdot \frac{(1-x^n)}{(1-x)} + \frac{1}{3} \epsilon x^n = \frac{1}{3} \epsilon.$$

We may conclude that

$$\left| \sum_{k=N}^{\infty} a_k x^k \right| \leqslant \frac{1}{3} \epsilon. \tag{5}$$

Next observe that, for $0 < x < 1$,

$$\left| \sum_{k=0}^{\infty} a_k x^k - \sum_{k=0}^{\infty} a_k \right| \leqslant \sum_{k=0}^{N-1} |a_n| (1-x^n) + \frac{1}{3} \epsilon + \frac{1}{3} \epsilon.$$

This follows from (4) and (5). But the *finite* sum on the right hand side tends to zero as $x \to 1 -$. We can therefore find a $\delta > 0$ such that, for any x satisfying $1 - \delta < x < 1$,

$$\sum_{k=0}^{N-1} |a_n| (1-x^n) < \frac{1}{3} \epsilon.$$

Given any $\epsilon > 0$, we have therefore found a $\delta > 0$ such that, for any x satisfying $1 - \delta < x < 1$,

$$\left| \sum_{k=0}^{\infty} a_k x^k - \sum_{k=0}^{\infty} \right| < \epsilon.$$

This concludes the proof.

20.10 Stirling's formula

The proof of proposition 17.2 is completed in exercises 17.4(1 and 2).

20.11 Properties of limits

Proposition 18.36 Suppose that $f: A \to \mathbb{R}^p$ and $g: A \to \mathbb{R}^q$, where A is a set in $\mathbb{R}^n$ containing all points $\mathbf{x} \in \mathbb{R}^n$ satisfying $0 < \|\mathbf{x} - \boldsymbol{\xi}\| < \Delta$. Let $f(\mathbf{x}) \to \mathbf{l}$ as $\mathbf{x} \to \boldsymbol{\xi}$ and $g(\mathbf{x}) \to \mathbf{m}$ as $\mathbf{x} \to \boldsymbol{\xi}$ and suppose that λ and μ are real numbers. Then

(ii) $\lambda f(\mathbf{x}) + \mu g(\mathbf{x}) \to \lambda \mathbf{l} + \mu \mathbf{m}$ (provided $p = q$)

(ii) $f(\mathbf{x})g(\mathbf{x}) \to \mathbf{lm}$ (provided $p = 1$ or $q = 1$)

(iii) $f(\mathbf{x})/g(\mathbf{x}) \to \mathbf{l}/\mathbf{m}$ (provided $q = 1$ and $\mathbf{m} \neq \mathbf{0}$)

as $\mathbf{x} \to \boldsymbol{\xi}$.

Proof It is perhaps easiest to prove the analogous results for sequences as for proposition 4.8 and then to appeal to the vector analogue of theorem 8.9.

Proposition 18.37 Suppose that $f: A \to \mathbb{R}^m$, where A is a set in $\mathbb{R}^n$ containing all points $\mathbf{x}$ satisfying $0 < \|\mathbf{x} - \boldsymbol{\xi}\| < \Delta$. Then $f(\mathbf{x}) \to \mathbf{l}$ as $\mathbf{x} \to \boldsymbol{\xi}$ if and only if the co-ordinates of $f(\mathbf{x})$ converge to the corresponding co-ordinates of $\mathbf{l}$ as $\mathbf{x} \to \boldsymbol{\xi}$.

Proof We give a direct proof. Write

$$f(\mathbf{x}) = (f^1(\mathbf{x}), f^2(\mathbf{x}), \ldots, f^m(\mathbf{x})).$$

(i) Suppose that $f^k(\mathbf{x}) \to l_k$ as $\mathbf{x} \to \boldsymbol{\xi}$ for each $k = 1, 2, \ldots, m$. Then

$$\|f(\mathbf{x}) - \mathbf{l}\| = \{(f^1(\mathbf{x}) - l_1)^2 + (f^2(\mathbf{x}) - l_2)^2 + \ldots + (f^k(\mathbf{x}) - l_k)^2\}^{1/2}$$
$$\to \{0 + 0 + \ldots + 0\}^{1/2} = 0 \text{ as } \mathbf{x} \to \boldsymbol{\xi}.$$

(ii) Suppose that $f(\mathbf{x}) \to \mathbf{l}$ as $\mathbf{x} \to \boldsymbol{\xi}$. Then, for each $k = 1, 2, \ldots, m$,

$$|f^k(\mathbf{x}) - l_k| \leq \|f(\mathbf{x}) - \mathbf{l}\| \to 0 \text{ as } \mathbf{x} \to \boldsymbol{\xi}.$$

Proposition 18.38 A function whose co-ordinates are all polynomials is continuous at every point. A function whose co-ordinates are all rational functions is continuous at every point at which it is defined.

Proof After proposition 18.37, we need only show that the co-ordinates are continuous. We begin with the polynomial case. We have that $x_k \to \xi_k$ as $x \to \xi$ and hence, from proposition 18.36(ii),

$$x_1^{m_1} x_2^{m_2} \ldots x_n^{m_n} \to \xi_1^{m_1} \xi_2^{m_2} \ldots \xi_n^{m_n} \quad \text{as} \quad x \to \xi$$

for any non-negative integres $m_1, m_2, \ldots, m_n$. Now consider a polynomial

$$P(x) = \sum a_{m_1, m_2, \ldots, m_n} x_1^{m_1} x_2^{m_2} \ldots x_n^{m_n}$$

where the sum extends over a finite set of values of $(m_1, m_2, \ldots, m_n)$. From proposition 18.36(i), we obtain that

$$P(x) \to \sum a_{m_1, m_2, \ldots, m_n} \xi_1^{m_1} \xi_2^{m_2} \ldots \xi_n^{m_n} \quad \text{as} \quad x \to \xi$$

and hence P is continuous at ξ.

In the case of a rational function, a similar argument is required which also uses proposition 18.36(iii).

20.12   Limits along paths

Proposition 18.44 Suppose that $f: A \to \mathbb{R}$, where A is a set in $\mathbb{R}^n$ containing all the points $x \in \mathbb{R}^n$ which satisfy $0 < \|x - \xi\| < \Delta$. Then

$$f(x) \to l \text{ as } x \to \xi$$

if and only if $f(x) \to l$ as $x \to \xi$ along all paths g with $g(0) = \xi$.

Proof The 'only if' part is an immediate consequence of the vector generalisation of theorem 8.17. For the 'if' part, we let $\langle x_k \rangle$ be any sequence of points in A such that $x_k \neq \xi$ ($k = 1, 2, \ldots$) and $x_k \to \xi$ as $k \to \infty$. We next define $g: [0, 1] \to \mathbb{R}^n$ so that $g(0) = \xi$ and, for each $k = 1, 2, \ldots,$

$$g(t) = (1 - \alpha)x_{k+1} + \alpha x_k \quad (1/(k + 1) \leqslant t \leqslant 1/k)$$

where α is given by $t = (1 - \alpha)(k + 1)^{-1} + \alpha k^{-1}$.

The function g is continuous on $[0, 1]$ and satisfies $g(1/k) = x_k$ ($k = 1, 2, \ldots$). Since $f(g(t)) \to l$ as $x \to \xi$, it follows that $f(x_k) \to l$ as $k \to \infty$. The result then follows from the vector analogue of theorem 8.9.

20.13   Chain rule

Proposition 19.24 Suppose that g is differentiable at ξ and f is differentiable at $\eta = g(\xi)$. Then $f \circ g$ is differentiable at ξ and

$$(f \circ g)'(\xi) = f'(\eta)g'(\xi).$$

Proof The following proof is a more detailed version of that given in §19.23. We have that

$$f(\eta + h) - f(\eta) - f'(\eta) = \epsilon_1(h)\|h\|$$

$$g(\xi + h) - g(\xi) - g'(\xi)h = \epsilon_2(h)\|h\|$$

where $\epsilon_1(h) \to 0$ as $h \to 0$ and $\epsilon_2(h) \to 0$ as $h \to 0$. In the second formula, we shall replace h by $k = g(\xi + h) - g(\xi)$. Since g is differentiable at ξ, it is continuous at ξ (theorem 19.13). Thus $k \to 0$ as $h \to 0$ by the vector analogue of theorem 8.17. Also, the function ϵ_1 is continuous at 0 (provided $\epsilon_1(0) = 0$) and it follows similarly that $\epsilon_1(k) \to 0$ as $h \to 0$.

We have to show that

$$f \circ g(\xi + h) - f \circ g(\xi) - f'(\eta)g'(\xi)h = \epsilon_0(h)\|h\|$$

where $\epsilon_0(h) \to 0$ as $h \to 0$. Since $g(\xi + h) = g(\xi) + k = \eta + k$, we may rewrite the left hand side as

$$f(\eta + k) - f(\eta) - f'(\eta)g'(\xi)h$$

$$= f(\eta + k) - f(\eta) - f'(\eta)k + f'(\eta)(k - g'(\xi)h)$$

$$= \epsilon_1(k)\|k\| + f'(\eta)(g(\xi + h) - g(\xi) - g'(\xi)h)$$

$$= \epsilon_1(k)\|k\| + f'(\eta)\epsilon_2(h)\|h\|.$$

A linear function is continuous everywhere and hence $f'(\eta)\epsilon_2(h) \to 0$ as $h \to 0$. We know that $\epsilon_1(k) \to 0$ as $h \to 0$. To complete the proof it is therefore only necessary to deal with the term $\|k\|$ in the preceding expression. We have that

$$\|k\| = \|g(\xi + h) - g(\xi)\|$$

$$= \|g'(\xi)h + \epsilon_2(h)\|h\| \|.$$

but by exercise 18.25 (6), there exists a constant K such that $\|g'(\xi)h\| \leqslant K\|h\|$. This concludes the proof.

20.14 Second derivatives

Proposition 19.33 If all the second order partial derivatives in the matrix (1) of §19.31 are continuous at ξ, then the $n \times n$ matrix $f''(\xi)$ is symmetric, i.e. for all i and j, $f_{ij}(\xi) = f_{ji}(\xi)$.

Proof We give the proof for $n = 2$ although the general case is entirely analogous. Consider

$$\phi(x) = \frac{f(x_1, x_2) - f(x_1, \xi_2) - f(\xi_1, x_2) + f(\xi_1, \xi_2)}{(x_1 - \xi_1)(x_2 - \xi_2)}$$

$$= \frac{F(x_1, x_2) - F(\xi_1, x_2)}{(x_1 - \xi_1)(x_2 - \xi_2)}$$

where F is defined by $F(x_1, x_2) = f(x_1, x_2) - f(x_1, \xi_2)$. By the mean value theorem, there exists a point u on the straight line segment joining (x_1, x_2) and (ξ_1, x_2) such that

$$\frac{F(x_1, x_2) - F(\xi_1, x_2)}{(x_1 - \xi_1)} = F_1(u_1, x_2)$$

and hence

$$\phi(x) = \frac{f_1(u_1, x_2) - f_1(u_1, \xi_2)}{(x_2 - \xi_2)}.$$

Again by the mean value theorem, there exists a point v on the line segment joining (u_1, x_2) and (u_1, ξ_2) such that

$$\phi(x) = \frac{f_1(u_1, x_2) - f_1(u_1, \xi_2)}{(x_2 - \xi_2)} = f_{21}(v).$$

Since the second partial derivatives are continuous at ξ, it follows that

$$\lim_{x \to \xi} \phi(x) = f_{21}(\xi).$$

But a precisely similar argument shows that

$$\lim_{x \to \xi} \phi(x) = f_{12}(\xi).$$

SOLUTIONS TO EXERCISES

- *Exercise 1.8*

1 There are three possibilities: $x = 0, x > 0, x < 0$. In the first case
$x^2 = 0$. In the second case rule III yields that $x^2 = x . x > 0 . x = 0$.
Rewrite the third possibility in the form $0 > x$ and apply rule IV. Then
$0 = 0 . x < x . x = x^2$. In each case $x^2 \geqslant 0$, i.e. $x^2 > 0$ *or* $x^2 = 0$.

(i) We are given that $0 < a < 1$. Since $a > 0$ we may apply rule III. Then
$0 = 0 . a < a . a < 1 . a = a$.

(ii) We are given that $b > 1$. Since $b > 0$ we may apply rule III. Then
$b^2 = b . b > 1 . b = b$.

2 Since $bB > 0$,

$$aB = \frac{a}{b} . bB < \frac{A}{B} . bB = Ab \quad \text{(rule III)}.$$

We deduce that

$$a(b + B) = ab + aB < ab + Ab = (a + A)b \quad \text{(rule II)}.$$

Since $(b + B)^{-1} > 0$ and $b^{-1} > 0$ (example 1.5), it follows from rule III
that

$$\frac{a}{b} < \frac{a + A}{b + B}.$$

The second half of the inequality to be demonstrated is obtained similarly.

3 Suppose that $a > b$ and $c > d$. Then, by rule II, $a + c > b + c$ and
$b + c > b + d$. Hence $a + c > b + d$ (rule I).

   If $b > 0$ and $d > 0$, we first observe that $c > d > 0$. Then, by rule
III, $ac > bc$ and $bc > bd$, Hence $ac > bd$ (rule I).

4 Substitute the values $a = 5, b = 4, c = 3, d = 1$ in inequalities (i) and
(ii). We obtain the *false* assertions

$$2 = 5 - 3 > 4 - 1 = 3$$

$$\frac{5}{3} > \frac{4}{1} \quad \text{i.e. } 5 > 12.$$

Note that the inequalities are false in spite of the fact that $b > 0$ and $d > 0$. On the other hand inequality (iii) is true when $b > 0$ and $d > 0$ (exercise 1.8(3)). However, if we take $a = c = -1$ and $b = d = -2$ in inequality (ii), we obtain the *false* assertion

$$1 = (-1)(-1) > (-2)(-2) = 4.$$

5 We have, for *any* $\epsilon > 0$, $b < a + \epsilon$ and $a - \epsilon < b$. Since $b < a + \epsilon$ for *any* $\epsilon > 0$, it follows that $b \leqslant a$ (example 1.7). Since $a < b + \epsilon$ for *any* $\epsilon > 0$, it follows that $a \leqslant b$. Hence $a = b$.

6 Take $x = (a + b)/2$.

● *Exercise 1.12*

1 If n is an even natural number it may be expressed in the form $n = 2k$. But then

$$x^n = x^{2k} = (x^k)^2 \geqslant 0$$

for all values of x (exercise 1.8(1)). Hence, if $y < 0$, $x^n = y$ has no solutions. The equation $x^n = 0$ has only the solution $x = 0$ since $x \neq 0$ implies that $x^n \neq 0$. If $y > 0$, our assumption about the existence of nth roots assures of the existence of a unique $x > 0$ such that $x^n = y$. But, since n is even, $x^n = y$ if and only if $(-x)^n = y$. Hence the equation has exactly two solutions, one positive and one negative.

  Next suppose that n is odd. If $y = 0$ there is no difficulty in showing that $x^n = y$ has exactly one solution. If $y > 0$, there is exactly one *positive* solution and this is the only solution because $x < 0$ implies $x^n < 0$ when n is odd. If $y < 0$, we use the fact that $z^n = -y$ has one and only one solution and hence the same is true of $(-x)^n = -y$, i.e. $x^n = y$.

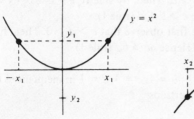
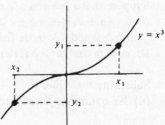

2 (i) $8^{2/3} = \{(2^3)^2\}^{1/3} = \{4^3\}^{1/3} = 4$

(ii) $27^{-4/3} = \left(\dfrac{1}{27^4}\right)^{1/3} = \left(\dfrac{1}{(3^3)^4}\right)^{1/3} = \left\{\left(\dfrac{1}{3^4}\right)^3\right\}^{1/3} = \dfrac{1}{81}$

(iii) $32^{6/5} = \{(2^5)^6\}^{1/5} = 2^6 = 64.$

3 We take for granted the truth of (i), (ii) and (iii) in the case when r and s are integers. To prove the results when r and s are any rational numbers, it is helpful to have available the following preliminary results in which m and n denote natural numbers:

(a) $(y^m)^{1/n} = (y^{1/n})^m$ (b) $(y^{1/m})^{1/n} = y^{1/mn}$ (c) $y^{1/n}z^{1/n} = (yz)^{1/n}.$

To prove (a), observe that

$$\{(y^{1/n})^m\}^n = (y^{1/n})^{mn} = \{(y^{1/n})^n\}^m = y^m$$

and hence $(y^{1/n})^m$ is the unique positive solution of $x^n = y^m$, i.e. $(y^{1/n})^m = (y^m)^{1/n}$. To prove (b), observe that

$$\{(y^{1/m})^{1/n}\}^{mn} = \{\{(y^{1/m})^{1/n}\}^n\}^m = (y^{1/m})^m = y$$

and hence $(y^{1/m})^{1/n}$ is the unique positive solution of $x^{mn} = y$. To prove (c), observe that

$$\{y^{1/n}z^{1/n}\}^n = (y^{1/n})^n(z^{1/n})^n = yz$$

and hence $y^{1/n}z^{1/n}$ is the unique positive solution of $x^n = yz$.

We may now prove (iii), (ii) and (i) in the general case when $r = p/m$ and $s = q/n$ (p and q integers).

(iii) $y^r z^r = (y^p)^{1/m}(z^p)^{1/m} = (y^p z^p)^{1/m} = \{(yz)^p\}^{1/m}$

$\qquad = (yz)^{p/m} = (yz)^r$

(ii) $(y^r)^s = \{[(y^p)^{1/m}]^q\}^{1/n} = \{[(y^p)^q]^{1/m}\}^{1/n} = (y^{pq})^{1/mn}$

$\qquad = y^{pq/mn} = y^{rs}$

(i) $y^{r+s} = y^{(pn+qm)/mn} = \{y^{pn} \cdot y^{qm}\}^{1/mn} = \{y^{pn}\}^{1/mn}\{y^{qm}\}^{1/mn}$

$\qquad = y^{p/m}y^{q/n} = y^r y^s.$

4. Write $ax^2 + bx + c = a(x - \alpha)(x - \beta)$. When $\alpha < x < \beta, x - \alpha > 0$ and $x - \beta < 0$. Hence $(x - \alpha)(x - \beta) < 0$. When $x < \alpha, x - \alpha < 0$ and $x - \beta < 0$. and therefore $(x - \alpha)(x - \beta) > 0$. When $x > \beta, x - \alpha > 0$ and $x - \beta > 0$ and therefore $(x - \alpha)(x - \beta) > 0$.

'Complete the square' as in §1.10 to obtain

$$ax^2 + bx + c = \{(2ax + b)^2 - (b^2 - 4ac)\}/4a.$$

Since $(2ax + b)^2 \geqslant 0,$

$$ax^2 + bx + c \geqslant c - \tfrac{1}{4}b^2 a^{-1}$$

with equality when $2ax + b = 0$, i.e. $x = -b/2a$.

5 Apply the Cauchy–Schwarz inequality using the numbers $\sqrt{a_1}, \sqrt{a_2},$
$\ldots, \sqrt{a_n}$ and $1/\sqrt{a_1}, \ldots, 1/\sqrt{a_n}$. We obtain

$$n^2 = \left(\sqrt{a_1} \cdot \frac{1}{\sqrt{a_1}} + \ldots + \sqrt{a_n} \cdot \frac{1}{\sqrt{a_n}} \right)^2$$

$$\leqslant (a_1 + a_2 + \ldots + a_n)\left(\frac{1}{a_1} + \frac{1}{a_2} + \ldots + \frac{1}{a_n} \right)$$

and the result follows.

$$6 \quad \sum_{k=1}^{n} (a_k + b_k)^2 = \sum_{k=1}^{n} a_k^2 + 2 \sum_{k=1}^{n} a_k b_k + \sum_{k=1}^{n} b_k^2$$

$$\leqslant \sum_{k=1}^{n} a_k^2 + 2\left(\sum_{k=1}^{n} a_k^2 \right)^{1/2} \left(\sum_{k=1}^{n} b_k^2 \right)^{1/2} + \sum_{k=1}^{n} b_k^2$$

(Cauchy–Schwarz inequality)

$$= \left\{ \left(\sum_{k=1}^{n} a_k^2 \right)^{1/2} + \left(\sum_{k=1}^{n} b_k^2 \right)^{1/2} \right\}^2$$

Consider, for example, the case $n = 3$ and suppose that (x_1, x_2, x_3),
(y_1, y_2, y_3) and (z_1, z_2, z_3) are the co-ordinates of the vertices of a triangle in three dimensional space. The length of the side joining the first
two vertices is given by

$$\left\{ \sum_{k=1}^{3} (x_k - y_k)^2 \right\}^{1/2} = \left\{ \sum_{k=1}^{3} (x_k - z_k + z_k - y_k)^2 \right\}^{1/2}$$

$$\leqslant \left\{ \sum_{k=1}^{3} (x_k - z_k)^2 \right\}^{1/2} + \left\{ \sum_{k=1}^{3} (z_k - y_k)^2 \right\}^{1/2}.$$

- *Exercise 1.20*

1 We first show that $|a| < b$ implies $-b < a < b$ and then that
$-b < a < b$ implies $|a| < b$.

(i) Suppose that $|a| < b$. From theorem 1.15, $a \leqslant |a|$ and $|a| \geqslant -a$.
Hence $a < b$ and $-a < b$, i.e. $a > -b$. It follows that $-b < a < b$.

(ii) Suppose that $-b < a < b$. Then $a < b$ and $-a < b$. Since, for each
a, $|a| = a$ or $|a| = -a$, it follows that $|a| < b$.

2 From theorem 1.18,

$$|c - d| \geqslant |c| - |d|$$

and

$$|c - d| = |d - c| \geqslant |d| - |c| = -\{|c| - |d|\}.$$

Thus, $-|c - d| \leqslant |c| - |d| \leqslant |c - d|$ and the result follows from exercise 1.20(1).

3 Only (iv) is not immediate. We have

$$d(x, y) = |x - y| = |x - z + z - y| \leqslant |x - z| + |z - y|$$

$$= d(x, z) + d(z, y).$$

4 Suppose that $r + s\sqrt{2} = t$, where t is rational. Then, provided $s \neq 0$,

$$\sqrt{2} = \frac{t - r}{s}.$$

But the right hand side of this equation is a rational number (why?) and so we have a contradiction.

5 We are given that $\alpha = r + s\sqrt{2}$ satisfies the equation $ax^2 + bx + c = 0$. Hence

$$a(r + s\sqrt{2})^2 + b(r + s\sqrt{2}) + c = 0$$

$$\{ar^2 + 2as^2 + br + c\} + \{2a + b\}s\sqrt{2} = 0.$$

Since a, b, c, s and r are rational, it follows that $(2a + b)s = 0$. Thus

$$a(r - s\sqrt{2})^2 + b(r - s\sqrt{2}) + c = \{ar^2 + 2as^2 + br + c\}$$

$$- (2a + b)s\sqrt{2} = 0$$

and so $\beta = r - s\sqrt{2}$ is also a root of the equation.

6 Suppose that $m^2 = 3n^2$. Then m^2 is divisible by 3 and hence m is divisible by 3. (Try $m = 3k + 1$ or $m = 3k + 2$.) Hence $m = 3k$. But then $9k^2 = 3n^2$, i.e. $3k^2 = n^2$ and so n is also divisible by 3.

Suppose that $m^3 = 2n^3$. Then m^3 is divisible by 2 and hence m is divisible by 2. (Try $m = 2k + 1$.) Hence $m = 2k$. But then $8k^3 = 2n^3$, i.e. $4k^3 = n^3$ and so n is also divisible by 2.

● *Exercise 2.10*

1 (i) False   (ii) true   (iii) true   (iv) false   (v) true.

2 By exercise 1.20(1), $|\xi - x| < \delta$ if and only if $-\delta < \xi - x < \delta$. This last inequality is equivalent to $\delta > x - \xi > -\delta$ which is, in turn,

equivalent to $\xi + \delta > x > \xi - \delta$. But $(\xi - \delta, \xi + \delta) =$
$\{x : \xi - \delta < x < \xi + \delta\}$.

3 (i) $(0, 1)$: bounded above; some upper bounds are 100, 2 and 1, the
smallest upper bound is 1; no maximum.

(ii) $(-\infty, 2]$: bounded above; some upper bounds are 100, 3 and 2; the
smallest upper bound is 2 and this is also a maximum.

(iii) $\{-1, 0, 2, 5\}$: bounded above; some upper bounds are 100, 6 and
5; the smallest upper bound is 5 and this is also a maximum.

(iv) $(3, \infty)$: unbounded above.

(v) $[0, 1]$: bounded above; some upper bounds are 100, 2 and 1; the
smallest upper bound is 1 and this is also a maximum.

4 (i) $(0, 1)$: bounded below; some lower bounds are $-10, -1$, and 0; the
largest lower bound is 0; no minimum.

(ii) $(-\infty, 2]$: unbounded below.

(iii) $\{-1, 0, 2, 5\}$: bounded below; some lower bounds are $-10, -2$
and -1; the largest lower bound is -1 and this is also a minimum.

(iv) $(3, \infty)$: bounded below; some lower bounds are $-10, 2$ and 3; the
largest lower bound is 3; no minimum.

(v) $[0, 1]$: bounded below; some lower bounds are $-10, -1$ and 0;
largest lower bound is 0 and this is also a minimum.

5 $\{3, 4\}$.

6 Take $y = \frac{1}{2}x$. No $m \in (0, \infty)$ can be a minimum because $\frac{1}{2}m$ is a smaller
element of the set.

● *Exercise 2.13*

1 Since $B = \sup S$ is an upper bound for $S, x \leqslant B$ for each $x \in S$. But
$S_0 \subset S$ means that, for each $x \in S_0$, it is true that $x \in S$. Hence $x \leqslant B$
for each $x \in S_0$. Thus B is an upper bound for S_0 and therefore at least
as large as the *smallest* upper bound $\sup S_0$, i.e. $\sup S_0 \leqslant B = \sup S$.

2 The proof is very similar to that of theorem 2.12. Let $B = \sup S$ and let
$T = \{x + \xi : x \in S\}$. Since $x \leqslant B$ for any $x \in S$, it is true that
$\xi + x \leqslant \xi + B$ for any $x \in S$ and hence that $\xi + B$ is an upper bound
for T. If C is the smallest upper bound for T, it follows that $C \leqslant \xi + B$.
    On the other hand, $y \leqslant C$ for any $y \in T$ and therefore $y - \xi \leqslant C - \xi$

for any $y \in T$. Since $S = \{y - \xi : y \in T\}$, it follows that $C - \xi$ is an upper bound for S and therefore that $B \leqslant C - \xi$.

We have shown that $C \leqslant \xi + B$ and $C \geqslant \xi + B$. Hence $C = \xi + B$.

3 Let $B = \sup S$ and let $T = \{-x : x \in S\}$. Since $x \leqslant B$ for any $x \in S$ it is true that $-x \geqslant -B$ for any $x \in S$ and hence that $-B$ is a lower bound for T. If C is the largest lower bound for T, it follows that $C \geqslant -B$.

On the other hand, $y \geqslant C$ for any $y \in T$ and therefore $-y \leqslant -C$ for any $y \in T$. Since $S = \{-y : y \in T\}$, it follows that $-C$ is an upper bound for S and therefore that $-C \geqslant B$.

We have shown that $C \geqslant -B$ and $C \leqslant -B$. Hence $C = -B$.

To obtain the required analogues, consider a non-empty set T which is bounded below and then apply theorem 2.12 and exercises 2.13(1) and (2) to the set $S = \{-x : x \in T\}$. We obtain

(i) $\inf_{x \in T} \xi x = \xi \inf_{x \in T} x \quad (\xi > 0)$

(ii) If $T_0 \subset T$, then $\inf T_0 \geqslant \inf T$

(iii) $\sup_{x \in T} (-x) = - \inf_{x \in T} x.$

4 (i) 1   (ii) 2   (iii) 1   (iv) 0.

5 (i) The set $D = \{|\xi - x| : x \in S\}$ has 0 as a lower bound. If $\xi \in S$, then 0 is a minimum for D. Exercise 4(iv) provides an example for which $d(\xi, S) = 0$ but $\xi \notin S$.

(ii) Let $\xi = \sup S$. Then $|\xi - x| = \xi - x$ for each $x \in S$. We therefore have to show that no $h > 0$ is a lower bound for the set $D = \{\xi - x : x \in S\}$. If this is false, we can find an $h > 0$ such that $\xi - x \geqslant h$ for all $x \in S$. But then $x \leqslant \xi - h$ for all $x \in S$ and hence $\xi - h$ is an upper bound for S smaller than the smallest upper bound.

If $\xi = \inf S$, consider instead $d(-\xi, T)$ where $T = \{-x : x \in S\}$.

(iii) Since I is an interval, $\xi \notin I$ implies that ξ is either an upper bound or a lower bound for I. Suppose the former. Let B be the smallest upper bound of I. Then $B \in I$ because I is closed. Given any $x \in I$,

$$|\xi - x| = \xi - x = \xi - B + B - x = \xi - B + |B - x|.$$

Hence, by exercise 2.13(2),

$$\inf_{x \in S} |\xi - x| = \xi - B + \inf_{x \in S} |B - x|$$

and therefore $d(\xi, S) = \xi - B + d(B, S)$. But $\xi - B \geqslant 0, d(B, S) \geqslant 0$ and $d(\xi, S) = 0$. It follows that $\xi = B$ and hence $\xi \in I$. Similarly if ξ is a lower bound for I.

If I is an open interval other than $\mathbb{R}$ or $\emptyset$ then ξ can be taken as its supremum or infimum (at least one of these must exist).

6 We may assume that the sets S and T have no elements in common (otherwise the problem is trivial). With reference to the given 'hint', the set T_0 is not empty because $t \in T_0$. The set T_0 is bounded below by s. Let $b = \inf T_0$.

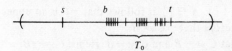

If $b \notin T$, then $b \in S$. But b is at zero distance from T_0 by exercise 2.13(5ii) and hence we have found a point of S at zero distance from T. If $b \in T$, then $b > s$ and the interval (s, b) is a non-empty subset of S. Hence b is a point of T at zero distance from S.

● *Exercise 3.6*

1 We have

$$\frac{n-1}{n} = 1 - \frac{1}{n} < 1$$

for all $n \in \mathbb{N}$ and therefore 1 is an upper bound for S. Suppose that $1 - h$, where $h > 0$, is a smaller upper bound. Then, for all $n \in \mathbb{N}$,

$$1 - \frac{1}{n} \leqslant 1 - h$$

$$\frac{1}{n} \geqslant h$$

$$n \leqslant \frac{1}{h}$$

and hence h^{-1} is an upper bound for $\mathbb{N}$. This is a contradiction and therefore 1 is the smallest upper bound for S.

The set S has no maximum. For no $n \in \mathbb{N}$ is it true that

$$\frac{n-1}{n} = 1.$$

2 Suppose that S is bounded above. Then it has a smallest upper bound B. Since BX^{-1} cannot be an upper bound, there exists an $n \in \mathbb{N}$ such that

$$X^n > BX^{-1}$$

$$X^{n+1} > B.$$

Hence B is not an upper bound for S and we have a contradiction.

It is clear the 0 is a lower bound for T. Suppose that h is a larger lower bound (i.e. $h > 0$). Then, for any $n \in \mathbb{N}$,

$$x^n \geqslant h$$

$$\left(\frac{1}{x}\right)^n \leqslant \frac{1}{h}.$$

It follows that the set $S = \{(1/x)^n : n \in \mathbb{N}\}$ is bounded above by h^{-1} which contradicts the first part of the question.

3 Let S be a non-empty set of integers which is bounded above by B. Since $\mathbb{N}$ is unbounded above, we can find an $n \in \mathbb{N}$ such that $n > B$. The set $T = \{n - x : x \in S\}$ is a non-empty set of natural numbers and hence has a minimum m (theorem 3.5). The integer $n - m$ is then the maximum of the set S.

Similarly if S is bounded below.

4 Since $\mathbb{N}$ is unbounded above, there exists a natural number n such that $n > (b - a)^{-1}$. Otherwise $(b - a)^{-1}$ would be an upper bound for $\mathbb{N}$. Let m be the smallest integer satisfying $m > an$. Then $m - 1 \leqslant an$. Hence

$$a < \frac{m}{n} \text{ and } \frac{m}{n} \leqslant a + \frac{1}{n} < a + (b - a) = b$$

as required.

5 Suppose that $m \in S$. Then $\frac{1}{2}(m + 1) \in S$ and $\frac{1}{2}(m + 1) > m$. Hence m cannot be a maximum. Similarly m cannot be a minimum.

It is obvious that S is bounded above by 1. Suppose that H $(0 < H < 1)$ were a smaller upper bound. By exercise 3.6(4) we could find a rational number r such that $H < r < 1$. But then r would be an element of S larger than H. Thus 1 is the smallest upper bound.

A similar argument shows that 0 is the largest lower bound.

6 We know that all numbers of the form $r\sqrt{2}$ are irrational provided r is rational and non-zero (exercise 1.20(4)). By exercise 3.16(4) we can find a rational number r such that

$$\frac{a}{\sqrt{2}} < r < \frac{b}{\sqrt{2}}$$

and hence $a < r\sqrt{2} < b$. (If r happens to be zero, locate a second rational number s satisfying $0 < s\sqrt{2} < b$.)

• *Exercise 3.11*

1 (i) Let $s_n = 1^2 + 2^2 + \ldots + n^2$ and let $P(n)$ be the proposition that $s_n = \frac{1}{6}n(n + 1)(2n + 1)$. Then $P(1)$ is true because $s_1 = 1$ and $\frac{1}{6}1(1 + 1)(2 + 1) = 1$. We next assume that $P(n)$ is true and try and deduce that $P(n + 1)$ is true.

$$
\begin{aligned}
s_{n+1} &= 1^2 + 2^2 + \ldots + n^2 + (n + 1)^2 = s_n + (n + 1)^2 \\
&= \tfrac{1}{6}n(n + 1)(2n + 1) + (n + 1)^2 \quad \text{(assuming } P(n) \text{ true)} \\
&= \tfrac{1}{6}(n + 1)\{n(2n + 1) + 6(n + 1)\} = \tfrac{1}{6}(n + 1)(2n^2 + 7n + 6) \\
&= \tfrac{1}{6}(n + 1)(n + 2)(2n + 3) = \tfrac{1}{6}m(m + 1)(2m + 1) \\
&\hspace{8cm} (m = n + 1).
\end{aligned}
$$

(ii) Let $s_n = 1^3 + 2^3 + \ldots + n^3$ and let $P(n)$ be the proposition that $s_n = \frac{1}{4}n^2(n + 1)^2$. Then $P(1)$ is true because $s_1 = 1$ and $\frac{1}{4}1^2(1 + 1)^2 = 1$. We next assume that $P(n)$ is true and try and deduce that $P(n + 1)$ is true.

$$
\begin{aligned}
s_{n+1} &= 1^3 + 2^3 + \ldots + n^3 + (n + 1)^3 = s_n + (n + 1)^3 \\
&= \tfrac{1}{4}n^2(n + 1)^2 + (n + 1)^3 \quad \text{(assuming } P(n) \text{ true)} \\
&= \tfrac{1}{4}(n + 1)^2(n^2 + 4n + 4) = \tfrac{1}{4}(n + 1)^2(n + 2)^2.
\end{aligned}
$$

2 Let $P(n)$ be the assertion

$$
1 + x + \ldots + x^n = \frac{1 - x^{n+1}}{1 - x}.
$$

Provided $x \neq 1$, this is clearly true when $n = 0$. Assuming $P(n)$ is true,

$$
\begin{aligned}
1 + x + x^2 + \ldots + x^n + x^{n+1} &= \frac{1 - x^{n+1}}{1 - x} + x^{n+1} \\
&= \frac{1 - x^{n+1} + x^{n+1} - x^{n+2}}{1 - x} \\
&= \frac{1 - x^{n+2}}{1 - x}
\end{aligned}
$$

and therefore $P(n + 1)$ is true.

The second formula of the question is obtained by writing $x = a/b$. (Note that, if $a = b$ or $b = 0$, the formula is obviously true.)

3 Suppose that $P(\xi) = 0$. Then

$$
\begin{aligned}
P(x) &= P(x) - P(\xi) \\
&= a_n(x^n - \xi^n) + a_{n-1}(x^{n-1} - \xi^{n-1}) + \ldots + a_1(x - \xi).
\end{aligned}
$$

By the previous question we have

$$x^k - \xi^k = (x - \xi)(x^{k-1} + x^{k-2}\xi + \ldots + x\xi^{k-2} + \xi^{k-1})$$

$$= (x - \xi)F_{k-1}(x)$$

where $F_{k-1}(x)$ is a polynomial of degree $k - 1$. It follows that

$$P(x) = (x - \xi)\{a_n F_{n-1}(x) + a_{n-1}F_{n-2}(x) + \ldots + a_1\} = (x - \xi)Q(x).$$

We apply this result to the second part of the question and obtain, in the first place, $P(x) = (x - \xi_1)Q_1(x)$ where $Q_1(x)$ is of degree $n - 1$. Then $0 = P(\xi_2) = (\xi_2 - \xi_1)Q_1(\xi_2)$. Since $\xi_2 \neq \xi_1$, it follows that $Q_1(\xi_2) = 0$ and hence that $Q_1(x) = (x - \xi_2)Q_2(x)$ where $Q_2(x)$ is of degree $n - 2$. Thus $P(x) = (x - \xi_1)(x - \xi_2)Q_2(x)$. Next consider ξ_3 and so on.

$$4 \quad \binom{n}{r} + \binom{n}{r-1} = \frac{n!}{r!(n-r)!} + \frac{n!}{(r-1)!(n-r+1)!}$$

$$= \frac{n!}{(r-1)!(n-r)!} \left\{ \frac{1}{r} + \frac{1}{n-r+1} \right\}$$

$$= \frac{n!}{(r-1)!(n-r)!} \frac{n+1}{r(n-r+1)}$$

$$= \frac{(n+1)!}{r!(n-r+1)!} = \binom{n+1}{r}.$$

Let $P(n)$ be the assertion that

$$(a + b)^n = \sum_{r=0}^{n} \binom{n}{r} a^{n-r}b^r.$$

Clearly $P(1)$ is true. Assuming $P(n)$ is true,

$$(a + b)^{n+1} = (a + b)(a + b)^n$$

$$= a\left\{ \sum_{r=0}^{n} \binom{n}{r} a^{n-r}b^r \right\} + b\left\{ \sum_{r=0}^{n} \binom{n}{r} a^{n-r}b^r \right\}$$

$$= \sum_{r=0}^{n} \binom{n}{r} a^{n-r+1}b^r + \sum_{r=0}^{n} \binom{n}{r} a^{n-r}b^{r+1}.$$

We replace r by s in the first sum and replace $r + 1$ by s in the second sum. Then

$$(a + b)^{n+1} = \sum_{s=0}^{n} \binom{n}{s} a^{n-s+1}b^s + \sum_{s=1}^{n+1} \binom{n}{s-1} a^{n-s+1}b^s$$

$$= a^{n+1} + \sum_{s=1}^{n} \left\{ \binom{n}{s} + \binom{n}{s-1} \right\} a^{n+1-s}b^{s} + b^{n+1}$$

$$= \sum_{s=0}^{n+1} \binom{n+1}{s} a^{n+1-s}b^{s}$$

as required.

5 Suppose there exists a natural number k for which $P(k)$ is false. Since the set $\{2^{n} : n \in \mathbb{N}\}$ is unbounded above, we can find an $M = 2^{N} > k$. Now put $S = \{n : P(n)$ is false and $n < M\}$. This is a non-empty set of natural numbers which is bounded above. From exercise 3.6(3) it has a maximum element m. The assertion $P(n)$ is true for $m < n \leqslant M$ and hence $P(m+1)$ is true. But $P(m+1)$ implies $P(m)$ and hence we have a contradiction.

6 From exercise 3.11(2),

$$a^{n} - b^{n} = (a-b)(a^{n-1} + a^{n-2}b + \ldots + ab^{n-2} + b^{n-1}).$$

Hence, if $a > b > 0$,

$$nb^{n-1} \leqslant \frac{a^{n} - b^{n}}{a-b} \leqslant na^{n-1}.$$

Therefore,

$$\frac{1}{na^{n-1}}(a^{n} - b^{n}) \leqslant a - b \leqslant \frac{1}{nb^{n-1}}(a^{n} - b^{n}).$$

If $x > y$, the result follows on writing $a = x^{1/n}$ and $b = y^{1/n}$. If $x < y$, write $a = y^{1/n}$ and $b = x^{1/n}$.

● *Exercise 4.6*

1 Let $\epsilon > 0$ be given. We must find a value of N such that, for any $n > N$,

$$\left| \frac{n^{2}-1}{n^{2}+1} - 1 \right| < \epsilon.$$

But

$$\left| \frac{n^{2}-1}{n^{2}+1} - 1 \right| = 1 - \frac{n^{2}-1}{n^{2}+1} = \frac{2}{n^{2}+1}$$

and so we simply need to find an N such that, for any $n > N$,

$$\frac{2}{n^{2}+1} < \epsilon.$$

i.e.

$n^2 + 1 > 2/\epsilon$.

Choose $N = (2/\epsilon)^{1/2}$. Then, for any $n > N$,

$n^2 + 1 > n^2 > N^2 = 2/\epsilon$.

We have shown that, given any $\epsilon > 0$, we can find an N (namely $(2/\epsilon)^{1/2}$) such that, for any $n > N$,

$$\left| \frac{n^2 - 1}{n^2 + 1} - 1 \right| < \epsilon.$$

Thus

$$\lim_{n \to \infty} \frac{n^2 - 1}{n^2 + 1} = 1.$$

2 Let $\epsilon > 0$ be given. We must find a value of N such that, for any $n > N$,

$$\left| \frac{1}{n^r} - 0 \right| < \epsilon$$

i.e.

$n^r > 1/\epsilon$.

We choose $N = (1/\epsilon)^{1/r}$. Then, for any $n > N$,

$n^r > N^r = 1/\epsilon$

and the result follows.

3 Let $\epsilon > 0$ be given. We must find a value of N such that, for any $n > N$,

$|\lambda x_n - \lambda l| < \epsilon$.

If $\lambda = 0$ there is nothing to prove. We may therefore assume $\lambda \neq 0$. We are given that $x_n \to l$ as $n \to \infty$. Since $\epsilon/|\lambda| > 0$ it follows that we can find an N such that, for any $n > N$,

$|x_n - l| < \epsilon/|\lambda|$

i.e.

$|\lambda x_n - \lambda l| < \epsilon$.

This completes the proof.

- *Exercise 4.20*

1 $\dfrac{n^3 + 5n^2 + 2}{2n^3 + 9} = \dfrac{1 + 5n^{-1} + 2n^{-3}}{2 + 9n^{-3}} \to \dfrac{1 + 0 + 0}{2 + 0}$ as $n \to \infty$.

2 First consider the case when $|x| < 1$. Then $x^n \to 0$ as $n \to \infty$. Hence

$$\frac{x+x^n}{1+x^n} \to \frac{x+0}{1+0} \text{ as } n \to \infty.$$

Next suppose that $x > 1$. Then $x^{-n} \to 0$ as $n \to \infty$. Hence

$$\frac{x+x^n}{1+x^n} = \frac{x.x^{-n}+1}{x^{-n}+1} \to \frac{0+1}{0+1} \text{ as } n \to \infty.$$

This leaves $x = 1$ and $x = -1$ giving

$$\frac{1+1^n}{1+1^n} = 1 \quad \text{and} \quad \frac{-1+(-1)^n}{1+(-1)^n}.$$

The latter expression is not even defined for odd n. Hence

$$\lim_{n \to \infty} \left\{\frac{x+x^n}{1+x^n}\right\} = \begin{cases} 1 & (x \geqslant 1) \\ x & (-1 < x < 1) \\ 1 & (x < -1). \end{cases}$$

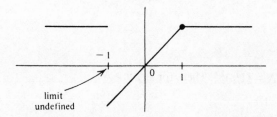

limit
undefined

3 We have

$$0 \leqslant \sqrt{(n+1)} - \sqrt{n} = \frac{(\sqrt{(n+1)} - \sqrt{n})(\sqrt{(n+1)} + \sqrt{n})}{(\sqrt{(n+1)} + \sqrt{n})}$$

$$= \frac{n+1-n}{\sqrt{(n+1)} + \sqrt{n}} < \frac{1}{\sqrt{n}} \to 0 \text{ as } n \to \infty$$

(see exercise 4.6(2)). It follows from the sandwich theorem that

$$\sqrt{(n+1)} - \sqrt{n} \to 0 \text{ as } n \to \infty.$$

4 Let N be the smallest natural number such that $N > x$. Then, for $n \geqslant N$,

$$\frac{x^n}{n!} \leqslant \frac{x}{1}.\frac{x}{2}.\frac{x}{N-1}.\frac{x}{N} \cdots \frac{x}{N}$$

$$= \frac{x^{N-1}}{(N-1)!}\left(\frac{x}{N}\right)^{n-N+1}.$$

Since $x/N < 1$, $(x/N)^n \to 0$ as $n \to \infty$ and the result follows from the sandwich theorem.

5 Suppose that $(1 + 1/n)^{\alpha+1}|x| > 1$ for all $n \in \mathbb{N}$. Then

$$\frac{1}{n} > \left\{\frac{1}{|x|}\right\}^{1/(\alpha+1)} - 1 > 0$$

for all $n \in \mathbb{N}$ and this is a contradiction (see example 3.4). Hence, for some $N \in \mathbb{N}$,

$$(1 + 1/N)^{\alpha+1}|x| \leqslant 1.$$

If $x \neq 0$, consider the expression

$$\left|\frac{(n+1)^{\alpha+1}x^{n+1}}{n^{\alpha+1}x^n}\right| = \left(1 + \frac{1}{n}\right)^{\alpha+1}|x|.$$

If $n \geqslant N$,

$$|(n+1)^{\alpha+1}x^{n+1}| \leqslant |n^{\alpha+1}x^n|.$$

It follows that, for $n \geqslant N$,

$$|n^{\alpha+1}x^n| \leqslant |N^{\alpha+1}x^N|.$$

We conclude that, for $n \geqslant N$,

$$|n^\alpha x^n| \leqslant \frac{1}{n}|N^{\alpha+1}x^N|$$

and thus $n^\alpha x^n \to 0$ as $n \to \infty$ by the sandwich theorem.

6 We wish to prove that

$$(n+1)^{1/(n+1)} \leqslant n^{1/n}.$$

This inequality is equivalent to

$$(n+1)^n \leqslant n^{n+1}$$

i.e.

$$\left(1 + \frac{1}{n}\right)^n \leqslant n. \tag{1}$$

But we know from example 4.19 that

$$\left(1 + \frac{1}{n}\right)^n < 3$$

for all $n \in \mathbb{N}$. Hence (1) holds provided $3 \leqslant n$.

● *Exercise 4.29*

1 (i) Let $\epsilon > 0$ be given. We have to find an N such that, for any $n > N$,

$$||x_n - l| - 0| < \epsilon.$$

But $\|x_n - l| - 0| = |x_n - l|$ and so what has to be proved is just the definition of $x_n \to l$ as $n \to \infty$.

(ii) By exercise 1.20(2),

$$\||x_n| - |l|\| \leqslant |x_n - l|$$

and hence $|x_n| \to |l|$ as $n \to \infty$ by the sandwich theorem.

2 For any $\epsilon > 0$, there exists an N such that, for any $n > N$,

$$|x_n - l| < \epsilon.$$

i.e.

$$l - \epsilon < x_n < l + \epsilon.$$

Since this is true for *any* $\epsilon > 0$, then it is true when $\epsilon = \frac{1}{2}l$ (for an appropriate value of N, say $N = N_1$). For $n > N_1$ we therefore have that

$$x_n > l - \tfrac{1}{2}l = \tfrac{1}{2}l.$$

3 (i) Let $H > 0$ be given. We have to find an N such that, for any $n > N$,

$$2^n > H.$$

Since the set $\{2^n : n \in \mathbb{N}\}$ is unbounded above (exercise 3.6(2)), there exists an $N \in \mathbb{N}$ such that $2^N > H$. If $n > N$, then

$$2^n > 2^N > H$$

and the result follows. (Alternatively one may appeal to the inequality $2^n \geqslant n$ $(n \geqslant 1)$ which is easily proved by induction.)

(ii) Let $H > 0$ be given. We have to find an N such that, for any $n > N$,

$$-\sqrt{n} < -H$$

i.e. $\sqrt{n} > H$

i.e. $n > H^2$.

We may therefore take $N = H^2$.

(iii) The sequence $\langle (-1)^n n \rangle$ cannot converge because it is unbounded. It cannot diverge to $+\infty$ because its odd terms are all negative. It cannot diverge to $-\infty$ because its even terms are all positive.

4 (i) Suppose that $x_n \to 0$ as $n \to \infty$. Let $H > 0$ be given. We have to find an N such that, for any $n > N$,

$$1/x_n > H.$$

But $H^{-1} > 0$. Since $x_n \to 0$ as $n \to \infty$, we can therefore find an N such that, for any $n > N$,

$$x_n = |x_n - 0| < H^{-1}$$

i.e. $1/x_n > H$.

(ii) To prove that $1/x_n \to + \infty$ as $n \to \infty$ implies that $x_n \to 0$ as $n \to \infty$, one simply has to reverse the above argument.

5 Let $H > 0$ be given. If $\langle x_n \rangle$ is unbounded above, there exists an N such that $x_N > H$. If $\langle x_n \rangle$ increases, then, for any $n \geqslant N, x_n \geqslant x_N > H$. It follows that $x_n \to + \infty$. Similarly in the case when $\langle x_n \rangle$ decreases and is unbounded below.

6 Suppose there is a value of $n \in \mathbb{N}$ for which no $x \in S$ can be found satisfying $|\xi - x| < 1/n$. Then $1/n$ is a lower bound for the set $D = \{ |\xi - x| : x \in S \}$ which contradicts the assertion that $d(\xi, S) = 0$. That $x_n \to \xi$ as $n \to \infty$ follows from $|\xi - x_n| < 1/n \ (n = 1, 2, \ldots)$ because of the sandwich theorem.

If $\xi = \sup S$, then from exercise 2.13(5ii) $d(\xi, S) = 0$. It therefore follows from the first part of the question that a sequence $\langle x_n \rangle$ of points of S can be found such that $x_n \to \xi$ as $n \to \infty$. (Note that the terms of this sequence are not necessarily distinct.)

If S is unbounded above, then, given any $n \in \mathbb{N}$, we can find an $x_n \in S$ such that $x_n > n$. Hence $x_n \to + \infty$ as $n \to \infty$.

● *Exercise 5.7*

1 Suppose that $n^{1/n} \to l$ as $n \to \infty$. It follows from theorem 5.2 that

$$(2n)^{1/2n} \to l \text{ as } n \to \infty.$$

But $2^{1/2n} \to 1$ as $n \to \infty$ (example 4.14). Thus $n^{1/2n} \to l$ as $n \to \infty$. Hence

$$n^{1/n} = n^{1/2n} n^{1/2n} \to l.l = l^2 \text{ as } n \to \infty.$$

It follows that $l = l^2$. Since $l \geqslant 1$, we deduce that $l = 1$.

2 We prove that $a < x_n < b$ by induction. It is given that $a < x_1 < b$. If we assume that $a < x_n < b$, it follows that

$$x_{n+1} - a = x_n^2 + k - a > a^2 - a + k = 0.$$

Similarly

$$x_{n+1} - b = x_n^2 + k - b < b^2 - b + k = 0.$$

Hence $a < x_{n+1} < b$. (Note that the condition $0 < k < \frac{1}{4}$ ensures that the quadratic equation $x^2 - x + k = 0$ has two real roots and that these are both positive.)

Next consider

$$x_{n+1} - x_n = x_n^2 - x_n + k.$$

But $x_n^2 - x_n + k < 0$ because $a < x_n < b$ (see exercise 1.12(4)). This completes the proof that $a < x_{n+1} < x_n < b$.

Since $\langle x_n \rangle$ decreases and is bounded below by a, it converges. Suppose that $x_n \to l$ as $n \to \infty$. Then $x_{n+1} \to l$ as $n \to \infty$. But $x_{n+1} = x_n^2 + k$ and so

$$l = l^2 + k.$$

This implies that $l = a$ or $l = b$. But b cannot be the infimum of the sequence $\langle x_n \rangle$ because it is not a lower bound. Hence $l = a$.

3 It is easily shown by induction that $k > x_n > 0$ $(n = 1, 2, \ldots)$. Observe that

$$x_{n+1} - x_{n-1} = \frac{k}{1 + x_n} - \frac{k}{1 + x_{n-2}}$$

$$= \frac{k(x_{n-2} - x_n)}{(1 + x_n)(1 + x_{n-2})}.$$

Hence $x_{n+1} - x_{n-1}$ has the same sign as $x_{n-2} - x_n$. It follows, using an induction argument that one of the sequences $\langle x_{2n} \rangle$ and $\langle x_{2n-1} \rangle$ increases and the other decreases. (In fact $\langle x_{2n-1} \rangle$ increases if $x_3 \geqslant x_1$ and decreases if $x_3 \leqslant x_1$.)

That both sequences converge follows from theorem 4.17. Suppose that $x_{2n} \to l$ as $n \to \infty$ and that $x_{2n-1} \to m$ as $n \to \infty$. Then

$$l = \frac{k}{1 + m}; \qquad m = \frac{k}{1 + l}$$

$$l + lm = k; \qquad m + lm = k.$$

It follows that $l = m$ and that $l^2 + l = k$ as required. The conclusion about $\langle x_n \rangle$ is that $x_n \to l$ as $n \to \infty$.

4 If $0 < a < b$, it is easily seen that the geometric mean $G = \sqrt{(ab)}$ (see example 3.10) and the harmonic mean $H = \{\frac{1}{2}(1/a + 1/b)\}^{-1}$ (see exercise 1.12(5)) satisfy

$$a < H < G < b.$$

We are given $\frac{1}{2} = x_1 < y_1 = 1$. On the assumption that $x_{n-1} < y_{n-1}$,

$$x_{n-1} < x_n < y_{n-1}$$

because x_n is the geometric mean of x_{n-1} and y_{n-1}. Further

$$x_n < y_n < y_{n-1}$$

because y_n is the harmonic mean of x_n and y_{n-1}. It follows by induction that $x_{n-1} < x_n < y_n < y_{n-1}$ $(n = 2, 3, \ldots)$.

The sequence $\langle x_n \rangle$ increases and is bounded above by $y_1 = 1$. The sequence $\langle y_n \rangle$ decreases and is bounded below by $x_1 = \frac{1}{2}$. Hence both sequences converge. Suppose that $x_n \to l$ as $n \to \infty$ and $y_n \to m$ as $n \to \infty$. Then

$$l^2 = lm$$

$$\frac{1}{m} = \frac{1}{2}\left(\frac{1}{l} + \frac{1}{m}\right).$$

Both of these equations yield $l = m$.

5 As in example 4.19 we apply the inequality of the geometric and arithmetic means with $a_1 = a_2 = \ldots = a_{n-1} = (1 + y/(n-1))$ and $a_n = 1$. Then

$$\left(1 + \frac{y}{n-1}\right)^{(n-1)/n} \leqslant \frac{(n-1)(1 + y(n-1)^{-1}) + 1}{n} = \left(1 + \frac{y}{n}\right).$$

Hence

$$\left(1 + \frac{y}{n-1}\right)^{n-1} \leqslant \left(1 + \frac{y}{n}\right)^{n}.$$

This analysis is only valid if $a_1, a_2, \ldots, a_n$ are non-negative numbers and hence we have only shown that the sequence increases when

$$1 + \frac{y}{n-1} \geqslant 0 \quad \text{i.e. } n \geqslant 1 - y.$$

As in example 4.19,

$$\left(1 + \frac{y}{n}\right)^{n} \leqslant 1 + |y| + \frac{|y|^2}{2!} + \ldots + \frac{|y|^n}{n!}. \tag{2}$$

Since there exists an N such that, for any $n > N$,

$$\frac{|y|^n}{n!} \leqslant \left(\frac{1}{2}\right)^{n}$$

we may obtain from (2)

$$\left(1 + \frac{y}{n}\right)^{n} \leqslant 1 + |y| + \ldots + \frac{|y|^N}{N!} + \left(\frac{1}{2}\right)^{N+1} + \ldots + \left(\frac{1}{2}\right)^{n}$$

$$\leqslant C + \frac{1 - (\frac{1}{2})^{n+1}}{1 - \frac{1}{2}} < C + 2$$

where C is a constant.

It follows that the sequence is convergent since it increases and is bounded above. Note that the limit is positive since $1 + y/n$ is positive when n is sufficiently large.

6 From the previous question we know that the sequence

$$\left\langle \left(1 - \frac{x^2}{n^2}\right)^{n^2} \right\rangle$$

converges to a positive limit l. It follows that positive numbers a and b can be found such that

$$a < \left(1 - \frac{x^2}{n^2}\right)^{n^2} < b$$

for sufficiently large values of n. Hence

$$a^{1/n} < \left(1 - \frac{x}{n}\right)^n \left(1 + \frac{x}{n}\right)^n < b^{1/n}.$$

The result then follows from the sandwich theorem and example 4.14.

● *Exercise 5.15*

1 For the sequence $\langle(-1)^n(1 + 1/n)\rangle$ the set $L = \{-1, 1\}$. Observe that

$$(-1)^{2n}(1 + 1/2n) \to 1 \text{ as } n \to \infty$$

$$(-1)^{2n-1}(1 + 1/(2n - 1)) \to -1 \text{ as } n \to \infty$$

and the sequence possesses no subsequences which tend to a limit other than 1 or -1. It follows that

(i) $\lim_{n \to \infty} \sup (-1)^n(1 + 1/n) = 1$ (ii) $\lim_{n \to \infty} \inf (-1)^n(1 + 1/n) = -1$.

Note that

$$\sup_{n \geqslant 1} (-1)^n(1 + 1/n) = (-1)^2(1 + \tfrac{1}{2}) = \tfrac{3}{2}$$

$$\inf_{n \geqslant 1} (-1)^n(1 + 1/n) = (-1)^1(1 + 1/1) = -2.$$

2 Observe that every rational number in the interval $(0, 1)$ occurs infinitely often as a term in the given sequence.

Let $x \in [0, 1]$. By exercise 3.6(4), a term r_{n_1} of the sequence can be found such that

$$x - 1 < r_{n_1} < x + 1.$$

A term r_{n_2} with $n_2 > n_1$ can then be found such that

$$x - \tfrac{1}{2} < r_{n_2} < x + \tfrac{1}{2}.$$

Continuing in this way we construct a subsequence $\langle r_{n_k} \rangle$ such that

$$x - \frac{1}{k} < r_{n_k} < x + \frac{1}{k}$$

and hence $r_{n_k} \to x$ as $k \to \infty$ by the sandwich theorem.

3 The hypothesis of the question implies that $\langle x_n \rangle$ contains a subsequence all of whose terms are at least as large as b. By the Bolzano–Weierstrass theorem, this subsequence in turn contains a convergent subsequence $\langle x_{n_r} \rangle$. Suppose that $x_{n_r} \to l$ as $r \to \infty$. Since $x_{n_r} \geqslant b$ $(r = 1, 2, \ldots)$, it follows from theorem 4.23 that $l \geqslant b$.

4 Suppose it is false that, for any $\epsilon > 0$, there exists an N such that, for any $n > N, x_n < \bar{l} + \epsilon$. Then for some $\epsilon > 0$ it is true that for each N we can find an $n > N$ such that $x_n \geqslant \bar{l} + \epsilon$. From exercise 5.15(3) we can then deduce the existence of a convergent subsequence with limit $l \geqslant \bar{l} + \epsilon$. This contradicts the definition of $\bar{l}$ (see §5.12).

The corresponding result for $\underline{l}$ is that for any $\epsilon > 0$, we can find an N such that, for any $n > N, x_n > \underline{l} - \epsilon$.

5 Suppose that $\underline{l} = \bar{l} = l$. Let $\epsilon > 0$ be given. By the previous question we can find an N_1 such that, for any $n > N_1$,

$$x_n < l + \epsilon. \tag{3}$$

We can find an N_2 such that, for any $n > N_2$,

$$l - \epsilon < x_n. \tag{4}$$

Take $N = \max \{N_1, N_2\}$. Provided $n > N$, inequalities (3) and (4) are true simultaneously. Hence, for any $n > N$,

$$l - \epsilon < x_n < l + \epsilon$$

i.e. $|x_n - l| < \epsilon$.

Thus $x_n \to l$ as $n \to \infty$.

That $x_n \to l$ as $n \to \infty$ implies $\bar{l} = \underline{l} = l$ is trivial.

To prove the last part of the question one only has to note that, if m is the limit of a convergent subsequence, then $\underline{l} \leqslant m \leqslant \bar{l}$.

6 The sequence $\langle M_n \rangle$ decreases because we are taking the supremum at each stage of a smaller and smaller set. Any lower bound for $\langle x_n \rangle$ is also a lower bound for $\langle M_n \rangle$. It follows from theorem 4.17 that $\langle M_n \rangle$ converges. Suppose that $M_n \to M$ as $n \to \infty$.

Suppose that $x_{n_r} \to l$ as $r \to \infty$. Then, for $n_r \geqslant n$,

$$x_{n_r} \leqslant M_n$$

and hence $l \leqslant M_n$ by theorem 4.23. From theorem 4.23 again it follows that $l \leqslant M$. This is true for each $l \in L$ and so $\bar{l} \leqslant M$.

From exercise 5.15(4) we know that, given any $\epsilon > 0$, we can find an n such that, for any $k \geqslant n$,

$$x_k < \bar{l} + \epsilon.$$

Thus $\bar{l} + \epsilon$ is an upper bound for the set $\{x_k : k \geqslant n\}$ and it follows that

$$M \leqslant M_n \leqslant \bar{l} + \epsilon.$$

Since this is true for *any* $\epsilon > 0$, $M \leqslant \bar{l}$ (see example 1.7).

We conclude that $M = \bar{l}$

i.e. $\displaystyle \lim_{n \to \infty} \left\{ \sup_{k \geqslant n} x_k \right\} = \limsup_{n \to \infty} x_n.$

- *Exercise 5.21*

1 If $n > m$,

$$|x_n - x_m| \leqslant |x_n - x_{n-1}| + |x_{n-1} - x_{n-2}| + \ldots + |x_{m+1} - x_m|$$

$$\leqslant \alpha^{n-1} + \alpha^{n-2} + \ldots + \alpha^m$$

$$= \alpha^m \left\{ \frac{1 - \alpha^{n-m}}{1 - \alpha} \right\} < \frac{\alpha^m}{1 - \alpha}.$$

Given $\epsilon > 0$, choose N so large that, for any $m > N$,

$$\frac{\alpha^m}{1 - \alpha} < \epsilon \quad \text{(Example 4.12)}.$$

The result then follows.

Exercise 4.20(3) provides the example $y_n = \sqrt{n} \ (n = 1, 2, \ldots)$.

2 Note first that $a \leqslant x_n \leqslant b \ (n = 1, 2, \ldots)$. Hence

$$\frac{a}{b} \leqslant \frac{x_{n+1}}{x_n} \leqslant \frac{b}{a}.$$

Next observe that

$$x_{n+2}^2 - x_{n+1}^2 = x_{n+1}x_n - x_{n+1}^2 = x_{n+1}(x_n - x_{n+1})$$

$$|x_{n+2} - x_{n+1}| = \frac{x_{n+1}}{x_{n+2} + x_{n+1}} |x_n - x_{n+1}|$$

$$\leqslant \frac{b}{a + b} |x_n - x_{n+1}|.$$

It follows that

$$|x_{n+1} - x_n| \leqslant \left(\frac{b}{a+b}\right)^{n-1} |x_2 - x_1|$$

and so we may proceed as in the previous question.

3 If $x_{n+2} = \frac{1}{2}(x_n + x_{n+1})$,

$$x_{n+2} + \frac{1}{2}x_{n+1} = x_{n+1} + \frac{1}{2}x_n = \ldots = x_2 + \frac{1}{2}x_1.$$

Put $l = \frac{2}{3}(x_2 + \frac{1}{2}x_1)$. Then

$$x_{n+1} + \frac{1}{2}x_n = \frac{3}{2}l$$

$$x_{n+1} - l = \frac{1}{2}(l - x_n).$$

Hence

$$|x_{n+1} - l| = \frac{1}{2}|x_n - l| = \ldots = \frac{1}{2^n}|x_1 - l| \to 0 \text{ as } n \to \infty.$$

It follows that $x_n \to l$ as $n \to \infty$.

If $x_{n+2} = \{x_{n+1}x_n\}^{1/2}$,

$$x_{n+2}^2 x_{n+1} = x_{n+1}^2 x_n = \ldots = x_2^2 x_1.$$

Put $l = \{x_2^2 x_1\}^{1/3}$. Then

$$x_{n+1}^2 x_n = l^3$$

and it follows that $x_n \to l$ as $n \to \infty$.

4 Let $\langle x_n \rangle$ be a sequence of points in $[a, b]$. Since $[a, b]$ is bounded, the sequence $\langle x_n \rangle$ is bounded. By the Bolzano–Weierstrass theorem, the sequence therefore has a convergent subsequence $\langle x_{n_r} \rangle$. Suppose that $x_{n_r} \to l$ as $r \to \infty$. Since $a \leqslant x_{n_r} \leqslant b$, it follows from theorem 4.23 that $a \leqslant l \leqslant b$. Hence $\langle x_{n_r} \rangle$ converges to a point of $[a, b]$.

5 Suppose that I is unbounded above. By exercise 4.29(6), we can find a sequence $\langle x_n \rangle$ of points of I such that $x_n \to +\infty$ as $n \to \infty$. But no subsequence of such a sequence can converge and so this is a contradiction. Hence I is bounded above. Let $b = \sup I$. By theorem 4.29(6) we can find a sequence $\langle x_n \rangle$ of points of I such that $x_n \to b$ as $n \to \infty$. *All* subsequences of $\langle x_n \rangle$ also converge to b (theorem 5.2). Hence $b \in I$. Similarly for $a = \inf I$.

6 Let $\langle x_n \rangle$ be a sequence of *distinct* points of S. Since $\langle x_n \rangle$ is bounded it contains a convergent subsequence $\langle x_{n_r} \rangle$. Suppose that $x_{n_r} \to \xi$ as $r \to \infty$. Then ξ is a cluster point of S.

● *Exercise 6.26*

1 We have

$$\frac{3n-2}{n(n+1)(n+2)} = -\frac{1}{n} + \frac{5}{n+1} - \frac{4}{n+2}.$$

It follows that

$$\sum_{n=1}^{N} \frac{3n-2}{n(n+1)(n+2)} = (-\tfrac{1}{1} + \tfrac{5}{2} - \tfrac{4}{3})$$

$$+ (-\tfrac{1}{2} + \tfrac{5}{3} - \tfrac{4}{4})$$

$$+ (-\tfrac{1}{3} + \tfrac{5}{4} - \tfrac{4}{5})$$

$$+ (-\tfrac{1}{4} + \tfrac{5}{5} - \tfrac{4}{6})$$

$$\cdots\cdots\cdots\cdots\cdots$$

$$+ \left(-\frac{1}{N-2} + \frac{5}{N-1} - \frac{4}{N}\right)$$

$$+ \left(-\frac{1}{N-1} + \frac{5}{N} - \frac{4}{N+1}\right)$$

$$+ \left(-\frac{1}{N} + \frac{5}{N+1} - \frac{4}{N+2}\right)$$

$$= -1 + \frac{5}{2} - \frac{1}{2} - \frac{4}{N+1} + \frac{5}{N+1} - \frac{4}{N+2}$$

$$\to 1 \text{ as } N \to \infty.$$

2 A convergent sequence is bounded (theorem 4.25). It follows that, for some H,

$$a_n \leqslant Hb_n \quad (n = 1, 2, \ldots).$$

Thus the convergence of $\sum_{n=1}^{\infty} b_n$ implies that of $\sum_{n=1}^{\infty} a_n$ by the comparison test.

Since $l > 0$, from exercise 4.29(2), there exists an N such that, for any $n > N$,

$$a_n > \tfrac{1}{2} l b_n$$

and hence the convergence of $\sum_{n=1}^{\infty} a_n$ implies that of $\sum_{n=1}^{\infty} b_n$ by the comparison test.

(i) Take $a_n = \dfrac{1}{2n}$ and $b_n = \dfrac{1}{n}$. Since $\displaystyle\sum_{n=1}^{\infty} \frac{1}{n}$ diverges (theorem 6.5), it it follows that $\displaystyle\sum_{n=1}^{\infty} \frac{1}{2n}$ diverges.

(ii) Take $a_n = \dfrac{1}{2n-1}$ and $b_n = \dfrac{1}{n}$. Again $\displaystyle\sum_{n=1}^{\infty} \dfrac{1}{2n-1}$ diverges.

(iii) Take $a_n = \dfrac{2}{n^2+3}$ and $b_n = \dfrac{1}{n^2}$. Since $\displaystyle\sum_{n=1}^{\infty} \dfrac{1}{n^2}$ converges (theorem

6.6), it follows that $\displaystyle\sum_{n=1}^{\infty} \dfrac{2}{n^2+3}$ converges.

3 We have

$$b_n = a_{n+1} + \ldots + a_{2n} = \sum_{k=1}^{2n} a_k - \sum_{k=1}^{n} a_k \to 0 \text{ as } n \to \infty.$$

Since the sequence $\langle a_n \rangle$ decreases,

$$b_n \geqslant n a_{2n} \geqslant 0.$$

It follows from the sandwich theorem that $2na_{2n} \to 0$ as $n \to \infty$.
Similarly $(2n-1)a_{2n-1} \to 0$ as $n \to \infty$.

4 The series diverges because its terms do not tend to zero (see example 5.11).

5 (i) We use the ratio test.

$$\left| \frac{a_{n+1}}{a_n} \right| = \frac{((n+1)!)^2}{(2(n+1))!} \cdot \frac{(2n)!}{(n!)^2} = \frac{(n+1)^2}{(2n+2)(2n+1)} \to \frac{1}{4} \text{ as } n \to \infty.$$

The series therefore converges.

(ii) Again the ratio test may be applied.

$$\left| \frac{a_{n+1}}{a_n} \right| = \frac{((n+1)!)^2}{(2(n+1))!} 5^{n+1} \frac{(2n)!}{(n!)^2} \frac{1}{5^n} \to \frac{5}{4} \text{ as } n \to \infty.$$

The series therefore diverges.

(iii) We use the nth root test

$$\left\{ \left(\frac{n}{n+1} \right)^{n^2} \right\}^{1/n} = \left\{ \left(1 + \frac{1}{n} \right)^n \right\}^{-1} \to \frac{1}{e} \text{ as } n \to \infty.$$

The series therefore converges because $e > 1$.

(iv) Again the nth root test may be applied.

$$\left\{ \left(\frac{n}{n+1} \right)^{n^2} 4^n \right\}^{1/n} \to \frac{4}{e} \text{ as } n \to \infty.$$

The series therefore diverges because $e < 4$.

(v) We have

$$\frac{\sqrt{(n+1)} - \sqrt{n}}{n} = \frac{1}{n} \frac{1}{(\sqrt{(n+1)} + \sqrt{n})} < \frac{1}{n^{3/2}} \quad (n = 1, 2, \ldots)$$

and hence the series converges by the comparison test (see theorem 6.6).

(vi) This series converges by theorem 6.13.

6 We have

$$t_{3n} = 1 + \frac{1}{3} - \frac{1}{2} + \frac{1}{5} + \frac{1}{7} - \frac{1}{4} + \ldots + \frac{1}{4n-3} + \frac{1}{4n-1} - \frac{1}{2n}$$

$$= \left\{ 1 - \frac{1}{2} + \frac{1}{3} - \frac{1}{4} + \frac{1}{5} - \ldots - \frac{1}{2n} \right\}$$

$$+ \left\{ \frac{1}{2n+2} + \frac{1}{2n+4} + \ldots + \frac{1}{4n} \right\}$$

$$= s_{2n} + \frac{1}{2} \left\{ \frac{1}{n+1} + \frac{1}{n+2} + \ldots + \frac{1}{2n} \right\}.$$

Here s_n denotes the nth partial sum of the given conditionally convergent series with sum s. Observe that

$$\left\{ 1 + \frac{1}{2} + \frac{1}{3} + \ldots + \frac{1}{2n} \right\} - \left\{ 1 - \frac{1}{2} + \frac{1}{3} - \ldots - \frac{1}{2n} \right\}$$

$$= 2 \left\{ \frac{1}{2} + \frac{1}{4} + \ldots + \frac{1}{2n} \right\}.$$

From this identity it follows that

$$\left\{ \frac{1}{n+1} + \frac{1}{n+2} + \ldots + \frac{1}{2n} \right\} = s_{2n}.$$

Hence

$$t_{3n} = s_{2n} + \tfrac{1}{2} s_{2n} \to \tfrac{3}{2} s \text{ as } n \to \infty.$$

● *Exercise 7.16*

1

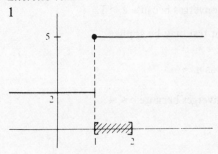

Each vertical line meets the graph in one and only one point. The range is $\{2, 5\}$. The image of the set $[1, 2]$ is $\{5\}$.

2

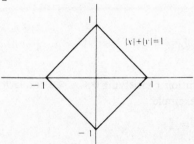

$|x| + |y| = 1$

(i) The equation does not define a function from $\mathbf{R}$ to itself. The value $x = 2$ is an example of an x for which no corresponding y exists.

(ii) The equation does not define a function from $[-1, 1]$ to $[-1, 1]$. The value $x = 0$ is an example of an x for which *two* corresponding values of y (namely $+1$ and -1) exist.

(iii) The equation *does* define a function from $[-1, 1]$ to $[0, 1]$. For each x satisfying $-1 \leqslant x \leqslant 1$ there is precisely one y with $0 \leqslant y \leqslant 1$ satisfying $|x| + |y| = 1$, i.e. $y = 1 - |x|$.

3 We have

$$f \circ g(x) = f(g(x)) = \frac{1 - g(x)}{1 + g(x)} = \frac{1 - 4x(1 - x)}{1 + 4x(1 - x)} \quad (0 \leqslant x \leqslant 1)$$

$$g \circ f(x) = g(f(x)) = 4f(x)(1 - f(x)) = 4\left(\frac{1 - x}{1 + x}\right)\left(\frac{2x}{1 + x}\right)$$
$$(0 \leqslant x \leqslant 1).$$

The two formulae give different results when $x = 1$ and hence define different functions.

In order that $f^{-1} : [0, 1] \to [0, 1]$ exist, it must be true that, for each y satisfying $0 \leqslant y \leqslant 1$, the equation

$$y = \frac{1 - x}{1 + x}$$

has a unique solution x satisfying $0 \leqslant x \leqslant 1$. This requirement is easily checked by solving the equation

$$y + yx = 1 - x$$

$$x = \frac{1 - y}{1 + y}.$$

The inverse function therefore exists and

$$f^{-1}(y) = \frac{1-y}{1+y} \quad (0 \leqslant y \leqslant 1).$$

(Note that $f = f^{-1}$).

The equation

$$y = 4x(1-x)$$

does not have a unique solution x satisfying $0 \leqslant x \leqslant 1$ for each y satisfying $0 \leqslant y \leqslant 1$. For example,

$$4.\tfrac{1}{4}(1 - \tfrac{1}{4}) = 4.\tfrac{3}{4}(1 - \tfrac{3}{4}) = \tfrac{3}{4}.$$

Thus g^{-1} does *not* exist.

4 The function f has no inverse function. When $y = -1$, for example, the equation $y = x^2$ has no real solution at all.

   The function g has an inverse function. For each $y \geqslant 0$ there exists precisely one $x \geqslant 0$ such that $y = x^2$ which we denote by $\sqrt{y}$ (see §1.9). Hence

$$g^{-1}(y) = \sqrt{y} \quad (y \geqslant 0).$$

5 We have $y = g(x)$ if and only if $x = g^{-1}(y)$.

   (i) $g^{-1} \circ g(x) = g^{-1}(g(x)) = g^{-1}(y) = x \quad (x \in A)$

   (ii) $g \circ g^{-1}(y) = g(g^{-1}(y)) = g(x) = y \quad (y \in B)$.

When g is as in question 4, these formulae reduce to

$$\sqrt{x^2} = x \quad (x \geqslant 0)$$

$$(\sqrt{y})^2 = y \quad (y \geqslant 0).$$

6 (i) We apply exercise 2.13. Put $T = \{f(x) : x \in S\}$. Then

$$\sup_{x \in S} \{f(x) + c\} = \sup_{y \in T} \{y + c\}$$

$$= c + \sup_{y \in T} y = c + \sup_{x \in S} f(x).$$

(ii) Let

$$H = \sup_{x \in S} f(x); \qquad K = \sup_{x \in S} g(x).$$

Then, for all $x \in S$,

$$f(x) + g(x) \leqslant H + K$$

and hence $H + K$ is an upper bound for the set

$\{f(x) + g(x) : x \in S\}$.

The result follows.

To show that equality need not hold, take $S = [0, 1]$ and let $f(x) = x \ (0 \leqslant x \leqslant 1)$ and $g(x) = 1 - x \ (0 \leqslant x \leqslant 1)$. Then $f(x) + g(x) = 1 \ (0 \leqslant x \leqslant 1)$. Thus

$$\sup_{x \in S} \{f(x) + g(x)\} = 1$$

but

$$\sup_{x \in S} f(x) + \sup_{x \in S} g(x) = 1 + 1 = 2.$$

- *Exercise 8.15*

1 Since rational functions are continuous at every point at which they are defined (theorem 8.13), we obtain

(i) $\displaystyle \lim_{x \to 1} \frac{x^2 + 4}{x^2 - 4} = \frac{1 + 4}{1 - 4} = -\frac{5}{3}$.

(ii) $\displaystyle \lim_{x \to 0} \frac{x^{73} + 5x^{42} + 9}{3x^{23} + 7} = \frac{0 + 5.0 + 9}{3.0 + 7} = \frac{9}{7}$.

2

$$f(x) = \begin{cases} 3 - x & (x > 1) \\ 1 & (x = 1) \\ 2x & (x < 1). \end{cases}$$

(i) $f(x) \to 2$ as $x \to 1-$. Given any $\epsilon > 0$, we must show how to find a $\delta > 0$ such that

$|f(x) - 2| < \epsilon$

provided that $1 - \delta < x < 1$. Since we are only concerned with values of x satisfying $x < 1$, we can replace $f(x)$ by $2x$. The condition $|f(x) - 2| < \epsilon$ then becomes $|2x - 2| < \epsilon$, i.e. $1 - \frac{1}{2}\epsilon < x < 1 + \frac{1}{2}\epsilon$. We therefore have to find a $\delta > 0$ such that

$1 - \frac{1}{2}\epsilon < x < 1 + \frac{1}{2}\epsilon$

provided that $1 - \delta < x < 1$. The choice $\delta = \frac{1}{2}\epsilon$ clearly suffices.

(ii) $f(x) \to 2$ as $x \to 1+$. Given any $\epsilon > 0$, we must show how to find a $\delta > 0$ such that

$$|f(x) - 2| < \epsilon$$

provided that $1 < x < 1 + \delta$. Since we are only concerned with values of x satisfying $x > 1$, we can replace $f(x)$ by $3 - x$. The condition $|f(x) - 2| < \epsilon$ then becomes $|3 - x - 2| < \epsilon$, i.e. $1 - \epsilon < x < 1 + \epsilon$. The choice $\delta = \epsilon$ therefore suffices.

It follows from proposition 8.4 that

$$\lim_{x \to 1} f(x)$$

exists and is equal to 2. Note that $f(1) = 1$ and so the function is not continuous at the point 1.

3 If $x \neq 0$,

$$f(x) = \frac{1 + 2x + x^2 - 1}{x} = 2 + x.$$

Since in considering a limit as $x \to 0$ we deliberately exclude consideration of the value of the function at $x = 0$, it follows that

$$\lim_{x \to 0} f(x) = \lim_{x \to 0} (2 + x) = 2.$$

4 Let $\epsilon > 0$ be given. We must find a $\delta > 0$ such that $|x - \xi| < \epsilon$ provided that $0 < |x - \xi| < \delta$. The choice $\delta = \epsilon$ clearly suffices.

Observe that

$$|f(x) - \xi| \leqslant |\xi - x|$$

and hence $f(x) \to \xi$ as $x \to \xi$ by the sandwich theorem (proposition 8.14).

5 If $\xi > 0$, let $0 < X < \xi < Y$ and take $\xi = y$ in exercise 3.11(6). If $X < x < Y$,

$$\frac{1}{n} Y^{1/n} Y^{-1} |x - \xi| \leqslant |x^{1/n} - \xi^{1/n}| \leqslant \frac{1}{n} X^{1/n} X^{-1} |x - \xi|$$

and the sandwich theorem applies.

To prove that $f(x) \to 0$ as $x \to 0+$, we must show that, given any $\epsilon > 0$, we can find a $\delta > 0$ such that

$$x^{1/n} = |x^{1/n} - 0| < \epsilon$$

provided that $0 < x < \delta$. The choice $\delta = \epsilon^n$ suffices.

6 For any $\epsilon > 0$, we can find a $\delta > 0$ such that

$$|f(x) - l| < \epsilon$$

provided $0 < |x - \xi| < \delta$. Since this is true for *any* $\epsilon > 0$, it is therefore true for $\epsilon = l$ (with an appropriate value of δ, say $\delta = h$). We then obtain

$$|f(x) - l| < l$$

i.e. $0 = l - l < f(x) < l + l$

provided that $0 < |x - \xi| < h$, i.e. $\xi - h < x < \xi + h$ and $x \neq \xi$.

(i) if $l < 0$, an $h > 0$ can be found such that $f(x) < 0$ provided that $\xi - h < x < \xi + h$ and $x \neq \xi$.

(ii) If $l = 0$, nothing of this sort can be said.

• *Exercise 8.20*

1 (i) Given $H > 0$, we must find a $\delta > 0$ such that

$$x^{-1} < -H \tag{5}$$

provided that $-\delta < x < 0$. If $x < 0$, (5) is equivalent to $1 > -Hx$ which is in turn equivalent to

$$-H^{-1} < x$$

and hence the choice $\delta = H^{-1}$ will suffice.

(ii) Given $H > 0$, we must find an X such that

$$x^2 > H$$

provided that $x > X$. The choice $X = \sqrt{H}$ suffices.

2 Let $\epsilon > 0$ be given. We must find an X such that

$$|f(g(x)) - l| < \epsilon$$

provided that $x > X$. Since $f(y) \to l$ as $y \to +\infty$, we can find a Y such that

$$|f(y) - l| < \epsilon \tag{6}$$

provided that

$$y > Y. \tag{7}$$

Since $g(x) \to +\infty$ as $x \to +\infty$, we can find an X such that, for any $x > X$,

$$g(x) > Y.$$

If $x > X$, it follows that (7) is satisfied with $y = g(x)$ and hence that (6) is satisfied with $y = g(x)$. But this is what we had to prove.

The problems of theorem 8.17 arise because of the possibility that $g(x) = \eta$ for some values of x. In the above problem η is replaced by $+\infty$ which is not a possible value for g.

3 Let $\epsilon > 0$ be given. We have to find an X such that, for any $x > X$,

$$|f(x^{-1}) - l| < \epsilon.$$

Since $f(y) \to l$ as $y \to 0+$, we can find a $\delta > 0$ such that

$$|f(y) - l| < \epsilon$$

provided that $0 < y < \delta$.
We choose $X = \delta^{-1}$. Then, if $x > X$,

$$y = x^{-1} < X^{-1} = \delta$$

and the result follows.

4 Suppose that $f(x) \to \lambda$ as $x \to \xi$. Then $f(x_n) \to \lambda$ as $n \to \infty$ and $f(y_n) \to \lambda$ as $n \to \infty$ (theorem 8.9). This is a contradiction.

5 Take $x_n = 1/n$ $(n = 1, 2, \ldots)$ and $y_n = \sqrt{2}/n$ $(n = 1, 2, \ldots)$. Then $f(x_n) = 1 \to 1$ as $n \to \infty$ and $f(y_n) = 0 \to 0$ as $n \to \infty$. Hence

$$\lim_{x \to 0} f(x)$$

does not exist by the previous question.

6 We have $0 \leqslant |xf(x)| \leqslant |x|$ and so the result follows from the sandwich theorem.

● *Exercise 9.17*

1

(i)

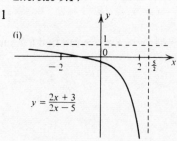

$$y = \frac{2x + 3}{2x - 5}$$

(ii)

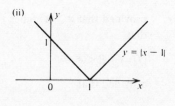

$$y = |x - 1|$$

Continuous on $(-2, 2)$ and on $[0, 1]$. Recall that a rational function is continuous wherever it is defined.

Continuous on $(-2, 2)$ and on $[0, 1]$. Note that the function has a 'corner' at $x = 1$.

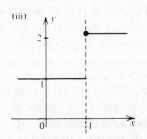

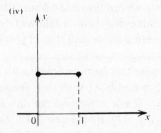

Not continuous on $(-2, 2)$ and not continuous on $[0, 1]$. We have $f(x) \to 1$ as $x \to 1 -$ but $f(1) = 2$.

Not continuous on $(-2, 2)$ but continuous on $[0, 1]$.

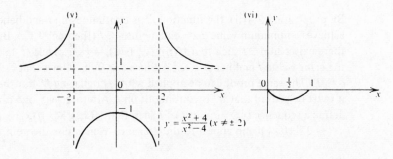

$$y = \frac{x^2 + 4}{x^2 - 4} \quad (x \neq \pm 2)$$

Continuous on $(-2, 2)$ and on $[0, 1]$.

Continuous on $(-2, 2)$ and on $[0, 1]$.

All of the functions except (v) are bounded on $(-2, 2)$: all but (i) and (ii) attain a maximum on $(-2, 2)$: all but (i) and (v) attain a minimum on $(-2, 2)$.

2 Let $g: I \to \mathbb{R}$ be defined by $g(x) = f(x) - x^2$. Then g is continuous on I. Given any point $\xi \in I$, let $\langle r_n \rangle$ be a sequence of rational numbers in I such that $r_n \to \xi$ as $n \to \infty$. Then $g(r_n) \to g(\xi)$ as $n \to \infty$ (proposition 9.6). But $g(r_n) = 0$ $(n = 1, 2, \ldots)$. It follows that

$$0 = g(\xi) = f(\xi) - \xi^2.$$

3 Let P be a polynomial of *odd* degree n. Let

$$P(x) = a_n x^n + a_{n-1} x^{n-1} + \ldots + a_1 x + a_0$$

$$= x^n \left\{ a_n + \frac{a_{n-1}}{x} + \ldots + \frac{a_1}{x^{n-1}} + \frac{a_0}{x^n} \right\}.$$

If $a_n > 0$, then $P(x) \to +\infty$ as $x \to +\infty$ and $P(x) \to -\infty$ as $x \to -\infty$.

Hence we can find a and b such that $P(a) < 0 < P(b)$. Since P is continuous everywhere, it follows from corollary 9.10 that there exists a ξ between a and b such that $P(\xi) = 0$. (Similarly if $a_n < 0$.)

4 Suppose that ξ can be found with $f(\xi) > 0$. Since $f(x) \to 0$ as $x \to +\infty$, we can find $b > \xi$ such that, for any $x > b$, $|f(x)| < f(\xi)$. Also, since $f(x) \to 0$ as $x \to -\infty$, we can find $a < \xi$ such that, for any $x < a$, $|f(x)| < f(\xi)$. By theorem 9.12, f attains a maximum on $[a, b]$. Suppose that $f(\eta) \geqslant f(x)$ ($x \in [a, b]$). Then $f(\eta) \geqslant f(\xi) > f(x)$ ($x \notin [a, b]$). Hence f attains a maximum on $\mathbb{R}$.

If a ξ can be found for which $f(\xi) < 0$, then a similar argument shows the existence of a minimum on $\mathbb{R}$.

If $f(x) = 0$ for all x, the result is trivial.

By proposition 9.4(ii), the function f^2 is continuous on I and hence achieves a minimum value c at some point $\xi \in I$ (theorem 9.12). But there exists an $\eta \in I$ such that $|f(\eta)| \leqslant \frac{1}{2} |f(\xi)| = \frac{1}{2}\sqrt{c}$. Thus f^2 takes the value $\frac{1}{4}c$ on I and hence $c \leqslant \frac{1}{4}c$. Since $c \geqslant 0$, it follows that $c = 0$. (The same proof can be applied with $|f|$ replacing f^2 but then it must be proved that $|f|$ is continuous on I. Alternatively, one can define a sequence $\langle x_n \rangle$ of points of I such that $|f(x_n)| \leqslant \frac{1}{2} |f(x_{n-1})| \leqslant \ldots \leqslant 2^{-n} |f(x_1)|$ and appeal to the Bolzano–Weierstrass theorem.)

6 That f is continuous on I follows from proposition 9.3. We have

$$|x_{n+1} - x_n| = |f(x_n) - f(x_{n-1})| \leqslant \alpha |x_n - x_{n-1}|.$$

Hence

$$|x_{n+1} - x_n| \leqslant \alpha^{n-1} |x_2 - x_1| \quad (n = 2, 3, \ldots).$$

As in exercise 5.21(1), we may conclude that $\langle x_n \rangle$ is a Cauchy sequence and hence that $\langle x_n \rangle$ converges. Suppose that $x_n \to l$ as $n \to \infty$. Because I is closed we have that $l \in I$. But f is continuous on I and therefore we may deduce from the equation

$$x_{n+1} = f(x_n)$$

that $l = f(l)$.

● *Exercise 10.11*

1 We have to consider

$$\frac{1}{h} \left\{ \frac{1}{1 + (x+h)^2} - \frac{1}{1 + x^2} \right\} = \frac{1}{h} \left\{ \frac{1 + x^2 - 1 - x^2 - 2xh - h^2}{(1 + (x+h)^2)(1 + x^2)} \right\}$$

$$= \frac{-2x - h}{(1 + (x + h)^2)(1 + x^2)} \rightarrow \frac{-2x}{(1 + x^2)^2} \text{ as } h \rightarrow 0.$$

2 Let $n = -m$ where $m \in \mathbb{N}$. Then, by theorem 10.9(iii),

$$Dx^n = D\frac{1}{x^m} = \frac{x^m \cdot 0 - 1 \cdot mx^{m-1}}{x^{2m}} = -mx^{-m-1} = nx^{n-1}.$$

3 If $h > 0$,

$$\frac{f(1 + h) - f(1)}{h} = \frac{2 + 2h - 2}{h} = 2 \rightarrow 2 \text{ as } h \rightarrow 0+.$$

If $h < 0$,

$$\frac{f(1 + h) - f(1)}{h} = \frac{(1 + h)^2 + 1 - 2}{h} = 2 + h \rightarrow 2 \text{ as } h \rightarrow 0-.$$

That $f'(1) = 2$ then follows from proposition 8.4.

On the other hand, if $h > 0$,

$$\frac{g(0 + h) - g(0)}{h} = \frac{h - 0}{h} \rightarrow 1 \text{ as } h \rightarrow 0+$$

but, if $h < 0$,

$$\frac{g(0 + h) - g(0)}{h} = \frac{-h - 0}{h} \rightarrow -1 \text{ as } h \rightarrow 0-.$$

4 By exercise 3.11(3), $P(x) = (x - \xi)R(x)$ where R is a polynomial of degree $n - 1$. Thus

$$P'(x) = R(x) + (x - \xi)R'(x) \quad \text{(theorem 10.9(ii))}.$$

Since $P'(\xi) = 0$ it follows that $R(\xi) = 0$ and therefore $R(x) = (x - \xi)Q(x)$ where Q is a polynomial of degree $n - 2$.

5 Put $g_l(x) = (x - \xi)^l$. Then

$$D^k g_l(x) = l(l - 1) \dots (l - k + 1)(x - \xi)^{l-k} \quad (k = 0, 1, 2, \dots, l).$$

It follows that $D^k g_l(\xi) = 0 \ (k = 0, 2, \dots, l - 1)$ and

$$D^l g_l(x) = l! \quad (x \in \mathbb{R}).$$

Hence $D^k g_l(x) = 0 \ (k = l + 1, l + 2, \dots)$. We now apply these results to the 'Taylor polynomial' P and obtain

$$D^k P(\xi) = 0 + 0 + \dots + \frac{1}{k!} f^{(k)}(\xi)D^k g_k(\xi) + \dots + 0$$

$$= f^{(k)}(\xi) \quad (k = 0, 1, \dots, n - 1).$$

6 We prove the result by induction. It certainly holds for $n = 1$. If

$$D^n fg = \sum_{j=0}^{n} \binom{n}{j} D^j f D^{n-j} g$$

then

$$D^{n+1} fg = \sum_{j=0}^{n} \binom{n}{j} \{D^{j+1} f D^{n-j} g + D^j f D^{n-j+1} g\}$$

and the proof continues as in the solution to exercise 3.11(4).

- *Exercise 10.15*

1 Let $y > x$. Then, by exercise 3.11(6) with $X = x$ and $Y = y$,

$$\frac{1}{n} y^{1/n} y^{-1} \leqslant \frac{y^{1/n} - x^{1/n}}{y - x} \leqslant \frac{1}{n} x^{1/n} x^{-1}.$$

It follows from the sandwich theorem that

$$\lim_{y \to x+} \frac{y^{1/n} - x^{1/n}}{y - x} = \frac{1}{n} x^{1/n} x^{-1}.$$

A similar argument shows that the left hand limit is the same.

2 Write $r = m/n$ where m is an integer and n a natural number. Then

$$D\{x^{m/n}\} = D\{(x^m)^{1/n}\} = \frac{1}{n} (x^m)^{1/n} x^{-m} m x^{m-1} = \frac{m}{n} x^{m/n} x^{-1}.$$

3 (i) $D\{1 + x^{1/27}\}^{3/5} = \frac{3}{5}\{1 + x^{1/27}\}^{-2/5} \cdot \frac{1}{27} x^{-26/27}$ $(x > 0)$.

(ii) $D\{x + (x + x^{1/2})^{1/2}\}^{1/2}$

$= \frac{1}{2}\{x + (x + x^{1/2})^{1/2}\}^{-1/2}\{1 + \frac{1}{2}(x + x^{1/2})^{-1/2}(1 + \frac{1}{2}x^{-1/2})\}$ $(x > 0)$.

4 $\dfrac{d}{dx} \{f(x^2)\} = f'(x^2)2x$; $\dfrac{d}{dx} \{f(x)\}^2 = 2f(x)f'(x)$.

When $x = 1$,

$$2f'(1) = 2f(1)f'(1)$$

and the result follows.

5 By theorem 10.13,

$$1 = Dx = D(f^{-1} \circ f)(x) = Df^{-1}(y) Df(x).$$

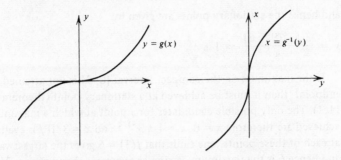

Observe that, for $h > 0$,

$$\frac{(0 + h)^{1/3} - 0^{1/3}}{h} = h^{-2/3} \to +\infty \text{ as } h \to 0+.$$

Thus g^{-1} is *not* differentiable at 0.

6 If x is rational, then $y = f(x) = x$ is rational and so

$$f \circ f(x) = f(f(x)) = f(y) = y = f(x) = x.$$

If x is irrational, then $y = f(x) = -x$ is irrational and so

$$f \circ f(x) = f(f(x)) = f(y) = -y = -f(x) = x.$$

A thoughtless application of theorem 10.13 would yield

$$Df(x) = \frac{1}{Df(y)} \quad (y = f(x)).$$

But the function f is not differentiable at any point – thus nothing can be deduced from theorem 10.13.

• *Exercise 11.8*

1 The stationary points are the roots of the equation $f'(x) = 0$. But $f(x) = x^3 - 3x^2 + 2x$ and so

$$f'(x) = 3x^2 - 6x + 2$$

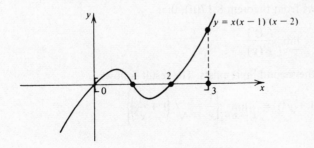

and hence the stationary points are given by

$$x = \frac{3 \pm \sqrt{(9-6)}}{3} = 1 \pm \frac{1}{\sqrt{3}}.$$

If the maximum on the compact interval $[0, 3]$ is not attained at an endpoint, then it must be achieved at a stationary point (theorem 11.2). The only possible candidates for a point at which a maximum is achieved are therefore $x = 0, x = 1 \pm 3^{-1/2}$ and $x = 3$. If f is evaluated at each of these points, one finds that $f(3) = 6$ gives the largest value and hence 6 is the maximum. A similar argument shows that $-2/3\sqrt{3}$ is the minimum and this is achieved at $x = 1 + 3^{-1/2}$.

2 Choose the constant h so that $F = f + hg$ satisfies $F(a) = F(b)$ and thus Rolle's theorem may be applied. We need

$$f(a) + hg(a) = f(b) + hg(b)$$

i.e.

$$h = -\frac{f(b) - f(a)}{g(b) - g(a)}.$$

We obtain the existence of a $\xi \in (a, b)$ such that

$$0 = F'(\xi) = f'(\xi) + hg'(\xi)$$

i.e.

$$h = -\frac{f'(\xi)}{g'(\xi)}$$

and the result follows.

3 Apply the previous result with $b = x$. We deduce the existence of a $\xi \in (a, x)$ such that

$$\frac{f'(\xi)}{g'(\xi)} = \frac{f(x)}{g(x)}.$$

Observe that ξ depends on the value of x, i.e. 'ξ is a function of x'. By exercise 8.15(4), we have $\xi \to a$ as $x \to a$. Moreover $\xi \neq a$ when $x > a$ and so it follows from theorem 8.17(ii), that

$$\lim_{x \to a+} \frac{f'(\xi)}{g'(\xi)} = \lim_{y \to a+} \frac{f'(y)}{g'(y)}$$

provided that the second limit exists. The result follows.

$$\lim_{y \to +\infty} y - \sqrt{(1 + y^2)} = \lim_{x \to 0+} \left\{ \frac{1}{x} - \sqrt{\left(1 + \frac{1}{x^2}\right)} \right\}$$

$$= \lim_{x \to 0+} \frac{1 - \sqrt{(1 + x^2)}}{x}$$

$$= \lim_{x \to 0+} \frac{-\frac{1}{2}(1 + x^2)^{-1/2} \cdot 2x}{1} = 0.$$

4 Given that $f'(x) = x^2$ for all x, it follows that

$$D\{f(x) - \tfrac{1}{3}x^3\} = f'(x) - x^2 = 0$$

for all x and hence there exists a constant c such that $f(x) - \tfrac{1}{3}x^3 = c$ for all x (theorem 11.7).

5 From Rolle's theorem we may deduce the existence of numbers $\eta_0, \eta_1, \ldots, \eta_{n-1}$ such that $\xi_0 < \eta_0 < \xi_1 < \eta_1 < \ldots < \xi_{n-1} < \eta_{n-1} < \xi_n$ and

$$f'(\eta_i) - P'(\eta_i) = 0 \quad (i = 0, 1, \ldots, n - 1).$$

Continuing in this way we can demonstrate the existence of a ξ such that

$$f^{(n)}(\xi) - P^{(n)}(\xi) = 0.$$

But a polynomial of degree $n - 1$ has the property that $P^{(n)}(x) = 0$ for all x.

6 Since $g(0) = g(1) = 0$, it follows that $f(0) = f(1) = 1$ and so Rolle's theorem may be applied on the interval $[0, 1]$. We obtain the existence of a $\xi \in (0, 1)$ such that $f'(\xi) = 0$. But then $f(\xi) = 1$ and so Rolle's theorem may be applied on the intervals $[0, \xi]$ and $[\xi, 1]$ and so on. This argument shows that between every pair of points x and y for which $f(x) = f(y) = 1$, we can find a point z for which $f(z) = 1$.

Suppose that $0 < a < b < 1$ and (a, b) contains no point x for which $f(x) = 1$. Let

$$A = \sup \{x : x \leqslant a \text{ and } f(x) = 1\}; B = \inf \{y : y \geqslant b \text{ and } f(y) = 1\}.$$

Because $a < b$, $A < B$. Also, the continuity of f implies that $f(A) = f(B) = 1$ [use exercise 4.29(6) and theorem 8.9]. But we know that a point C can be found between A and B for which $f(C) = 1$ and this is a contradiction.

Let $x \in [0, 1]$. Since every open subinterval of $[0, 1]$ contains a point ξ for which $f(\xi) = 1$, we can find a sequence ξ_n of points of $[0, 1]$ with $f(\xi_n) = 1$ such that $\xi_n \to x$ as $n \to \infty$. This implies that $f(x) = 1$ because f is continuous.

- *Exercise 11.11*

1 Let $f: \mathbb{R} \to \mathbb{R}$ be defined by $f(x) = x^n$ and apply Taylor's theorem in the form

$$f(y) = f(\xi) + \frac{1}{1!}(y - \xi)f'(\xi) + \ldots + \frac{1}{n!}(y - \xi)^n f^{(n)}(\xi) + E_{n+1}.$$

Put $y = x + 1$ and $\xi = 1$. One then has to observe that, for $k = 0, 1, \ldots, n$,

$$\frac{1}{k!} f^{(k)}(\xi) = \frac{n(n-1) \ldots (n-k+1)\xi^{n-k}}{k!} = \binom{n}{k}$$

and that

$$E_{n+1} = \frac{1}{(n+1)!} x^{n+1} f^{(n+1)}(\eta) = 0.$$

2 By Taylor's theorem

$$(1+4)^{1/2} = 2 + \tfrac{1}{2} \cdot 4^{-1/2} + \frac{1}{2!} \cdot \tfrac{1}{2}(\tfrac{1}{2}-1)4^{-3/2} + \frac{1}{3!} \cdot \tfrac{1}{2}(\tfrac{1}{2}-1)(\tfrac{1}{2}-2)\eta^{-5/2}$$

where η lies between 4 and 5. Hence

$$\sqrt{5} = 2\tfrac{15}{64} + \tfrac{1}{16}\eta^{-5/2}.$$

But

$$\eta^{5/2} > 4^{5/2} = 2^5$$

and the result follows.

3 By Taylor's theorem, for any $x \in I$,

$$f(x) = f(\xi) + \frac{(x-\xi)}{1!} f'(\xi) + \ldots + \frac{(x-\xi)^{n-1}}{(n-1)!} f^{(n-1)}(\xi) + E_n.$$

Hence, with the given information,

$$f(x) - f(\xi) = \frac{1}{n!}(x-\xi)^n f^{(n)}(\eta) \qquad (8)$$

for some η between x and ξ. Since $f^{(n)}$ is continuous at ξ, it follows from exercise 8.15(6), that, if x is taken sufficiently close to ξ, then $f^{(n)}(\eta)$ will have the same sign as $f^{(n)}(\xi)$ (provided that $f^{(n)}(\xi) \neq 0$).

The problem therefore reduces to examining the sign of the right hand side of (8) under the various cases listed.

(i) $f^{(n)}(\xi) > 0$ and n even. Then the right hand side of (8) is positive for x sufficiently close to ξ and hence f has a *local minimum* at ξ.

(ii) $f^{(n)}(\xi) > 0$ and n odd. If x is sufficiently close to ξ, then the right hand side of (8) is positive if $x > \xi$ and negative if $x < \xi$.

(iii) $f^{(n)}(\xi) < 0$ and n even. We obtain a *local maximum*.

(iv) $f^{(n)}(\xi) < 0$ and n odd. This case is similar to (ii) – see the diagram below.

(i)
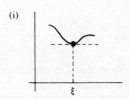

(ii)
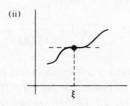

(iii)
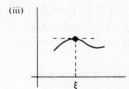

(iv)
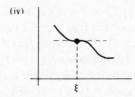

Of the three functions given, (v) falls into class (ii), (vi) falls into class (i) and (vii) into class (iii).

- *Exercise 12.12*

1 We have $f(a) \leqslant f(x) \leqslant f(b)$ $(a \leqslant x \leqslant b)$.

2 Observe that

$$f'(x) = \frac{1}{n}(x + 1)^{1/n}(x + 1)^{-1} - \frac{1}{n}x^{1/n}x^{-1} \leqslant 0 \quad (x > 0)$$

because $n \geqslant 1$ and hence

$$\left(1 + \frac{1}{x}\right)\left(1 + \frac{1}{x}\right)^{-1/n} \geqslant 1 \quad (x > 0).$$

3 If x and y lie in I and $y > x$, then $f(y) \geqslant f(x)$. Hence

$$f'(x) = \lim_{y \to x} \frac{f(y) - f(x)}{y - x} = \lim_{y \to x+} \frac{f(y) - f(x)}{y - x} \geqslant 0.$$

The function $f: \mathbb{R} \to \mathbb{R}$ defined by $f(x) = x^3$ is *strictly* increasing on $\mathbb{R}$ but $f'(0) = 0$.

4 Since f is continuous, $f(\mathbb{R})$ is an interval (theorem 9.9). Since $f(x) \to +\infty$ as $x \to +\infty$ and $f(x) \to -\infty$ as $x \to -\infty$, it follows that $f(\mathbb{R}) = \mathbb{R}$.

We have

$$f'(x) = 1 + 3x^2 > 0 \quad (x \in \mathbb{R})$$

and hence theorem 12.10 assures us of the existence of an inverse function $f^{-1} \colon \mathbb{R} \to \mathbb{R}$.

The unique value of x satisfying $-1 = y = f(x) = 1 + x + x^3$ is obviously $x = -1$. Hence

$$Df^{-1}(-1) = Df^{-1}(y) = \frac{1}{Df(x)} = \frac{1}{Df(-1)} = \frac{1}{4}.$$

5 Suppose that ξ is not an endpoint of I. To show that f is continuous at ξ, we need to prove that $f(\xi -) = f(\xi +)$ (see corollary 12.5).

If $f(\xi -) < f(\xi +)$ then we can find λ such that $f(\xi -) < \lambda < f(\xi +)$ and $\lambda \neq f(\xi)$. Choose a and b in I so that $a < \xi < b$. Then $f(a) < \lambda < f(b)$ and so we can find an $x \in I$ such that $f(x) = \lambda$ and hence $f(\xi -) < f(x) < f(\xi +)$. Since $f(x) \neq f(\xi)$ this is inconsistent with corollary 12.5 and hence $f(\xi -) = f(\xi +)$.

A slightly different argument is required if ξ is an endpoint.

The function $f \colon (0, 2) \to \mathbb{R}$ defined by

$$f(x) = \begin{cases} x & (0 < x \leqslant 1) \\ x - 1 & (1 < x \leqslant 2) \end{cases}$$

shows that the requirement that f be increasing cannot be abandoned (although it might be replaced by some new condition).

6 For each x satisfying $0 \leqslant x \leqslant 1$, $0 < f(x) \leqslant M(x)$. It follows that

$$\phi(x) = \lim_{n \to \infty} \left\{ \frac{f(x)}{M(x)} \right\}^n$$

exists for each x satisfying $0 \leqslant x \leqslant 1$ and that $\phi(x) = 0$ or $\phi(x) = 1$.

If f increases on $[0, 1]$, then $f(x) = M(x)$ for each x satisfying $0 \leqslant x \leqslant 1$ and hence $\phi(x) = 1 \ (0 \leqslant x \leqslant 1)$. Thus ϕ is continuous.

It is more difficult to deduce that f is increasing given that ϕ is continuous. We show to begin with that, if ϕ is continuous on $[0, 1]$, then $\phi(x) = 1 \ (0 \leqslant x \leqslant 1)$.

We know that $\phi([0, 1]) \subset \{0, 1\}$. But, by theorem 9.9, the image of an interval under a continuous function is another interval. Hence either $\phi([0, 1]) = \{0\}$ or $\phi([0, 1]) = \{1\}$. But the former case is impossible because $\phi(0) = 1$. Thus $\phi(x) = 1 \ (0 \leqslant x \leqslant 1)$.

Let $0 \leqslant x < y \leqslant 1$. We know that $f(x) = M(x)$ and $f(y) = M(y)$. Obviously $M(y) \geqslant M(x)$ and it follows that f increases on $[0, 1]$.

● *Exercise 12.21*

1 We have $f(x) = (x-1)(x-2)(x-3) = x^3 - 6x^2 + 11x - 6$. Hence

$$f'(x) = 3x^2 - 12x + 11$$

$$f''(x) = 6x - 12 = 6(x-2).$$

If follows immediately that f is convex for $x \geqslant 2$ and concave for $x \leqslant 2$. The roots of the quadratic equation $3x^2 - 12x + 11$ are

$$x = 2 \pm 3^{-1/2}$$

and it follows that f decreases on the interval $[2 - 3^{-1/2}, 2 + 3^{-1/2}]$ and increases on the intervals $(-\infty, 2 - 3^{-1/2}]$ and $[2 + 3^{-1/2}, \infty)$.

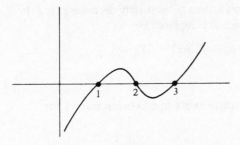

2 Take $x_1 = x, x_3 = y$ and $x_2 = \alpha x + \beta y$ in (1) of §12.13. Since $\alpha + \beta = 1$,

$$x_2 = \alpha x_1 + (1 - \alpha)x_3$$

and

$$\alpha = \frac{x_3 - x_2}{x_3 - x_1}; \qquad \beta = \frac{x_2 - x_1}{x_3 - x_1}.$$

Thus (1) of §12.13 is equivalent to

$$(x_3 - x_1)f(x_2) \leqslant (x_3 - x_2)f(x_1) + (x_2 - x_1)f(x_3). \tag{9}$$

Consider now the inequality

$$\frac{f(x_2) - f(x_1)}{x_2 - x_1} \leqslant \frac{f(x_3) - f(x_1)}{x_3 - x_1}.$$

This is equivalent to

$$(x_3 - x_1)f(x_2) - (x_3 - x_1)f(x_1) \leqslant (x_2 - x_1)f(x_3) - (x_2 - x_1)f(x_1)$$

i.e.

$$(x_3 - x_1)f(x_2) \leqslant (x_3 - x_1 - x_2 + x_1)f(x_1) + (x_2 - x_1)f(x_3)$$

which is (9). A similar argument establishes the inequality

$$\frac{f(x_3) - f(x_1)}{x_3 - x_1} \leqslant \frac{f(x_3) - f(x_2)}{x_3 - x_2}.$$

3 Let $X \in J$ and $Y \in J$. Set $X = f(x)$ and $Y = f(y)$. From (1) of §12.13,

$$f(\alpha x + \beta y) \leqslant \alpha f(x) + \beta f(y) = \alpha X + \beta Y$$

provided that $\alpha > 0, \beta > 0$ and $\alpha + \beta = 1$. Since f is strictly increasing on I, f^{-1} is strictly increasing on J. Thus

$$\alpha f^{-1}(X) + \beta f^{-1}(Y) = \alpha x + \beta y \leqslant f^{-1}(\alpha X + \beta Y)$$

and hence f^{-1} in concave on J.

If f is strictly decreasing on I, then f^{-1} is strictly decreasing on J. In this case the last step in the proof is replaced by

$$\alpha f^{-1}(X) + \beta f^{-1}(Y) = \alpha x + \beta y \geqslant f^{-1}(\alpha X + \beta Y)$$

and so f^{-1} is convex on J.

4 By the mean value theorem, there exists an η between x and ξ for which

$$\frac{f(x) - f(\xi)}{x - \xi} = f'(\eta).$$

Because f is convex, its derivative increases. Thus $f'(\eta) \geqslant f'(\xi)$ if $x > \xi$ and $f'(\eta) \leqslant f'(\xi)$ if $x < \xi$. Hence

$$f(x) - f(\xi) \geqslant f'(\xi)(x - \xi).$$

Geometrically the result asserts that the graph of the function lies above any tangent drawn to the graph.

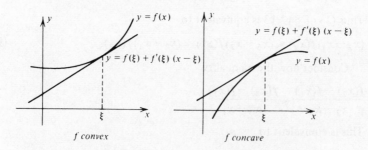

f convex   *f concave*

If f is concave the inequality should be reversed.

The geometry of the Newton–Raphson process is indicated below.

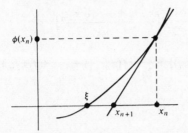

Given that $x_n \geqslant \xi$, $\phi(x_n) \geqslant 0$ and $\phi'(x_n) > 0$. (We cannot have $\phi'(x) = 0$ for any x because ϕ' is increasing and hence we would have $\phi'(y) = 0$ $(y \leqslant x)$). Thus $x_{n+1} \leqslant x_n$. On the other hand, by the inequality of the first part of the question,

$$\phi(x_{n+1}) - \phi(x_n) \geqslant \phi'(x_n)(x_{n+1} - x_n) = -\phi(x_n)$$

and hence $\phi(x_{n+1}) \geqslant 0$. It follows that $\xi \leqslant x_{n+1} \leqslant x_n$.

Since $\langle x_n \rangle$ is decreasing and bounded below, it converges. Say $x_n \to l$ as $n \to \infty$. Then $\phi(x_n) \to \phi(l)$ as $n \to \infty$ because ϕ is continuous and $\phi'(x_n) \to \phi'(l+)$ as $n \to \infty$ because ϕ' is increasing. We obtain

$$l = l - \frac{\phi(l)}{\phi'(l+)}$$

and hence $\phi(l) = 0$. Thus $l = \xi$.

5 Suppose that $f'(\xi) > 0$. Then, using question 4,

$$f(x) \geqslant f(\xi) + f'(\xi)(x - \xi) \to +\infty \text{ as } x \to +\infty.$$

Similarly, if $f'(\xi) < 0$,

$$f(x) \geqslant f(\xi) + f'(\xi)(x - \xi) \to +\infty \text{ as } x \to -\infty.$$

It follows that $f'(\xi) = 0$ for all values of ξ and therefore f is constant (theorem 11.7).

6 Let $P(n)$ be the assertion that

$$f\left(\frac{x_1 + x_2 + \ldots + x_n}{n}\right) \leqslant \frac{1}{n}\{f(x_1) + \ldots + f(x_n)\} \tag{10}$$

for any $x_1, x_2, \ldots, x_n$ in the interval I. That $P(2)$ holds is given. We assume that $P(2^n)$ holds and seek to establish $P(2^{n+1})$. Write $m = 2^n$. Then $2m = 2^{n+1}$ and

$$f\left(\frac{x_1 + \ldots + x_{2m}}{2m}\right) = f\left(\frac{1}{2}\left\{\frac{1}{m}(x_1 + \ldots + x_m) + \frac{1}{m}(x_{m+1} + \ldots + x_{2m})\right\}\right)$$

$$\leqslant \frac{1}{2} f\left(\frac{x_1 + \ldots + x_m}{m}\right) + \frac{1}{2} f\left(\frac{x_{m+1} + \ldots + x_{2m}}{m}\right)$$

$$\leqslant \frac{1}{2m} \{f(x_1) + \ldots + f(x_m)\} + \frac{1}{2m} \{f(x_{m+1}) + \ldots + f(x_{2m})\}$$

$$= \frac{1}{2m} \{f(x_1) + \ldots + f(x_{2m})\}.$$

We now assume that $P(n)$ holds and seek to deduce that $P(n-1)$ holds. Write

$$X = \frac{x_1 + x_2 + \ldots + x_{n-1}}{n-1}.$$

Then

$$f\left(\frac{x_1 + x_2 + \ldots + x_{n-1}}{n-1}\right) = f(X) = f\left(\frac{(n-1)X + X}{n}\right)$$

$$= f\left(\frac{x_1 + x_2 + \ldots + x_{n-1} + X}{n}\right)$$

$$\leqslant \frac{1}{n} \{f(x_1) + \ldots + f(x_{n-1}) + f(X)\}.$$

Hence

$$\left(1 - \frac{1}{n}\right) f(X) \leqslant \frac{1}{n} \{f(x_1) + f(x_2) + \ldots + f(x_{n-1})\}$$

and $P(n-1)$ follows.

Let α be a rational number satisfying $0 < \alpha < 1$. We may write $\alpha = m/n$. If $\beta = 1 - \alpha$, then $\beta = (n-m)/n$. Apply inequality (10) with $x_1 = x_2 = \ldots = x_m = x$ and $x_{m+1} = x_{m+2} = \ldots = x_n = y$. Then

$$f(\alpha x + \beta y) \leqslant \alpha f(x) + \beta f(y). \tag{11}$$

If α is irrational, consider a sequence $\langle \alpha_n \rangle$ of rational numbers such that $\alpha_n \to \alpha$ as $n \to \infty$. Then $\beta_n = 1 - \alpha_n \to 1 - \alpha = \beta$ as $n \to \infty$. We have

$$f(\alpha_n x + \beta_n y) \leqslant \alpha_n f(x) + \beta_n f(y) \quad (n = 1, 2, \ldots).$$

Since f is continuous, consideration of the limit shows that (11) holds even when α is irrational.

- *Exercise 13.18*

1 If $n \in \mathbb{N}$, we have for any x

$$D\left\{\frac{1}{n+1} x^{n+1}\right\} = x^n$$

and hence the result follows from theorem 13.14.

If n is replaced by a rational number $\alpha \neq -1$, the same argument applies. The restriction $0 < a < b$ is required because x^α need not be defined for $x \leqslant 0$ (e.g. if $\alpha = -2$, x^α is not defined for $x = 0$).

If $\alpha = -1$, the argument fails. (Why?)

2 The integrand is not defined when $x = 1$ and, in fact, has an unpleasant 'singularity' at this point.

3 As in example 13.17, we obtain

$$\frac{1}{n^{\alpha+1}}\{0^\alpha + 1^\alpha + \ldots + (n-1)^\alpha\}$$

$$\leqslant \int_0^1 x^\alpha \, dx \leqslant \frac{1}{n^{\alpha+1}}\{1^\alpha + 2^\alpha + \ldots + n^\alpha\}.$$

It follows that

$$0 \leqslant \frac{1}{n^{\alpha+1}}\{1^\alpha + 2^\alpha + \ldots + n^\alpha\} - \frac{1}{\alpha+1} \leqslant \frac{1}{n}$$

and the result follows.

- *Exercise 13.26*

1 Suppose that, for some $\xi \in [a, b]$, $g(\xi) > 0$. Then, because g is continuous on $[a, b]$, a subinterval $[c, d]$ containing ξ can be found for which

$$g(t) > \tfrac{1}{2}g(\xi) \quad (c \leqslant t \leqslant d)$$

(proof?). But then

$$\int_a^b g(t) \, dt > \int_c^d g(t) \, dt \geqslant \tfrac{1}{2}g(\xi)(d-c) > 0.$$

2 We integrate by parts. Then

$$\int_a^b x f''(x) \, dx = [x f'(x)]_a^b - \int_a^b f'(x) \, dx$$

$$= [x f'(x)]_a^b - [f(x)]_a^b$$

and the result follows.

3 Observe that $F'(t) = f(t) \geqslant \{F(t)\}^{1/2}$ (theorem 13.12). Hence

$$(x-1) = \int_1^x 1 \, dt \leqslant \int_1^x \{F(t)\}^{-1/2} F'(t) \, dt \quad \text{(theorem 13.23)}$$

$$= [2\{F(t)\}^{1/2}]_1^x \quad \text{(theorem 13.14)}$$

$$= 2\{F(x)\}^{1/2} \leqslant 2f(x).$$

4 By the continuity property (theorem 9.12), the function f attains a maximum M and a minimum m on the set $[a, b]$. Hence, by theorem 13.23,

$$m \int_a^b g(t)\, dt \leqslant \int_a^b f(t)g(t)\, dt \leqslant M \int_a^b g(t)\, dt. \tag{12}$$

We now appeal to the intermediate value theorem (corollary 9.10). The function F defined on $[a, b]$ by

$$F(x) = f(x) \int_a^b g(t)\, dt$$

is continuous on $[a, b]$ and hence a $\xi \in [a, b]$ can be found such that

$$F(\xi) = \int_a^b f(t)g(t)\, dt$$

because of (12).

   If $g(t) = 1$ $(a \leqslant t \leqslant b)$ and F is a primitive for f on $[a, b]$, then the result reduces to

$$F(b) - F(a) = F'(\xi)(b - a)$$

which is the mean value theorem (i.e. theorem 11.6) for the function F.

5 We have

$$\int_a^b f(t)g(t)\, dt = [f(t)G(t)]_a^b - \int_a^b f'(t)G(t)\, dt.$$

Since $f'(t) \geqslant 0$ $(a \leqslant t \leqslant b)$ we may apply the previous question and deduce the existence of $\xi \in [a, b]$ such that

$$\int_a^b f(t)g(t)\, dt = [f(t)G(t)]_a^b - G(\xi) \int_a^b f'(t)\, dt$$

$$= [f(t)G(t)]_a^b - G(\xi)[f(t)]_a^b$$

$$= f(b)\{G(b) - G(\xi)\} - f(a)\{G(a) - G(\xi)\}$$

$$= f(a) \int_a^\xi g(t)\, dt + f(b) \int_\xi^b g(t)\, dt.$$

6 We integrate the given integral by parts. Then

$$E_n = \frac{1}{(n-1)!} \int_\xi^x (x - t)^{n-1} f^{(n)}(t)\, dt$$

$$= \frac{1}{(n-1)!} \left\{ [(x-t)^{n-1} f^{(n-1)}(t)]_{\xi}^{x} \right.$$

$$\left. + \int_{\xi}^{x} (n-1)(x-t)^{n-2} f^{(n-1)}(t)\, dt \right\}$$

$$= -\frac{1}{(n-1)!} (x-\xi)^{n-1} f^{(n-1)}(\xi) + E_{n-1}$$

. . . .

$$= -\frac{1}{(n-1)!} (x-\xi)^{n-1} f^{(n-1)}(\xi) - \ldots - (x-\xi) f'(\xi) + E_1.$$

But

$$E_1 = \int_{\xi}^{x} f'(t)\, dt = f(x) - f(\xi)$$

and therefore

$$f(x) = f(\xi) + (x-\xi) f'(\xi) + \ldots + \frac{1}{(n-1)!} (x-\xi)^{n-1} f^{(n-1)}(\xi) + E_n$$

as required. Observe that, by question 4 above,

$$E_n = \frac{1}{(n-1)!} \int_{\xi}^{x} (x-t)^{n-1} f^{(n)}(t)\, dt$$

$$= \frac{1}{(n-1)!} f^{(n)}(\eta) \int_{\xi}^{x} (x-t)^{n-1}\, dt = \frac{(x-\xi)^n}{n!} f^{(n)}(\eta)$$

for some value of η between x and ξ.

- *Exercise 13.34*

1 (i) $\displaystyle\int_0^{1-\delta} \frac{dx}{\sqrt{(1-x)}} = [-2(1-x)^{1/2}]_0^{1-\delta} = 2 - 2\delta^{1/2} \to 2$ as $\delta \to 0+$.

(ii) $\displaystyle\int_0^{X} \frac{x^2\, dx}{(1+x^3)^2} = [-\tfrac{1}{3}(1+x^3)^{-1}]_0^{X}$

$$= \frac{1}{3} - \frac{1}{3} \frac{1}{(1+X^3)} \to \frac{1}{3} \text{ as } X \to +\infty.$$

(iii) Since

$$\frac{(1+x+x^3)(1+x^2)}{1+x+x^5} \to 1 \text{ as } x \to +\infty$$

we can find an H such that

$$\frac{1 + x + x^3}{1 + x + x^5} \leqslant \frac{H}{1 + x^2} \quad (x \geqslant 0).$$

The existence of the improper integral then follows from proposition 13.29 (see example 13.30).

(iv) $\displaystyle\int_0^{1-\delta} \frac{dx}{(1 - x)^{3/2}} = [2(1 - x)^{-1/2}]_0^{1-\delta} = \frac{2}{\sqrt{\delta}} - 2 \to +\infty$ as $\delta \to 0 +$.

(v) Put $1 - x = y$. Then

$$\int_0^{1-\delta} \frac{dx}{1 - x} = \int_1^{\delta} \frac{-dy}{y} = \int_{\delta}^1 \frac{dy}{y}$$

and so (v) reduces to (vi).

(vi) Put $y = 1/z$. Then

$$\int_{\delta}^1 \frac{dy}{y} = \int_{1/\delta}^1 -\frac{z \, dz}{z^2} = \int_1^{1/\delta} \frac{dz}{z} \to +\infty \text{ as } \delta \to 0 +$$

by example 13.33.

2 $\displaystyle I = \int_{1/2-y}^{1/2+y} \frac{2x - 1}{x(1 - x)} dx = \int_{1/2-y}^{1/2} \frac{2x - 1}{x(1 - x)} + \int_{1/2}^{1/2+y} \frac{2x - 1}{x(1 - x)} dx.$

Put $x = \frac{1}{2} - u$ in the first integral and $x = \frac{1}{2} + u$ in the second integral. Then

$$I = -\int_0^y \frac{2u}{(\frac{1}{2} - u)(\frac{1}{2} + u)} du + \int_0^y \frac{2u}{(\frac{1}{2} + u)(\frac{1}{2} - u)} du = 0.$$

However, the improper integral does *not* exist. We have

$$\int_{1/2-y}^{1/2} \frac{2x - 1}{x(1 - x)} dx = -2 \int_0^{2y} \frac{v \, dv}{(1 - v)(1 + v)}$$

$$= -\int_0^{2y} \left\{ \frac{1}{1 - v} - \frac{1}{1 + v} \right\} dv$$

$$\to -\infty \text{ as } y \to \tfrac{1}{2} - \quad \text{(exercise 13.34(1v))}.$$

On the other hand

$$\int_{1/2}^{1/2+z} \frac{2x - 1}{x(1 - x)} dx = +2 \int_0^{2y} \frac{v \, dv}{(1 - v)(1 + v)} \to +\infty \text{ as } z \to \tfrac{1}{2} -.$$

3 After theorem 13.32 it is enough to establish the existence of the improper integral

$$\int_1^{\to \infty} \frac{dx}{x^\alpha}.$$

But

$$\int_1^X \frac{dx}{x^\alpha} = \left[\frac{1}{1-\alpha} x^{1-\alpha} \right]_1^X = \frac{1}{\alpha-1} \left\{ 1 - \frac{1}{X^{\alpha-1}} \right\} \to \frac{1}{\alpha-1} \text{ as } X \to +\infty$$

provided that $\alpha > 1$.

- *Exercise 14.3*

1 We have

$$\log 2 = \int_1^2 \frac{dt}{t} \geqslant \tfrac{1}{2}(2-1) > 0.$$

Observe that

$$\log (2^n) = n \log 2 \to +\infty \text{ as } n \to \infty$$

and hence the logarithm is unbounded above on $(0, \infty)$. Since it is strictly increasing, it follows that $\log x \to +\infty$ as $x \to +\infty$. Also

$$\log (2^{-n}) = -n \log 2 \to -\infty \text{ as } n \to \infty.$$

Thus $\log x \to -\infty$ as $x \to 0+$.

2 Since the logarithm is concave,

$$\log y - \log 1 \leqslant (y-1)$$

for all $y > 0$ (see exercise 12.21(4)).
   Take $y = x^s$. Then, using theorem 14.2(ii),

$$s \log x = \log (x^s) \leqslant x^s - 1 \leqslant x^s.$$

(i) Given $r > 0$, put $s = \tfrac{1}{2}r$ in the above inequality. Then

$$x^{-r} \log x = x^{-r/2}(x^{-s} \log x) \leqslant x^{-r/2}/s \to 0 \text{ as } x \to +\infty.$$

(ii) Put $y = x^{-1}$ in (i).

3 We have $F'(x) = \log x + x \cdot x^{-1} - 1 = \log x$ and hence F is a primitive for the logarithm. By theorem 13.14,

$$\int_0^1 \log (1+x) \, dx = \int_1^2 \log t \, dt = [t \log t - t]_1^2$$

$$= 2 \log 2 - 1 = \log (4/e).$$

As in example 13.17,

$$\sum_{k=1}^n \frac{1}{n} \log \left(1 + \frac{(k-1)}{n} \right) \leqslant \int_0^1 \log (1+x) \, dx \leqslant \sum_{k=1}^n \frac{1}{n} \log \left(1 + \frac{k}{n} \right).$$

Hence

$$\log \left\{1 \cdot \left(1 + \frac{1}{n}\right)\left(1 + \frac{2}{n}\right) \ldots \left(1 + \frac{n-1}{n}\right)\right\}^{1/n} \leqslant \log \left(\frac{4}{e}\right)$$

$$\leqslant \log \left\{\left(1 + \frac{1}{n}\right) \ldots \left(1 + \frac{n}{n}\right)\right\}^{1/n}$$

$$\log \left\{\frac{(n+1)(n+2)\ldots(2n-1)}{n^{n-1}}\right\}^{1/n} \leqslant \log \left(\frac{4}{e}\right)$$

$$\leqslant \log \left\{\frac{(n+1)\ldots(2n)}{n^n}\right\}^{1/n}$$

$$\log \left\{\frac{n(2n)!}{2n \cdot n!}\right\}^{1/n} \frac{1}{n} \leqslant \log \left(\frac{4}{e}\right) \leqslant \log \left\{\frac{(2n)!}{n!}\right\}^{1/n} \frac{1}{n}.$$

It follows that

$$0 \leqslant \log \frac{1}{n} \left\{\frac{(2n)!}{n!}\right\}^{1/n} - \log \left(\frac{4}{e}\right) \leqslant \frac{1}{n} \log 2 \to 0 \text{ as } n \to \infty.$$

4 Consider the function $\phi : (0, \infty) \to \mathbb{R}$ defined by

$$\phi(x) = \log x - \alpha x.$$

If $\alpha \leqslant 0$, $\phi(x) \to -\infty$ as $x \to 0+$ and $\phi(x) \to +\infty$ as $x \to +\infty$ (exercise 14.3(2)). It then follows from the continuity of ϕ that a solution of the equation $\phi(x) = 0$ exists.

The case $\alpha > 0$ is somewhat more interesting. It remains true that $\phi(x) \to -\infty$ as $x \to 0+$ but, for this case, $\phi(x) \to -\infty$ as $x \to +\infty$ (exercise 14.3(2)). Consider the derivative

$$\phi'(x) = \frac{1}{x} - \alpha.$$

From this identity it follows that ϕ is strictly increasing on $(0, 1/\alpha]$ and strictly decreasing on $[1/\alpha, \infty)$. Thus

$$\phi(x) \leqslant \phi \left(\frac{1}{\alpha}\right) = \log \frac{1}{\alpha} - 1 \tag{13}$$

with equality if and only if $x = 1/\alpha$. Because ϕ is continuous on $(0, \infty)$, solutions of $\phi(x) = 0$ exist if and only if the maximum value $\phi(1/\alpha)$ of ϕ is non-negative, i.e.

$$\log (1/\alpha) \geqslant 1 = \log e. \tag{14}$$

Since the logarithm is strictly increasing, (14) holds if and only if $1/\alpha > e$, i.e. $e\alpha \leqslant 1$.

The inequality $e \log x \leqslant x$ $(x > 0)$ is simply (13) with $\alpha = 1/e$.

We prove that the sequence $\langle x_n \rangle$ is decreasing and bounded below by *e.* We have

$$x_{n+1} = e \log x_n \leqslant x_n$$

and so $\langle x_n \rangle$ decreases. If $x_n > e$, then

$$x_{n+1} = e \log x_n > e \log e = e$$

and so the fact that $\langle x_n \rangle$ is bounded below by e follows by induction.

Suppose that $x_n \to \xi$ as $n \to \infty$. Since $x_{n+1} = e \log x_n$ and the logarithm is continuous,

$$\xi = e \log \xi.$$

But this equation holds if and only if $\xi = e$.

5 Apply theorem 13.32 with $f(x) = 1/x$. Then

$$\Delta_n = 1 + \frac{1}{2} + \frac{1}{3} + \ldots + \frac{1}{n} - \int_1^n \frac{dt}{t}$$

and $\Delta_n \to \gamma$ as $n \to \infty$.

As in the solution to exercise 6.26(6),

$$1 - \frac{1}{2} + \frac{1}{3} - \frac{1}{4} + \ldots + \frac{1}{2n-1} - \frac{1}{2n} = \frac{1}{n+1} + \frac{1}{n+2} + \ldots + \frac{1}{2n}$$

$$= \left\{ 1 + \frac{1}{2} + \ldots + \frac{1}{2n} \right\} - \left\{ 1 + \frac{1}{2} + \ldots + \frac{1}{n} \right\}$$

$$= \{ \log 2n + \Delta_{2n} \} - \{ \log n + \Delta_n \}$$

$$= \log 2 + \Delta_{2n} - \Delta_n$$

$$\to \log 2 + \gamma - \gamma \text{ as } n \to \infty.$$

6 (i) $D \{ \log x \}^s = s \{ \log x \}^{s-1} 1/x$ (ii) $D \{ \log \log x \} = \frac{1}{\log x} \cdot \frac{1}{x}$.

If $r \neq 1$, take $s - 1 = -r$ in (i). Then

$$\int_2^X \frac{1}{x (\log x)^r} \, dx = \left[\frac{1}{1-r} (\log x)^{1-r} \right]_2^X \to \begin{cases} +\infty & (r < 1) \\ \dfrac{(\log 2)^{1-r}}{r-1} & (r > 1) \end{cases}$$

as $X \to +\infty$. If $r = 1$, we use (ii). Then

$$\int_2^X \frac{1}{x \log x} \, dx = [\log \log x]_2^X \to +\infty \text{ as } x \to +\infty.$$

The appropriate conclusions about the given series follow from theorem 13.32.

● *Exercise 14.5*

1 (i) Put $X = \exp x$ and $Y = \exp y$. By theorem 14.2(i),

$$\log XY = \log X + \log Y = x + y.$$

Hence

$$\exp(x + y) = \exp\{\log XY\} = XY = (\exp x)(\exp y).$$

Alternatively, consider

$$D\left\{\frac{\exp(x + y)}{\exp x}\right\} = \frac{\exp(x + y).\exp x - \exp x.\exp(x + y)}{(\exp x)^2} = 0$$

for a fixed value of y. We obtain

$$\frac{\exp(x + y)}{\exp x} = c \quad (x \in \mathbb{R})$$

for some constant c. The choice $x = 0$ shows that $c = \exp y$.

(ii) Put $X = \exp x$ in theorem 14.2(ii).

$$\log X^r = r \log X = r \log(\exp x) = rx.$$

Hence

$$\exp(rx) = \exp(\log X^r) = X^r = (\exp x)^r.$$

2 (i) By exercise 14.3(2i), for any $s > 0$,

$$X^{-s} \log X \to 0 \text{ as } X \to +\infty.$$

Hence

$$(\log X)^{-r}X = \{(\log X)X^{-1/r}\}^{-r} \to +\infty \text{ as } X \to +\infty.$$

It only remains to put $X = \exp x$ (see exercise 8.20(2)).

(ii) By exercise 14.3(2ii), for any $s > 0$,

$$X^s \log X \to 0 \text{ as } X \to 0+.$$

$$(\log X)^r X = \{(\log X)X^{1/r}\}^r \to 0 \text{ as } X \to 0+.$$

Put $X = \exp x$ again and the result follows.

3 (i) $\displaystyle\lim_{x \to 0} \frac{\exp x - 1}{x} = \lim_{x \to 0} \frac{\exp x}{1} = \exp 0 = 1.$

(ii) $\displaystyle\lim_{x \to 0} \frac{\log(1 + x)}{x} = \lim_{x \to 0} \frac{(1 + x)^{-1}}{1} = \frac{1}{1 + 0} = 1.$

4 We have

$H'(t) = h(t) \geqslant H(t)$.

Hence

$$\int_0^x \frac{H'(t)}{H(t)} \, dt \geqslant \int_0^x 1 \cdot dt = x \quad (x > 0).$$

Thus

$$[\log H(t)]_0^x \geqslant x.$$

But $H(0) = 1$ and it follows that

$$\log H(x) \geqslant x$$

i.e. $h(x) \geqslant H(x) \geqslant \exp x$.

5 We have

$$f'(x) = 18.3x^2 \exp(18x^3)$$

$$f''(x) = 18.6x \exp(18x^3) + (18.3)^2 x^4 \exp(18x^3).$$

The sign of $f''(x)$ is the same as that of

$$\phi(x) = 2x + 18.3x^4 = 2x(1 + 27x^3).$$

Observe that $\phi(0) = \phi(-\tfrac{1}{3}) = 0$. Since

$$\phi'(x) = 2 + 8.27x^3 = 2 + (6x)^3,$$

it follows that ϕ is strictly increasing for $6x > -2^{1/3}$ and strictly decreasing for $6x < -2^{1/3}$.

Thus ϕ is non-negative on the intervals $(-\infty, -\tfrac{1}{3}]$ and $[0, \infty)$ and so f is convex on these intervals (theorem 12.19). But ϕ is non-positive on $[-\tfrac{1}{3}, 0]$ and so f is concave on this interval.

6 Since f increases on $[0, \infty)$

$$f(x) \leqslant f(k) \quad (k - 1 \leqslant x \leqslant k)$$

$$f(k) \leqslant f(x) \quad (k \leqslant x \leqslant k + 1).$$

Hence

$$\int_{k-1}^k f(x) \, dx \leqslant f(k) \leqslant \int_k^{k+1} f(x) \, dx.$$

But

$$\int_0^n f(x) \, dx = \sum_{k=1}^n \int_{k-1}^k f(x) \, dx; \quad \int_1^{n+1} f(x) \, dx = \sum_{k=1}^n \int_k^{k+1} f(x) \, dx.$$

Apply the result with $f(x) = \log x$.

$$[x \log x - x]_0^n \leqslant \sum_{k=1}^n \log k \leqslant [x \log x - x]_1^{n+1} \tag{15}$$

$$n \log n - n \leqslant \log n! \leqslant (n+1) \log (n+1) - (n+1) + 1$$

$$n \log n - \log n! \leqslant n \leqslant (n+1) \log (n+1) - \log n!$$

$$\log \left\{ \frac{n^n}{n!} \right\} \leqslant n \leqslant \log \left\{ \frac{(n+1)^{n+1}}{n!} \right\}$$

$$\frac{n^n}{n!} \leqslant \exp n \leqslant \frac{(n+1)^{n+1}}{n!}.$$

(Note that (15) involves the use of an improper integral on the left hand side. From exercise 14.3(2), $x \log x \to 0$ as $x \to 0+$.)

- *Exercise 14.7*

1 (i) $a^{x+y} = \exp \{(x+y) \log a\}$

$\qquad = \exp \{x \log a + y \log a\}$

$\qquad = \exp (x \log a) \exp (y \log a) \quad$ (exercise 14.5(1i))

$\qquad = a^x a^y.$

(ii) $(ab)^x = \exp \{x \log (ab)\}$

$\qquad = \exp \{x \log a + x \log b\} \quad$ (theorem 14.2(i))

$\qquad = \exp (x \log a) \exp (x \log b) \quad$ (exercise 14.5(1i))

$\qquad = a^x b^x.$

(iii) $a^{-x} = \exp \{-x \log a\}$

$\qquad = \dfrac{1}{\exp (x \log a)} \quad$ (exercise 14.5(1ii))

$\qquad = \dfrac{1}{a^x}.$

(iv) Note that, since $a^x = \exp (x \log a)$, it follows that $x \log a = \log (a^x)$. Hence

$$a^{xy} = \exp \{xy \log a\} = \exp \{y \log (a^x)\} = (a^x)^y.$$

2 We have

$$D\{\exp (x \log a)\} = \log a . \exp (x \log a) = (\log a) a^x.$$

3 By theorem 8.9 and exercise 14.5(3ii),

$$n \log (1 + xn^{-1}) = x \frac{\log (1 + xn^{-1})}{xn^{-1}} \to x \text{ as } n \to \infty.$$

Since the exponential function is continuous at every point, it follows that

$$\left(1 + \frac{x}{n}\right)^n = \exp\left\{n \log \left(1 + \frac{x}{n}\right)\right\} \to \exp x \text{ as } n \to \infty.$$

4 By exercise 14.3(2i), $n^{-1} \log n \to 0$ as $n \to \infty$. Hence,

$$n^{1/n} = \exp\left\{\frac{1}{n} \log n\right\} \to \exp 0 = 1 \text{ as } n \to \infty.$$

5 (i) Consider

$$\int_1^{\to +\infty} e^{-x^2/2} \, dx.$$

Take $\phi(x) = e^{-x/2}$ in proposition 13.29. We know that

$$\int_1^X e^{-x/2} \, dx = [-2e^{-x/2}]_1^X \to 2e^{-1/2} \text{ as } X \to \infty$$

and

$$e^{-x^2/2} \leqslant e^{-x/2} \quad (x \geqslant 1).$$

A similar argument (with $\phi(x) = e^{x/2}$) establishes the existence of

$$\int_{\to -\infty}^{-1} e^{-x^2/2} \, dx.$$

It follows that the improper integral

$$\int_{\to -\infty}^{\to \infty} e^{-x^2/2} \, dx$$

exists.

(ii) See §17.3.

(iii) We have

$$\int_0^X \frac{x}{1 + x^2} \, dx = [\tfrac{1}{2} \log (1 + x^2)]_0^X$$

$$= \tfrac{1}{2} \log (1 + X^2) \to +\infty \text{ as } X \to +\infty.$$

Hence the improper integral

$$\int_{\to -\infty}^{\to +\infty} \frac{x \, dx}{1 + x^2}$$

does not exist, even though

$$\int_{-X}^{X} \frac{x\,dx}{1+x^2} = 0 \to 0 \text{ as } X \to +\infty.$$

6 (i) The trick is to differentiate the given equation with respect to x keeping y constant, and then with respect to y keeping x constant. For each x and y, we obtain

$$f'(x) = f'(x+y) = f'(y)$$

and it follows that

$$f'(x) = c \quad (x \in \mathbb{R})$$

for some constant c and hence that

$$f(x) = cx + d.$$

Since $f(0) = 2f(0)$, we must have $d = 0$.

(ii) As in (i) we obtain

$$f'(x)f(y) = f(x)f'(y)$$

and therefore

$$\frac{f'(x)}{f(x)} = c \quad (x \in \mathbb{R}).$$

Hence

$$\log f(x) = cx + d.$$

Since $f(0) = f(0)^2$, $d = 0$ and hence

$$f(x) = e^{cx} \quad (x \in \mathbb{R}).$$

(iii) As in (i) we obtain

$$\frac{f'(x)}{y} = \frac{f'(y)}{x}$$

and therefore

$$f'(x) = \frac{c}{x} \quad (x > 0).$$

Hence

$$f(x) = c \log x + d \quad (x > 0).$$

Since $f(1) = 2f(1)$, $d = 0$.

(iv) As in (i) we obtain

$$\frac{f'(x)f(y)}{y} = \frac{f'(y)f(x)}{x}$$

and therefore

$$\frac{f'(x)}{f(x)} = \frac{c}{x} \quad (x > 0)$$

$$\log f(x) = c \log x + d.$$

Since $f(1) = f(1)^2$, $d = 0$ and hence

$$f(x) = x^c \quad (x > 0).$$

- *Exercise 15.6*

1 (i) The interval of convergence is $\mathbb{R}$ because

$$|a_n|^{1/n} = \frac{1}{n} \to 0 \text{ as } n \to \infty.$$

(ii) The interval of convergence is $\{0\}$ because

$$\left|\frac{a_{n+1}}{a_n}\right| = \frac{(n+1)!}{n!} = (n+1) \to +\infty \text{ as } n \to \infty.$$

(iii) The interval of convergence is $\mathbb{R}$. This is most easily seen by comparing the series with the Taylor series expansion for $\exp x$.

(iv) As (iii).

(v) The interval of convergence is $(-4, 4)$. The radius of convergence is equal to 4 because

$$\left|\frac{a_{n+1}}{a_n}\right| = \frac{(n+1)^2}{(2n+2)(2n+1)} = \frac{(n+1)}{2(2n+1)} \to \frac{1}{4} \text{ as } n \to \infty.$$

Note that the series cannot converge at the endpoints $x = 4$ and $x = -4$ of the interval of convergence because $\langle a_n 4^n \rangle$ is strictly increasing and hence cannot tend to zero. We have

$$\frac{a_{n+1}4^{n+1}}{a_n 4^n} = \frac{(n+1)}{(n+\frac{1}{2})} > 1 \quad (n = 1, 2, \ldots).$$

(vi) The interval of convergence is $[-1, 1)$. The radius of convergence is equal to 1 because

$$\left|\frac{a_{n+1}}{a_n}\right| = \frac{\sqrt{n}}{\sqrt{(n+1)}} \to 1 \text{ as } n \to \infty.$$

The series does not converge for $x = 1$.
The series converges for $x = -1$ by theorem 6.13.

2 To employ exercise 13.26(6), we need the nth derivative of the function $f: (-1, \infty) \to \mathbb{R}$ defined by $f(x) = \log(1 + x)$. We have

$$f'(x) = \frac{1}{1 + x}$$

$$f''(x) = -\frac{1}{(1 + x)^2}$$

. . . .

$$f^{(n)}(x) = (n - 1)! \frac{(-1)^{n-1}}{(1 + x)^n}.$$

(A formal proof would require an induction argument.) These derivatives are evaluated at $x = 0$ and from exercise 13.26(6)

$$\log(1 + x) = f(0) + \frac{x}{1!}f'(0) + \ldots + \frac{x^{n-1}}{(n-1)!}f^{(n-1)}(0) + E_n$$

$$= x - \frac{x^2}{2} + \frac{x^3}{3} - \ldots + \frac{(-1)^{n-2}}{(n-1)}x^{n-1} + E_n$$

where

$$E_n = \frac{1}{(n-1)!}\int_0^x (x - t)^{n-1}f^{(n)}(t)\,dt$$

$$= \frac{1}{(n-1)!}\int_0^x (x - t)^{n-1}(n - 1)!\frac{(-1)^{n-1}}{(1 + t)^n}\,dt.$$

These formulae hold for any $x > -1$. We are concerned with the values of x for which $E_n \to 0$ as $n \to \infty$. We therefore have to consider

$$|E_n| = \left|\int_0^x \frac{(x - t)^{n-1}}{(1 + t)^n}\,dt\right|.$$

If $0 \leqslant x \leqslant 1$,

$$|E_n| = \int_0^x \frac{(x - t)^{n-1}}{(1 + t)^n}\,dt \leqslant \int_0^x (x - t)^{n-1}\,dt$$

$$= \frac{x^n}{n} \to 0 \text{ as } n \to \infty.$$

If $-1 < x < 0$,

$$|E_n| = \int_x^0 \frac{(t - x)^{n-1}}{(1 + t)^n}\,dt.$$

It is easily seen that

$$0 \leqslant \frac{t-x}{1+t} \leqslant -x \quad (-1 < x \leqslant t \leqslant 0)$$

and it follows that, for $-1 < x < 0$,

$$|E_n| \leqslant \int_x^0 (-x)^{n-1} \frac{dt}{1+t}$$

$$= (-x)^{n-1}[-\log(1+x)] \to 0 \text{ as } n \to \infty.$$

The series expansion for $\log 2$ is now obtained simply by setting $x = 1$.

3 To obtain the form of the remainder as given in theorem 11.10, we have to write $x = \eta$ in the formula for $f^{(n)}(x)$ in the previous question. This yields

$$E_n = \frac{(-1)^{n-1}}{n} \left(\frac{x}{1+\eta}\right)^n$$

where η lies between 0 and x. If $0 \leqslant x \leqslant 1$, there is no problem. Since $1 + \eta \geqslant 1$,

$$|E_n| \leqslant \frac{1}{n} x^n \to 0 \text{ as } n \to \infty.$$

If $-1 < x < 0$, we can assert no more than

$$|E_n| \leqslant \frac{1}{n} \left(\frac{-x}{1+x}\right)^n.$$

But

$$\frac{-x}{1+x} \geqslant 1 \quad (-1 < x \leqslant -\tfrac{1}{2})$$

and hence we are only able to show that the remainder term tends to zero for values of x satisfying $-\tfrac{1}{2} < x \leqslant 1$.

4 We have

$$R_n = \frac{1}{(n+1)!} + \frac{1}{(n+2)!} + \dots$$

$$= \frac{1}{(n+1)!} \left\{ 1 + \frac{1}{(n+2)} + \frac{1}{(n+2)(n+3)} + \dots \right\}$$

$$< \frac{1}{(n+1)!} \left\{ 1 + \frac{1}{(n+2)} + \frac{1}{(n+2)^2} + \dots \right\}$$

$$= \frac{1}{(n+1)!} \left\{ \frac{1}{1 - (n+2)^{-1}} \right\}$$

$$= \frac{1}{(n+1)!} \frac{n+2}{n+1}.$$

From the power series expansion for e^x with $x = 1$, we know that

$$0 < e - \left\{1 + \frac{1}{1!} + \frac{1}{2!} + \ldots + \frac{1}{n!}\right\} = R_n.$$

We have to take $n = 4$. Observe that

$$R_4 < \frac{1}{5!} \frac{6}{5} = \frac{1}{120} \cdot \frac{6}{5} = \frac{1}{100}.$$

5 Suppose that $e = m/n$ where m and n are natural numbers. Then

$$I = en! - \left\{1 + \frac{1}{1!} + \frac{1}{2!} + \ldots + \frac{1}{n!}\right\} n!$$

is a natural number. But, by the previous question,

$$0 < I = n! R_n < \frac{n!}{(n+1)!} \frac{(n+2)}{(n+1)}$$

$$= \frac{(n+2)}{(n+1)^2} \leq \frac{3}{4}$$

provided that n is a natural number. But there are no natural numbers in the interval $(0, \frac{3}{4})$ and hence we have a contradiction. (Note that

$$\frac{n+2}{(n+1)^2} = \frac{1}{(n+1)^2} + \frac{1}{(n+1)}$$

and hence is obviously decreasing.)

6 We prove by induction that, for any $x \neq 0$,

$$f^{(n)}(x) = P_n\left(\frac{1}{x}\right) e^{-1/x^2} \tag{16}$$

where P_n is a polynomial of degree $3n$. For $n = 0$, there is nothing to prove. Given that (16) holds,

$$f^{(n+1)}(x) = P_n'\left(\frac{1}{x}\right)\left(-\frac{1}{x^2}\right) e^{-1/x^2} + P_n\left(\frac{1}{x}\right) \cdot \frac{2}{x^3} e^{-1/x^2}$$

$$= \left\{P_n'\left(\frac{1}{x}\right)\left(-\frac{1}{x^2}\right) + P_n\left(\frac{1}{x}\right) \frac{2}{x^3}\right\} e^{-1/x^2}.$$

Observe that the polynomial P_{n+1} defined by $P_{n+1}(y) = -y^2 P_n'(y) + 2y^3 P_n(y)$ has degree $3n + 3$ provided that P_n has degree $3n$. This completes the induction argument.

We next consider the existence of the derivatives $f^{(n)}(0)$. Again we proceed by induction. We show that $f^{(n)}(0) = 0$ ($n = 0, 1, 2, \ldots$). This is given for $n = 0$. On the assumption that $f^{(n-1)}(0)$ exists and is zero,

$$\frac{f^{(n-1)}(x) - f^{(n-1)}(0)}{x} = \frac{1}{x} P_{n-1}\left(\frac{1}{x}\right) e^{-1/x^2}.$$

It follows that

$$\lim_{x \to 0+} \left\{ \frac{f^{(n-1)}(x) - f^{(n-1)}(0)}{x} \right\} = \lim_{y \to +\infty} y\, P_{n-1}(y)\, e^{-y^2} = 0$$

by exercise 14.5(2) (exponentials drown powers). The same argument also applies with the left hand limit. It follows that $f^{(n)}(0)$ exists and is zero as required.

Since $f^{(n)}(0) = 0$ ($n = 0, 1, 2, \ldots$), the Taylor series expansion of f about the point 0, is

$$0 + 0 + 0 + 0 + 0 + \ldots$$

This converges to $f(x)$ only when $x = 0$.

- *Exercise 15.10*

1 The radius of convergence R of the power series

$$f(x) = \sum_{n=0}^{\infty} \frac{\alpha(\alpha - 1) \ldots (\alpha - n + 1)}{n!} x^n$$

is given by

$$\frac{1}{R} = \lim_{n \to \infty} \frac{|\alpha(\alpha - 1) \ldots (\alpha - n)|}{(n+1)!} \frac{n!}{|\alpha(\alpha - 1) \ldots (\alpha - n + 1)|}$$

$$= \lim_{n \to \infty} \frac{|\alpha - n|}{n+1} = 1.$$

For $|x| < 1$ we may therefore apply proposition 15.8 and obtain

$$f'(x) = \sum \frac{\alpha(\alpha - 1) \ldots (\alpha - n + 1)}{n!} n x^{n-1}.$$

Hence,

$$(1 + x) f'(x)$$

$$= \sum_{n=1}^{\infty} \frac{\alpha(\alpha - 1) \ldots (\alpha - n + 1)}{(n-1)!} x^{n-1} + \sum_{n=1}^{\infty} \frac{\alpha(\alpha - 1) \ldots (\alpha - n + 1)}{(n-1)!} x^n$$

$$= \alpha + \sum_{n=1}^{\infty} \left\{ \frac{\alpha(\alpha - 1) \ldots (\alpha - n)}{n!} + \frac{\alpha(\alpha - 1) \ldots (\alpha - n + 1)}{(n-1)!} \right\} x^n$$

$$= \alpha + \sum_{n=1}^{\infty} \frac{\alpha(\alpha-1)\dots(\alpha-n)}{(n-1)!} \left\{ \frac{1}{n} + \frac{1}{\alpha-n} \right\} x^n$$

$$= \alpha \left\{ 1 + \sum_{n=1}^{\infty} \frac{\alpha(\alpha-1)\dots(\alpha-n+1)}{n!} x^n \right\}$$

$$= \alpha f(x)$$

It follows that, for $|x| < 1$,

$$D\{(1+x)^{-\alpha}f(x)\} = -\alpha(1+x)^{-\alpha-1}f(x) + (1+x)^{-\alpha}f'(x) = 0$$

and therefore

$$f(x) = c(1+x)^\alpha \quad (|x|<1)$$

for some constant c. But $f(0) = 1$ and hence $c = 1$.

2 The radius of convergence R is given by

$$\frac{1}{R} = \lim_{n \to \infty} \frac{(n+1)^2}{n^2} = 1.$$

We know that

$$\sum_{n=0}^{\infty} x^n = \frac{1}{1-x} \quad (|x|<1).$$

Hence, using proposition 15.8,

$$\sum_{n=1}^{\infty} nx^{n-1} = \frac{1}{(1-x)^2} \quad (|x|<1)$$

$$\sum_{n=2}^{\infty} n(n-1)x^{n-2} = \frac{2}{(1-x)^3} \quad (|x|<1).$$

Therefore,

$$\sum_{n=0}^{\infty} n^2 x^n = \sum_{n=1}^{\infty} \{n(n-1)+n\}x^n$$

$$= x^2 . \sum_{n=2}^{\infty} n(n-1)x^{n-2} + x . \sum_{n=1}^{\infty} nx^{n-1}$$

$$= \frac{2x^2}{(1-x)^3} + \frac{x}{(1-x)^2}$$

$$= \frac{2x^2 + x(1+x)}{(1-x)^3} = \frac{x(1+x)}{(1-x)^3} \quad (|x|<1).$$

3 The power series

$$F(y) = \sum_{n=0}^{\infty} \frac{a_n}{(n+1)} y^{n+1}$$

has the same radius of convergence as the given power series because

$$\limsup_{n \to \infty} \left| \frac{a_n}{(n+1)} \right|^{1/n} = \limsup_{n \to \infty} |a_n|^{1/n}.$$

By proposition 15.8, if $x \in I$ but is not an endpoint,

$$F'(x) = \sum_{n=0}^{\infty} \frac{a_n}{(n+1)} \cdot (n+1)x^n = f(x).$$

Hence

$$F(y) - F(0) = \int_0^y f(x)\, dx.$$

Observe that $F(0) = 0$.

4 The function $f: (-\infty, 1) \to \mathbb{R}$ defined by

$$f(x) = -\frac{\log(1-x)}{x} \tag{17}$$

has the Taylor series expansion

$$f(x) = 1 + \frac{x}{2} + \frac{x^2}{3} + \frac{x^3}{4} + \dots \quad (-1 \leqslant x < 1).$$

(Note that formula (17) does not make sense when $x = 0$, but it is natural to define $f(0) = 1$.)

If $0 < y < 1$ we obtain from the previous question

$$\int_0^y f(x)\, dx = y + \frac{y^2}{2^2} + \frac{y^3}{3^3} + \dots$$

and hence

$$\lim_{y \to 1-} \int_0^1 f(x)\, dx = \lim_{y \to 1-} \sum_{n=1}^{\infty} \frac{y^n}{n^2}.$$

But the final power series converges for $y = 1$, and from proposition 15.8 we know that the sum of a power series is continuous on its interval of convergence. Thus

$$-\int_0^1 \frac{\log(1-x)}{x}\, dx = \sum_{n=1}^{\infty} \frac{1}{n^2}.$$

5 We consider the power series

$$g(x) = \sum_{n=0}^{\infty} (a_n - b_n)(x - \xi)^n.$$

If $g(x) = 0$ $(x \in I)$, then, using proposition 15.8,

$$0 = g(\xi) = a_0 - b_0$$

$$0 = g'(\xi) = 1(a_1 - b_1)$$

$$0 = g''(\xi) = 2(a_2 - b_2)$$

$$0 = g'''(\xi) = 3.2(a_3 - b_3)$$

$$\dots$$

and the result follows.

6 By proposition 15.8,

$$f'(x) = \sum_{n=0}^{\infty} n a_n x^{n-1}.$$

From the previous question it follows that

$$a_{n-1} = n a_n \quad (n = 1, 2, \dots)$$

as required. But then

$$a_1 = \frac{a_0}{1}$$

$$a_2 = \frac{1}{2} a_1 = \frac{1}{2.1} a_0$$

$$a_3 = \frac{1}{3} a_2 = \frac{1}{3.2.1} a_0$$

and, in general,

$$a_n = \frac{a_0}{n!} \quad (n = 1, 2, \dots).$$

Thus

$$f(x) = a_0 \sum_{n=0}^{\infty} \frac{x^n}{n!} = a_0 e^x.$$

● *Exercise 16.3*

1 (i) $\cos 0 = 1 - \dfrac{0^2}{2!} + \dfrac{0^4}{4!} - \dots = 1$

(ii) $\sin 0 = 0 - \dfrac{0^3}{3!} + \dfrac{0^5}{5!} - \dots = 0$

(iii) $\cos(-x) = 1 - \dfrac{(-x)^2}{2!} + \dfrac{(-x)^4}{4!} - \dots = \cos x$

(iv) $\sin(-x) = (-x) - \dfrac{(-x)^3}{3!} + \dfrac{(-x)^5}{5!} - \ldots = -\sin x.$

We appeal to proposition 15.8. Then

$$D\cos x = D\left\{1 - \frac{x^2}{2!} + \frac{x^4}{4!} - \ldots\right\} = -\frac{2x}{2!} + \frac{4x^3}{4!} - \frac{6x^5}{6!} + \ldots$$

$$= -x + \frac{x^3}{3!} - \frac{x^5}{5!} + \ldots = -\sin x.$$

$$D\sin x = D\left\{x - \frac{x^3}{3!} + \frac{x^5}{5!} - \ldots\right\} = 1 - \frac{3x^2}{3!} + \frac{5x^4}{5!} - \ldots$$

$$= 1 - \frac{x^2}{2!} + \frac{x^4}{4!} - \ldots = \cos x.$$

2 We have

$$g'(x) = \cos(x+y) - \cos x \cos y + \sin x \sin y = h(x)$$

$$h'(x) = -\sin(x+y) + \sin x \cos y + \cos x \sin y = -g(x).$$

Hence

$$D\{g(x)^2 + h(x)^2\} = 2g(x)g'(x) + 2h(x)h'(x)$$

$$= 2g(x)h(x) - 2h(x)g(x) = 0$$

for all x. It follows that

$$g(x)^2 + h(x)^2 = g(0)^2 + h(0)^2 = 0$$

for all x. Since $\{g(x)\}^2 \geqslant 0$ and $\{h(x)\}^2 \geqslant 0$, we may conclude that, for all values of x,

$$g(x) = 0$$

and

$$h(x) = 0.$$

3 (i) $1 = \cos(x - x) = \cos x \cdot \cos(-x) - \sin x \cdot \sin(-x).$

(ii) $0 \leqslant \cos^2 x \leqslant \cos^2 x + \sin^2 x = 1$. Hence $|\cos x| \leqslant 1$. Similarly $|\sin x| \leqslant 1.$

(iii) $\sin 2x = \sin x \cdot \cos x + \cos x \cdot \sin x$

(iv) $\cos 2x = \cos x \cdot \cos x - \sin x \cdot \sin x.$

4 (i) $\displaystyle\lim_{x \to 0} \frac{\sin x}{x} = \lim_{x \to 0} \frac{\cos x}{1} = \cos 0 = 1.$

(ii) $\displaystyle\lim_{x \to 0} \frac{1 - \cos x}{x^2} = \lim_{x \to 0} \frac{\sin x}{2x} = \frac{1}{2}.$

5 By the mean value theorem, there exists ξ between 0 and x such that

$$\frac{\sin x - \sin 0}{x - 0} = \cos \xi$$

and we know, from question 3 above, that $|\cos \xi| \leqslant 1$.

The convergence of $\sum_{n=1}^{\infty} \sin (1/n^2)$ may be deduced from the comparison test because

$$\left| \sin \left(\frac{1}{n^2} \right) \right| \leqslant \frac{1}{n^2} \quad (n = 1, 2, \ldots).$$

Alternatively, we may appeal to exercise 6.26(2). From Question 4(i) above

$$\lim_{n \to \infty} \frac{\sin (1/n^2)}{1/n^2} = 1; \qquad \lim_{n \to \infty} \frac{\sin (1/n)}{1/n} = 1.$$

The latter result shows that $\sum_{n=1}^{\infty} \sin (1/n)$ diverges.

6 From exercise 13.26(5) it follows that, for some ξ between m and n,

$$\int_m^n \frac{\sin t}{t} \, dt = \frac{1}{m} \int_m^\xi \sin t \, dt + \frac{1}{n} \int_\xi^n \sin t \, dt$$

$$= \frac{1}{m} \{ \cos m - \cos \xi \} + \frac{1}{n} \{ \cos \xi - \cos n \}.$$

Hence

$$\left| \int_m^n \frac{\sin t}{t} \, dt \right| \leqslant \frac{4}{m}.$$

Let $\epsilon > 0$ be given and choose $N = 4/\epsilon$. Then for any $n > m > N$,

$$\left| \int_1^n \frac{\sin t}{t} \, dt - \int_1^m \frac{\sin t}{t} \, dt \right| \leqslant \frac{4}{m} < \frac{4}{N} = \epsilon$$

and it follows that we are dealing with a Cauchy sequence. Hence

$$\lim_{n \to \infty} \int_1^n \frac{\sin t}{t} \, dt$$

exists (theorem 5.19).

● *Exercise 16.5*

1 From the discussion of §16.4 we know that $\cos x \geqslant 0 \; (-\tfrac{1}{2}\pi \leqslant x \leqslant \tfrac{1}{2}\pi)$. Since $\sin (x + \tfrac{1}{2}\pi) = \cos x$, it follows immediately that

$\sin x \geqslant 0 \; (0 \leqslant x \leqslant \pi)$. But

$$D(\cos x) = -\sin x$$

and thus the cosine function decreases on $[0, \pi]$. Because $\cos(x + \pi) = -\cos x$, we may also deduce that the cosine function increases on $[\pi, 2\pi]$. Further,

$$D^2(\cos x) = -\cos x.$$

Thus $D^2(\cos x) \leqslant 0 \; (-\frac{1}{2}\pi \leqslant x \leqslant \frac{1}{2}\pi)$ and so the cosine function is concave on $[-\frac{1}{2}\pi, \frac{1}{2}\pi]$. Because $\cos(x + \pi) = -\cos x$, $D^2(\cos x) \geqslant 0$ $(\frac{1}{2}\pi \leqslant x \leqslant \frac{3}{2}\pi)$ and so the cosine function is convex on $[\frac{1}{2}\pi, \frac{3}{2}\pi]$.

2   We have

$$\tan(x + \pi) = \frac{\sin(x + \pi)}{\cos(x + \pi)} = \frac{-\sin x}{-\cos x} = \tan x$$

(provided that $\cos x \neq 0$). Also

$$D \tan x = \frac{\cos x . \cos x - \sin x \, (-\sin x)}{\cos^2 x} = \frac{1}{\cos^2 x}$$

(provided that $\cos x \neq 0$). Since $\cos x$ does not vanish on $(-\frac{1}{2}\pi, \frac{1}{2}\pi)$, it follows that $\tan x$ is strictly increasing on $(-\frac{1}{2}\pi, \frac{1}{2}\pi)$. We have $\cos x \to 0$ as $x \to \frac{1}{2}\pi -$ and $\sin x \to 1$ as $x \to \frac{1}{2}\pi -$. Since the sine and cosine functions are positive on $(0, \frac{1}{2}\pi)$, it follows that

$$\tan x = \frac{\sin x}{\cos x} \to +\infty \text{ as } x \to \tfrac{1}{2}\pi -.$$

Similarly for $\tan x \to -\infty$ as $x \to -\frac{1}{2}\pi +$.

3   We have

$$D \sin x = \cos x > 0 \quad (-\tfrac{1}{2}\pi < x < \tfrac{1}{2}\pi)$$

and hence the sine function is strictly increasing on $[-\frac{1}{2}\pi, \frac{1}{2}\pi]$. Since $\sin(-\frac{1}{2}\pi) = -\sin(\frac{1}{2}\pi)$ and $\sin(\frac{1}{2}\pi) = 1$, it follows that the image of $[-\frac{1}{2}\pi, \frac{1}{2}\pi]$ is $[-1, 1]$.

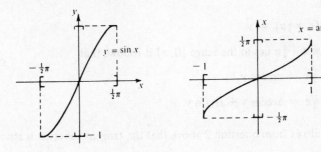

$$D(\arcsin y) = \frac{1}{D \sin x} = \frac{1}{\cos x}$$

where $x = \arcsin y$, i.e. $\sin x = y$. Since $\cos^2 x + \sin^2 x = 1$, we have

$$\cos x = \pm \sqrt{(1 - \sin^2 x)} = \pm \sqrt{(1 - y^2)}.$$

We take the *positive* sign because we know that the arcsine function increases. Thus

$$D(\arcsin y) = \frac{1}{\sqrt{(1 - y^2)}} \quad (-1 < y < 1).$$

We have

$$D \cos x = -\sin x < 0 \quad (0 < x < \pi).$$

Also $\cos (0) = 1$ and $\cos (\pi) = -1$. Hence the image of $[0, \pi]$ under the cosine function is $[-1, 1]$.

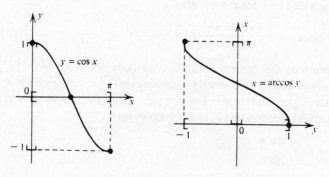

$$D(\arccos y) = \frac{1}{D \cos x} = \frac{1}{-\sin x} = \frac{-1}{\sqrt{(1 - y^2)}}.$$

We take the negative sign because the arccosine function decreases.
  Note that

$$\cos (x + \tfrac{1}{2}\pi) = -\sin x = \sin (-x).$$

  If x lies in the range $[-\tfrac{1}{2}\pi, \tfrac{1}{2}\pi]$ we may write $-x = \arcsin y$. But then

$$\cos (x + \tfrac{1}{2}\pi) = y.$$

Since $x + \tfrac{1}{2}\pi$ lies in the range $[0, \pi]$ it follows that

$$x + \tfrac{1}{2}\pi = \arccos y.$$

i.e. $\tfrac{1}{2}\pi = \arccos y + \arcsin y.$

4 It follows from question 2 above that the tangent function is strictly

increasing on $(-\tfrac{1}{2}\pi, \tfrac{1}{2}\pi)$ and $\tan x \to +\infty$ as $x \to \tfrac{1}{2}\pi-$ and $\tan x \to -\infty$ as $x \to -\tfrac{1}{2}\pi+$. Thus the image of $(-\tfrac{1}{2}\pi, \tfrac{1}{2}\pi)$ is $\mathbb{R}$.

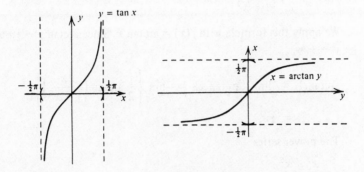

$$D \arctan y = \frac{1}{D \tan x} = \cos^2 x \quad (y = \tan x).$$

But $\cos^2 x + \sin^2 x = 1$ and so $1 + \tan^2 x = 1/\cos^2 x$. Thus

$$D \arctan y = \frac{1}{1 + \tan^2 x} = \frac{1}{1 + y^2}.$$

Finally,

$$\int_0^X \frac{dx}{1 + x^2} = [\arctan x]_0^X = \arctan X \to \frac{\pi}{2} \text{ as } X \to +\infty.$$

5 Differentiate with respect to x keeping y constant and then with respect to y keeping x constant. If $xy < 1$,

$$f'(x) = f'\left(\frac{x+y}{1-xy}\right)\left\{\frac{1(1-xy) - (x+y)(-y)}{(1-xy)^2}\right\}.$$

Hence

$$f'(x) = f'\left(\frac{x+y}{1-xy}\right)\left\{\frac{1+y^2}{(1-xy)^2}\right\}.$$

Also

$$f'(y) = f'\left(\frac{x+y}{1-xy}\right)\left\{\frac{1+x^2}{(1-xy)^2}\right\}.$$

Thus $f'(x)(1 + x^2) = f'(y)(1 + y^2)$ $(xy < 1)$. Hence

$$f'(x) = \frac{c}{1+x^2} \quad (x \in \mathbb{R}).$$

Therefore $f(x) = c \arctan x + d$ $(x \in \mathbb{R})$. But $2f(0) = f(0)$ and so $d = 0$.

If $x = y$, we obtain

$$f(x) = \frac{1}{2} f\left(\frac{2x}{1-x^2}\right) \quad (x > 1).$$

We apply this formula with $f(x) = \arctan x$. Since $\arctan y \to \frac{1}{2}\pi$ as $y \to +\infty$,

$$\arctan 1 = \lim_{x \to 1-} \{\arctan x\} = \lim_{x \to 1-} \left\{\frac{1}{2}\arctan\left(\frac{2x}{1-x^2}\right)\right\}$$

$$= \tfrac{1}{4}\pi.$$

The power series

$$x - \frac{x^3}{5} + \frac{x^5}{5} - \ldots$$

is easily seen to have interval of convergence $(-1, 1]$ (for $x = 1$, use theorem 6.13). We apply proposition 15.8. Then, for $|x| < 1$,

$$D\left\{\arctan x - x + \frac{x^3}{3} - \frac{x^5}{5} + \ldots\right\}$$

$$= \frac{1}{1+x^2} - 1 + x^2 - x^4 + \ldots$$

$$= \frac{1}{1+x^2} - \frac{1}{1+x^2} = 0.$$

It follows that

$$\arctan x = C + x - \frac{x^3}{3} + \frac{x^5}{5} - \ldots \quad (|x| < 1).$$

But $\arctan 0 = 0$ and so $C = 0$.

We know from proposition 15.8 that the sum of a power series is continuous on its interval of convergence. Hence

$$\tfrac{1}{4}\pi = \arctan 1 = \lim_{x \to 1-} \{\arctan x\}$$

$$= \lim_{x \to 1-} \left\{x - \frac{x^3}{3} + \frac{x^5}{5} - \ldots\right\}$$

$$= 1 - \tfrac{1}{3} + \tfrac{1}{5} - \ldots$$

6 (i) Consider the sequence $\langle x_n \rangle$ defined by

$$x_n = \{\tfrac{1}{2}\pi + 2n\pi\}^{-1}.$$

Then $x_n \to 0$ as $n \to \infty$. But

$$f(x_n) = \sin\left(\tfrac{1}{2}\pi + 2n\pi\right) = 1 \neq f(0) \text{ as } n \to \infty.$$

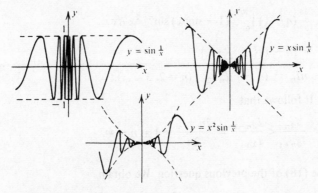

(ii) We have $|g(x)| \leqslant |x|$ and hence $g(x) \to 0$ as $x \to 0$ by the sandwich theorem. However,

$$\frac{g(0 + x_n) - g(0)}{x_n} \to 1 \text{ as } n \to \infty$$

$$\frac{g(0 - x_n) - g(0)}{-x_n} \to -1 \text{ as } n \to \infty.$$

(iii) We have $|h(x)| \leqslant x^2$. Hence

$$\left| \frac{h(x) - h(0)}{x} - 0 \right| \leqslant |x| \to 0 \text{ as } x \to 0$$

and hence $h'(0) = 0$ by the sandwich theorem.

- *Exercise 17.5*

1 We have $0 < \sin x < 1$ $(0 < x < \tfrac{1}{2}\pi)$. Hence

$$0 < \sin^{n+1}x < \sin^n x$$

and therefore

$$0 < I_{n+1} < I_n \quad (n = 0, 1, 2, \ldots).$$

To obtain the recurrence relation $nI_n = (n-1)I_{n-2}$, we integrate by parts. For $n \geqslant 2$,

$$I_n = \int_0^{\pi/2} \sin^n x$$

$$= \int_0^{\pi/2} \sin x \cdot \sin^{n-1}x \, dx$$

$$= [-\cos x \cdot \sin^{n-1}x]_0^{\pi/2} + \int_0^{\pi/2} \cos x \cdot (n-1)\sin^{n-2}x \cdot \cos x \, dx$$

$$= (n-1)\int_0^{\pi/2} (1 - \sin^2 x)\sin^{n-2}x\,dx$$

$$= (n-1)\{I_{n-2} - I_n\}.$$

i.e. $nI_n = (n-1)I_{n-2}$ $(n = 2, 3, \ldots)$. $\qquad\qquad$ (18)

It follows that

$$1 < \frac{I_{2n}}{I_{2n+1}} < \frac{I_{2n-1}}{I_{2n+1}} = \frac{2n+1}{2n} \to 1 \text{ as } n \to \infty.$$

2 Use (18) of the previous question. We obtain

$$I_{2n} = \frac{(2n-1)}{2n}I_{2n-2} = \frac{(2n-1)(2n-3)}{2n(2n-2)}I_{2n-4}$$

$$= \frac{(2n-1)(2n-3)\ldots 1}{2n(2n-2)\ldots 2}I_0$$

$$= \frac{2n(2n-1)(2n-2)(2n-3)\ldots 2.1}{(2n)^2(2n-2)^2\ldots 2^2}\cdot\int_0^{\pi/2}dx$$

$$= \frac{(2n)!}{(2^n n!)^2}\frac{\pi}{2} \quad (n = 0, 1, 2, \ldots).$$

$$I_{2n+1} = \frac{2n}{2n+1}I_{2n-1} = \frac{2n(2n-2)}{(2n+1)(2n-1)}I_{2n-3}$$

$$= \frac{2n(2n-2)\ldots 2}{(2n+1)(2n-1)\ldots 3}I_1$$

$$= \frac{(2n)^2(2n-2)^2\ldots 2^2}{(2n+1)(2n)(2n-1)\ldots 3.2}\cdot\int_0^{\pi/2}\sin x\,dx$$

$$= \frac{(2^n n!)^2}{(2n+1)!}[-\cos x]_0^{\pi/2} = \frac{(2^n n!)^2}{(2n+1)!} \quad (n = 0, 1, 2, \ldots).$$

It follows that

$$\frac{I_{2n}}{I_{2n+1}} = \frac{(2n)!}{(2^n n!)^2}\cdot\frac{\pi}{2}\cdot\frac{(2n+1)!}{(2^n n!)^2} = \frac{\pi}{2}(2n+1)\frac{\{(2n)!\}^2}{\{2^n n!\}^4}$$

$$\sim \frac{\pi}{2}(2n+1)\frac{\{C(2n)^{2n+1/2}e^{-2n}\}^2}{\{2^n Cn^{n+1/2}e^{-n}\}^4}$$

$$= \frac{\pi}{2}(2n+1)C^{-2}\frac{2^{4n}2.n^{4n+1}e^{-4n}}{2^{4n}.n^{4n+2}e^{-4n}}$$

$$= \pi\left(\frac{2n+1}{n}\right)C^{-2} \to 2\pi C^{-2} \text{ as } n \to \infty.$$

By the previous question, $2\pi C^{-2} = 1$, i.e. $C = \sqrt{(2\pi)}$.

3 The fact that $\langle d_n - (12n)^{-1} \rangle$ increases was demonstrated in §17.3. As in §17.3,

$$d_n - d_{n+1} = f(x) = \frac{x^2}{3} + \frac{x^4}{5} + \frac{x^6}{7} + \ldots \quad \left(x = \frac{1}{2n+1}\right).$$

Hence

$$d_n - d_{n+1} > \frac{x^2}{3} = \frac{1}{3}\frac{1}{(2n+1)^2} = \frac{1}{12n^2 + 12n + 3}.$$

But

$$\frac{1}{12n+1} - \frac{1}{12(n+1)+1} = \frac{12n + 13 - 12n - 1}{(12n+1)(12n+13)}$$

$$= \frac{12}{(12n+1)(12n+13)}.$$

It follows that

$$\left\{d_n - \frac{1}{12n+1}\right\} - \left\{d_{n+1} - \frac{1}{12(n+1)+1}\right\}$$

$$> \frac{(12n+1)(12n+13) - 12(12n^2 + 12n + 3)}{3(12n+1)(12n+13)(2n+1)^2}.$$

The numerator is equal to

$$12n(14-12) + (13-36) = 24n - 23 > 0 \quad (n = 1, 2, \ldots)$$

and therefore the sequence $\langle d_n - (12n+1)^{-1} \rangle$ decreases.

We know that $d_n \to d$ as $n \to \infty$ and therefore

$$d_n - \frac{1}{12n} < d < d_n - \frac{1}{12n+1} \quad (n = 1, 2, \ldots).$$

Taking exponentials and noting that $C = \exp d = \sqrt{(2\pi)}$, we obtain

$$e^{-1/12n} < \frac{\sqrt{(2\pi)} \cdot n^{n+1/2} e^{-n}}{n!} < e^{-1/(12n+1)}$$

as required.

We may deduce that

$$n!(1 - e^{-1/(12n+1)}) < n! - \sqrt{(2\pi)}n^{n+1/2} e^{-n} < n!(1 - e^{-1/12n}).$$

Since $e^x - 1$ is approximately equal to x for small values of x (Why?), the maximum possible error in approximating to $n!$ is about

$$\frac{1}{12n} n!.$$

When $n = 100$, this is small compared with $n!$ (though, of course, it will be a very large number compared with those usually encountered).

4 Consider, for $0 < \delta < \Delta$, the value of

$$\int_\delta^\Delta t^{x-1} e^{-t} dt = \left[\frac{t^x}{x} e^{-t} \right]_\delta^\Delta + \int_\delta^\Delta \frac{t^x}{x} e^{-t} dt$$

$$= \frac{1}{x} \{ \Delta^x e^{-\Delta} - \delta^x e^{-\delta} \} + \frac{1}{x} \int_\delta^\Delta t^x e^{-t} dt.$$

Observe that $\Delta^x e^{-\Delta} \to 0$ as $\Delta \to +\infty$ (exponentials drown powers) and $\delta^x e^{-\delta} \to 0$ as $\delta \to 0+$. It follows that

$$\Gamma(x) = \frac{1}{x} \Gamma(x + 1)$$

as required.

5 Let $0 < \alpha < a \leqslant x \leqslant y \leqslant b < \beta$ and let $0 < \delta < \Delta$. Then

$$\left| \int_\delta^\Delta t^{x-1} e^{-t} dt - \int_\delta^\Delta t^{y-1} e^{-t} dt \right| \leqslant \int_\delta^\Delta |t^{x-1} - t^{y-1}| e^{-t} dt. \qquad (19)$$

We may apply the mean value theorem and obtain

$$\frac{t^{x-1} - t^{y-1}}{x - y} = (\log t) t^{\xi-1} \qquad (20)$$

for some ξ between x and y.

For any $r > 0$, $t^{-r} \log t \to 0$ as $t \to +\infty$ and $t^r \log t \to 0$ as $t \to 0+$. It follows from (20) that we can find an H such that

$$|t^{x-1} - t^{y-1}| \leqslant H\{t^{\alpha-1} + t^{\beta-1}\}|x - y|$$

and hence it follows from (19) that

$$|\Gamma(x) - \Gamma(y)| \leqslant H|x - y|\{\Gamma(\alpha) + \Gamma(\beta)\}$$

and the continuity of the gamma function then follows from the sandwich theorem.

6 We show that $\log \Gamma(\frac{1}{2}x + \frac{1}{2}y) \leqslant \frac{1}{2} \log \Gamma(x) + \frac{1}{2} \log \Gamma(y)$ and appeal to exercise 12.21(6). By the Schwarz inequality (theorem 13.25), if $0 < \delta < \Delta$,

$$\left\{ \int_\delta^\Delta t^{(x+y-2)/2} e^{-t} dt \right\}^2 = \left\{ \int_\delta^\Delta (t^{(x-1)/2} e^{-t/2})(t^{(y-1)/2} e^{-t/2}) dt \right\}^2$$

$$\leqslant \left\{ \int_\delta^\Delta t^{x-1} e^{-t} dt \right\} \left\{ \int_\delta^\Delta t^{y-1} e^{-t} dt \right\}.$$

Thus

$$\left\{ \Gamma \left(\frac{x+y}{2} \right) \right\}^2 \leqslant \{\Gamma(x)\}\{\Gamma(y)\}$$

and the result follows.

- *Exercise 17.8*

1 Make the change of variable $-t = \log u$ in the integral

$$\int_\delta^\Delta t^{x-1} e^{-t} \, dt.$$

2 We use the inequality

$$(y + n)^y n! = (y + n)^y \Gamma(n + 1) \leqslant \Gamma(y + n + 1) \leqslant (n + 1)^y \Gamma(n + 1)$$

$$= (n + 1)^y n! \tag{21}$$

established in the proof of theorem 17.7 for $0 \leqslant y < 1$ and $n \in \mathbb{N}$.

Given x, we let $n + 1$ be the largest natural number satisfying $n + 1 \leqslant x$ and write $x = y + n + 1$. Then $0 \leqslant y < 1$.

Consider

$$\frac{\Gamma(x)}{\sqrt{(2\pi)} x^x x^{-1/2} e^{-x}} \leqslant \frac{(n + 1)^y n!}{\sqrt{(2\pi)} x^x x^{-1/2} e^{-x}} \sim \frac{(n + 1)^y \sqrt{(2\pi)} n^n n^{1/2} e^{-n}}{\sqrt{(2\pi)} x^x x^{-1/2} e^{-x}}$$

$$= \frac{(n + 1)^y n^n n^{1/2} e^{-n}}{(y + n + 1)^{y+n} (y + n + 1)^{1/2} e^{-y-n-1}}$$

$$= \left\{ \frac{n + 1}{y + n + 1} \right\}^y \left\{ \frac{n}{y + n + 1} \right\}^{1/2} \left\{ 1 + \frac{y + 1}{n} \right\}^{-n} e^{y+1}$$

$$\rightarrow 1 . 1 . \frac{1}{e^{y+1}} e^{y+1} = 1 \text{ as } n \rightarrow \infty.$$

A similar argument for the left hand side of inequality (21) completes the proof.

3 We have

$$\lim_{z \to 0} \left\{ \frac{\log(1 + z) - z}{z^2} \right\} = \lim_{z \to 0} \left\{ \frac{(1 + z)^{-1} - 1}{2z} \right\}$$

$$= \lim_{z \to 0} \left\{ \frac{-(1 + z)^{-2}}{2} \right\} = -\frac{1}{2}.$$

Consider

$$\sqrt{u}\, f_u(u + x\sqrt{u}) = \frac{\sqrt{u}}{\Gamma(u)} (u + x\sqrt{u})^{u-1} e^{-u-x\sqrt{u}}$$

$$\sim \frac{\sqrt{u}(u + x\sqrt{u})^{u-1} e^{-u-x\sqrt{u}}}{\sqrt{(2\pi)} u^u u^{-1/2} e^{-u}}$$

$$= \frac{1}{\sqrt{(2\pi)}} \cdot \left(1 + \frac{x}{\sqrt{u}}\right)^{u-1} e^{-x\sqrt{u}}$$

$$= \frac{1}{\sqrt{(2\pi)}} \left(1 + \frac{x}{\sqrt{u}}\right)^{-1} \exp\left\{u \log\left(1 + \frac{x}{\sqrt{u}}\right) - x\sqrt{u}\right\}.$$

Write $z = x/\sqrt{u}$. Then

$$u \log\left(1 + \frac{x}{\sqrt{u}}\right) - x\sqrt{u} = \frac{x^2}{z^2} \log(1 + z) - \frac{x^2}{z}$$

$$= x^2 \left\{\frac{\log(1 + z) - z}{z^2}\right\}$$

$$\to -\tfrac{1}{2}x^2 \text{ as } z \to 0$$

and the result follows.

4 We use proposition 13.29 to check that the improper integral exists for $x > 0$ and $y > 0$. The inequalities

$$t^{x-1}(1 - t)^{y-1} < t^{x-1} \quad (0 < t < 1)$$

$$t^{x-1}(1 - t)^{y-1} < (1 - t)^{y-1} \quad (0 < t < 1)$$

suffice for this purpose. The fact that $B(x, y)$ is a continuous function of x (and of y) is proved in the same manner as exercise 17.5(5). The fact that its logarithm is convex is proved in the same manner as exercise 17.5(6).

5 After the previous question, we need only show that $f(1) = 1$ and $f(x + 1) = xf(x)$. Now

$$f(1) = \frac{\Gamma(1 + y)}{\Gamma(y)} B(1, y) = y \int_0^{\to 1} (1 - t)^{y-1} \, dt = 1.$$

Also, if $0 < \alpha < \beta < 1$

$$\int_\alpha^\beta t^x (1 - t)^{y-1} \, dt = \left[-t^x \frac{1}{y}(1 - t)^y\right]_\alpha^\beta + \int_\alpha^\beta xt^{x-1} \frac{1}{y}(1 - t)^y \, dt.$$

It follows that $B(x + 1, y) = xy^{-1}B(x, y + 1)$. But

$$B(x, y + 1) = \int_{\to 0}^{\to 1} t^{x-1}(1 - t)^y \, dt = \int_{\to 0}^{\to 1} t^{x-1}(1 - t)^{y-1}(1 - t) \, dt$$

$$= B(x, y) - B(x + 1, y).$$

Hence

$$B(x + 1, y) = \frac{x}{y} \{B(x, y) - B(x + 1, y)\}$$

$$\left(1 + \frac{x}{y}\right) B(x + 1, y) = \frac{x}{y} B(x, y)$$

$$B(x + 1, y) = \frac{x}{x + y} B(x, y).$$

It follows that

$$f(x + 1) = \frac{\Gamma(x + 1 + y)}{\Gamma(y)} B(x + 1, y)$$

$$= (x + y) \frac{\Gamma(x + y)}{\Gamma(y)} \frac{x}{x + y} B(x, y) = x f(x).$$

Since f satisfies all the conditions of theorem 17.7, it follows that $f(x) = \Gamma(x)$ $(x > 0)$ and therefore

$$B(x, y) = \frac{\Gamma(x)\Gamma(y)}{\Gamma(x + y)}.$$

6 Take $x = y = \frac{1}{2}$ in the previous question. Now

$$B(\tfrac{1}{2}, \tfrac{1}{2}) = \frac{\Gamma(\tfrac{1}{2})\Gamma(\tfrac{1}{2})}{\Gamma(1)} = \Gamma(\tfrac{1}{2})^2.$$

On the other hand

$$B(\tfrac{1}{2}, \tfrac{1}{2}) = \int_{\to 0}^{\to 1} \frac{dt}{\sqrt{\{t(1 - t)\}}} = \int_{\to 0}^{\to \pi/2} \frac{2 \sin \theta \cos \theta \, d\theta}{\sin \theta \sqrt{(1 - \sin^2 \theta)}} = \pi.$$

Thus $\Gamma(\tfrac{1}{2}) = \sqrt{\pi}$.
Write $t = x^2/2$. Then

$$\int_{0}^{\to \infty} e^{-x^2/2} \, dx = \int_{0}^{\to \infty} (2t)^{-1/2} e^{-t} \, dt = \frac{1}{\sqrt{2}} \Gamma(\tfrac{1}{2}) = \sqrt{\frac{\pi}{2}}.$$

Hence

$$\int_{\to -\infty}^{\to \infty} e^{-x^2/2} \, dx = 2 \cdot \sqrt{\frac{\pi}{2}} = \sqrt{(2\pi)}.$$

● *Exercise 18.20*

1 (i) $x + y = (0, 1, 0) + (1, 1, 0) = (1, 2, 0)$

(ii) $x - y = (0, 1, 0) - (1, 1, 0) = (-1, 0, 0)$

(iii) $2x = 2(0, 1, 0) = (0, 2, 0)$

(iv) $\|x\| = \{0^2 + 1^2 + 0^2\}^{1/2} = 1$

(v) $\|x - y\| = \{(-1)^2 + 0^2 + 0^2\}^{1/2} = 1$

(vi) $\langle x, y \rangle = (0 \times 1) + (1 \times 1) + (0 \times 0) = 1.$

The vector x is of length 1 because $\|x\| = 1$. The distance between x and y is 1 because $\|x - y\| = 1$. If θ is the angle between x and y then $\theta = \pi/4$ because

$$\cos\theta = \frac{\langle x, y \rangle}{\|x\|.\|y\|} = \frac{1}{\sqrt{2}}.$$

2   Take $a = d$ and $b = c - d$ in the triangle inequality. Then $\|c\| = \|a + b\| \leqslant \|a\| + \|b\| = \|d\| + \|c - d\|$ and so $\|c - d\| \geqslant \|c\| - \|d\|$. Similarly $\|c - d\| \geqslant \|d\| - \|c\|$. Hence

$$- \|c - d\| \leqslant \|c\| - \|d\| \leqslant \|c - d\|$$

and the result follows from exercise 1.20(1).

3   We have that

$$\tfrac{1}{4}\{\|x + y\|^2 - \|x - y\|^2\} = \tfrac{1}{4}\{\langle x + y, x + y \rangle - \langle x - y, x - y \rangle\}$$

$$= \tfrac{1}{4}\{(\|x\|^2 + 2\langle x, y \rangle + \|y\|^2) - (\|x\|^2 - 2\langle x, y \rangle + \|y\|^2)\}$$

$$= \langle x, y \rangle.$$

4   The angle θ which u makes with the x_1-axis satisfies

$$\cos\theta = \frac{\langle i, u \rangle}{\|i\|.\|u\|} = (1 \times u_1) + (0 \times u_2) + (0 \times u_3) = u_1$$

because $\|i\| = \|u\| = 1.$

5.   The line through a and b is the same as the line through a in the direction $b - a$. The latter has parametric equation

$$x = a + t(b - a)$$

$$= (1 - t)a + tb.$$

If $t \geqslant 0$, then $x = (1 - t)a + tb = a + t(b - a)$ lies on the side of a in which $b - a$ points. If $t \leqslant 1$, then $s = 1 - t \leqslant 0$ and so $x = (1 - t)a + tb = sa + (1 - s)b = b + s(a - b)$ lies on the side of b in which $a - b$ points.

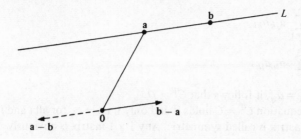

6 Referring to the diagrams of §18.15, it is clear the the shortest distance
   from **0** to the hyperplane is equal to $|\alpha|$ where $\alpha\mathbf{u}$ is a point on the
   hyperplane. But the latter condition means that $\langle\alpha\mathbf{u}, \mathbf{u}\rangle = c$ and hence
   $\alpha = c$ because $\|\mathbf{u}\| = 1$.

● *Exercise 18.25*

1 (i) A column vector is an $n \times 1$ matrix. When transposed this becomes a
       $1 \times n$ matrix which is a row vector. Similarly the transpose of a $1 \times n$
       matrix is an $n \times 1$ matrix.

   (ii) $\langle \mathbf{x}, \mathbf{y} \rangle = x_1 y_1 + x_2 y_2 + \ldots + x_n y_n$

$$x^T y = (x_1, x_2, \ldots, x_n)\begin{pmatrix} y_1 \\ y_2 \\ \vdots \\ y_n \end{pmatrix} = x_1 y_1 + x_2 y_2 + \ldots + x_n y_n$$

$$y^T x = (y_1, y_2, \ldots, y_n)\begin{pmatrix} x_1 \\ x_2 \\ \vdots \\ x_n \end{pmatrix} = y_1 x_1 + y_2 x_2 + \ldots + y_n x_n$$

$$yx^T = \begin{pmatrix} y_1 \\ y_2 \\ \vdots \\ y_n \end{pmatrix}(x_1, x_2, \ldots, x_n) = \begin{pmatrix} y_1 x_1 & y_1 x_2 \ldots y_1 x_n \\ y_2 x_1 & y_2 x_2 \ldots y_2 x_n \\ \vdots & \\ y_n x_1 & y_n x_2 \ldots y_n x_n \end{pmatrix}$$

The final object is an $n \times n$ matrix which cannot therefore be the same
as the previous objects which are real numbers (unless $n = 1$).

2 The element a_{ij} in the ith row and the jth column of A is the element
   in the jth column and ith row of A^T. Let $C = AB$ and $D = B^T A^T$
   Then

$$c_{ji} = \sum_{k=1}^{n} a_{jk}b_{ki}$$

$$d_{ij} = \sum_{k=1}^{n} b_{ki}a_{jk}.$$

Since $c_{ji} = d_{ij}$, it follows that $C^T = D$.

The equation $C^T = C$ holds if and only if $c_{ij} = c_{ji}$ for all i and j. Such a matrix is called symmetric. Any 1×1 matrix is obviously symmetric.

3 Suppose that $g : \mathbb{R}^n \to \mathbb{R}^m$ is affine. Then there exists an $m \times n$ matrix and an $m \times 1$ column vector c such that $y = g(x)$ if and only if $y = Lx + c$. Given any ξ, let $\eta = c + L\xi$. Then $Lx + c = L(x - \xi) + \eta$ and hence $g(x) = f(x - \xi) + \eta$ where $f : \mathbb{R}^n \to \mathbb{R}^m$ is the linear function for which $y = f(x)$ if and only if $y = Lx$.

If $g(x) = f(x - \xi) + \eta$ where f is linear, then $y = g(x)$ if and only if $y = Lx + c$ provided $c = \eta - L\xi$.

4 Take $z = (x_1, x_2, \ldots, x_n, y_1, y_2, \ldots, y_m)$ and $l_1 = (-l_{11}, -l_{12}, \ldots, -l_{1n}, 1, 0, \ldots, 0)$. Then the first equation of equations (2) of §18.22 takes the form

$$\langle l_1, z \rangle = c_1$$

which is the equation of a hyperplane in $\mathbb{R}^{n+m}$. Hence the set of points z satisfying equations (2) of §18.22 (i.e. the graph of the affine function g) consists of those points which are common to m hyperplanes. This is how we define a flat.

5 Suppose that $f(\alpha x + \beta y) = \alpha f(x) + \beta f(y)$ for all scalars α and β and all x and y in $\mathbb{R}^n$. Take L to be the matrix proposed in the hint. Then

$$f(x) = f(x_1 e_1 + x_2 e_2 + \ldots + x_n e_n)$$

$$= x_1 f(e_1) + x_2 f(e_2) + \ldots + x_n f(e_n)$$

$$= \begin{pmatrix} | & | & & | \\ f(e_1) & f(e_2) & \ldots & f(e_n) \\ | & | & & | \end{pmatrix} \begin{pmatrix} x_1 \\ x_2 \\ \vdots \\ x_n \end{pmatrix} = Lx.$$

If f is linear, then $f(\alpha x + \beta y) = \alpha f(x) + \beta f(y)$ because $L(\alpha x + \beta y) = \alpha Lx + \beta Ly$.

6 Suppose that $\mathbf{y} = f(\mathbf{x})$ if and only if $y = L\mathbf{x}$. We have that

$$|y_i| = \left| \sum_{j=1}^{n} l_{ij} x_j \right|$$

$$\leqslant \left\{ \sum_{j=1}^{n} l_{ij}^2 \right\}^{1/2} \left\{ \sum_{j=1}^{n} x_j^2 \right\}^{1/2}$$

$$= \left\{ \sum_{j=1}^{n} l_{ij}^2 \right\}^{1/2} \|\mathbf{x}\|.$$

Hence

$$\|\mathbf{y}\| = \left\{ \sum_{i=1}^{m} y_i^2 \right\}^{1/2} \leqslant \left\{ \sum_{i=1}^{m} \sum_{j=1}^{n} l_{ij}^2 \right\}^{1/2} \|\mathbf{x}\|.$$

If g is affine, then $g(\mathbf{x}) + \mathbf{c}$ for some constant vector $\mathbf{c}$ where f is linear. Hence

$$\|g(\mathbf{x}) - g(\boldsymbol{\xi})\| = \|f(\mathbf{x}) - f(\boldsymbol{\xi})\|$$

$$= \|f(\mathbf{x} - \boldsymbol{\xi})\| \leqslant K\|\mathbf{x} - \boldsymbol{\xi}\|.$$

● *Exercise 18.45*

1 (i) We have that

$$\left(\frac{1}{k}, \frac{k-1}{k}, \frac{k^2-1}{k^2} \right) = \left(\frac{1}{k}, 1 - \frac{1}{k}, 1 - \frac{1}{k^2} \right) \to (0, 1, 1) \text{ as } k \to \infty$$

by theorem 18.28
(ii) We have that

$$\left(\frac{x_1^2 - x_2^2}{x_1^2 + x_2^2}, \frac{x_1 - x_2}{x_1 + x_2} \right) \to \left(\frac{1-1}{1+1}, \frac{1-1}{1+1} \right) = (0, 0) \text{ as } (x_1, x_2) \to (1, 1)$$

by proposition 18.38.
(iii) We have that

$$(\sin(x_1 x_2), \cos(x_1 + x_2)) \to (\sin 0, \cos 0) = (0, 1) \text{ as } (x_1, x_2) \to (0, 0)$$

by proposition 18.37.

2 Consider

$$|\langle \mathbf{x}_k, \mathbf{y}_k \rangle - \langle \mathbf{l}, \mathbf{m} \rangle| = |\langle \mathbf{x}_k, \mathbf{y}_k \rangle - \langle \mathbf{l}, \mathbf{y}_k \rangle + \langle \mathbf{l}, \mathbf{y}_k \rangle - \langle \mathbf{l}, \mathbf{m} \rangle|$$

$$\leqslant |\langle \mathbf{x}_k - \mathbf{l}, \mathbf{y}_k \rangle| + |\langle \mathbf{l}, \mathbf{y}_k - \mathbf{m} \rangle|$$

$$\leqslant \|\mathbf{x}_k - \mathbf{l}\|.\|\mathbf{y}_k\| + \|\mathbf{l}\|.\|\mathbf{y}_k - \mathbf{m}\|$$

$$\to 0.\|\mathbf{m}\| + \|\mathbf{l}\|.0 \text{ as } k \to \infty.$$

3 The sequence of first co-ordinates has a convergent subsequence by the Bolzano–Weierstrass theorem for real numbers. Look at the correspond-ing subsequence of second co-ordinantes. This also has a convergent

subsequence by the Bolzano–Weierstrass theorem. We now have a sub-sub-sequence of the original sequence for which the first and second co-ordinates converge. Proceed similarly to the other co-ordinates.

4   Let $\epsilon > 0$ be given. We have to find a $\delta > 0$ such that

$$\left| \frac{x_1^3}{x_1^2 + x_2^2} - 0 \right| < \epsilon \tag{1}$$

provided that $0 < \|\mathbf{x} - \mathbf{0}\| < \delta$. Observe that

$$|x_1| < \{x_1^2 + x_2^2\}^{1/2} = \|\mathbf{x}\|.$$

Hence

$$\left| \frac{x_1^3}{x_1^2 + x_2^2} \right| \leqslant \frac{\|\mathbf{x}\|^3}{\|\mathbf{x}\|^2} = \|\mathbf{x}\|.$$

It is therefore enough to find a $\delta > 0$ such that $\|\mathbf{x}\| < \epsilon$ provided $0 < \|\mathbf{x}\| < \delta$. Clearly the choice $\delta = \epsilon$ suffices.

For R to be continuous at $(0, 0)$ we require that $R(x_1, x_2) \to R(0, 0)$ as $(x_1, x_2) \to (0, 0)$. We therefore need $R(0, 0) = 0$.

5   Along straight line paths in region A we have that f approaches 1. Along straight line paths in region B, f approaches -1.

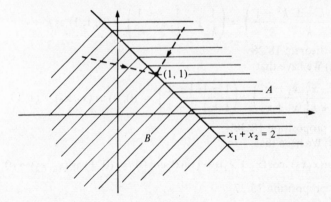

The limit

$$\lim_{(x_1, x_2) \to (1, 1)} f(x_1, x_2)$$

does not exist because, if it did, f would approach its value along *all* paths ending at $(1, 1)$. The function is continuous at all points not on the line $x_1 + x_2 = 2$.

6   A suitable f is given by

$$f(x_1\ x_2) = \frac{x_1 x_2^3}{x_1^2 + x_2^6} \quad ((x_1, x_2) \neq (0, 0)).$$

This function tends to different limits as (x_1, x_2) approaches $(0, 0)$ along different paths given by

$$\left.\begin{array}{l} x_1 = at^3 \\ x_2 = at. \end{array}\right\}$$

- **Exercise 19.11**

1 As in example 19.2

$$\begin{aligned} D_{\mathbf{u}}f(\boldsymbol{\xi}) &= \lim_{t \to 0} \frac{f(\boldsymbol{\xi} + t\mathbf{u}) - f(\boldsymbol{\xi})}{t} \\ &= \lim_{t \to 0} \frac{(1 + 3t/5)^2(2 + 4t/5) + (2 + 4t/5)^3(1 + 3t/5) - 2 - 8}{t} \\ &= \lim_{t \to 0} \frac{2 + 12t/5 + 4t/5 + o(t) + 8 + 48t + 24t/5 + o(t) - 10}{t} \\ &= \frac{1}{5}(12 + 4 + 48 + 24) = \frac{88}{5} \end{aligned}$$

where $o(t)$ denotes a term which tends to zero when divided by t. Note that this problem is easier after theorem 19.15 from which it follows that

$$\begin{aligned} D_{\mathbf{u}}f(\mathbf{x}) = f'(\mathbf{x})\mathbf{u} &= (2x_1x_2 + x_2^3, x_1^2 + 3x_2^2x_1)\begin{pmatrix} u_1 \\ u_2 \end{pmatrix} \\ &= (2x_1x_2 + x_2^3)u_1 + (x_1^2 + 3x_2^2x_1)u_2. \end{aligned}$$

Taking $\mathbf{x} = (1, 2)$ and $\mathbf{u} = (3/5, 4/5)$, we obtain that

$$D_{\mathbf{u}}f(\boldsymbol{\xi}) = (4 + 8)\tfrac{3}{5} + (1 + 12)\tfrac{4}{5} = \tfrac{88}{5}.$$

2   (i) $f_1(x_1, x_2) = 1$ $f_2(x_1, x_2) = 2.$

   (ii) $f_1(x_1, x_2) = 2x_1x_2^5$ $f_2(x_1, x_2) = 5x_1^2x_2^4.$

   (iii) $f_1(x_1, x_2) = (x_2 + e^{x_1x_2})^{-2}2(x_2 + e^{x_1x_2})(x_2e^{x_1x_2})$

$$= 2x_2e^{x_1x_2}(x_2 + e^{x_1x_2})^{-1}$$

$f_2(x_1, x_2) = (x_2 + e^{x_1x_2})^{-2}2(x_2 + e^{x_1x_2})(1 + x_1e^{x_1x_2})$

$$= 2(1 + x_1e^{x_1x_2})(x_2 + e^{x_1x_2})^{-1}.$$

   (iv) $f_1(x_1, x_2) = -\sin(x_2\sin x_1)\{x_2\cos x_1\}$

$f_2(x_1, x_2) = -\sin(x_2\sin x_1)\{\sin x_1\}.$

3 We need to solve the equations

$$f_1(x_1, x_2) = 4x_1^3x_2 - 2x_1x_2^3 = 0$$
$$f_2(x_1, x_2) = x_1^4 - x_1^2 3x_2^2 = 0$$

simultaneously. The solutions satisfy

$(x_1 = 0$ or $x_2 = 0$ or $2x_1^2 = x_2^2)$ and $(x_1 = 0$ or $x_1^2 = 3x_2^2)$

which we may rewrite as

$(x_1 = 0$ and $x_1 = 0)$ or $(x_1 = 0$ and $x_1^2 = 3x_2^2)$ or $(x_2 = 0$ and $x_1 = 0)$
or $(x_2 = 0$ and $x_1^2 = 3x_2^2)$ or $(2x_1^2 = x_2^2$ and $x_1 = 0)$
or $(2x_1^2 = x_2^2$ and $x_1^2 = 3x_2^2)$.

Thus all points of the form $(0, x_2)$ are solutions and no others.

4 We need to solve the equations

$$f_1(x_1, x_2, x_3) = x_2x_3 = 0$$
$$f_2(x_1, x_2, x_3) = x_1x_3 = 0$$
$$f_3(x_1, x_2, x_3) = x_1x_2 = 0$$

$$\frac{f(0 + t\mathbf{u}) - f(0)}{t} = \frac{t^2 u_1^2 t u_2}{t(t^4 u_1^4 + t^2 u_2^2)}$$

$$= \frac{u_1^2 u_2}{t^2 u_1^4 + u_2^2} \to \frac{u_1^2}{u_2} \text{ as } t \to 0$$

(unless $u_2 = 0$ in which case the limit is zero). Thus directional derivatives exist in all directions at $\mathbf{0}$ but f is not continuous at $\mathbf{0}$ because

$$\lim_{(x_1, x_2) \to (0, 0)} f(x_1, x_2)$$

does not exist. This may be proved as in example 18.43.

- *Exercise 19.26*

1 A partial derivative of a polynomial is again a polynomial and hence is continuous at every point. A function whose co-ordinates are all rational functions is differentiable at every point at which it is defined.

2 In each case the function is differentiable because its partial derivatives are continuous at the point in question.

(i) $f_1(x_1, x_2) = 2x_1 + 2x_2 \qquad f_2(x_1, x_2) = 2x_1 + 3x_2^2$

$$f'(1, 2) = (6, 14).$$

The tangent flat has equation

$$y - 13 = (6, 14) \begin{pmatrix} x_1 - 1 \\ x_2 - 2 \end{pmatrix} = 6(x_1 - 1) + 14(x_2 - 2).$$

(ii) $f_1^1(x_1, x_2) = -x_2 \sin(x_1 x_2) \qquad f_2^1(x_1, x_2) = -x_1 \sin(x_1 x_2)$

$f_1^2(x_1, x_2) = x_2 \cos(x_1 x_2) \qquad f_2^2(x_1, x_2) = x_1 \cos(x_1 x_2)$

$$f'(\pi, 1) = \begin{pmatrix} 0 & 0 \\ -1 & -\pi \end{pmatrix}.$$

The tangent flat has equation

$$\begin{pmatrix} y_1 + 1 \\ y_2 - 0 \end{pmatrix} = \begin{pmatrix} 0 & 0 \\ -1 & -\pi \end{pmatrix} \begin{pmatrix} x_1 - \pi \\ x_2 - 1 \end{pmatrix}$$

i.e.

$$\left. \begin{array}{l} y_1 + 1 = 0 \\ \quad y_2 = -(x_1 - \pi) - \pi(x_2 - 1). \end{array} \right\}$$

(iii) $f_1^1(x_1, x_2) = 1$ $f_2^1(x_1, x_2) = 0$

$\qquad f_1^2(x_1, x_2) = 1$ $f_2^2(x_1, x_2) = 1$

$\qquad f_1^3(x_1, x_2) = x_2$ $f_2^3(x_1, x_2) = x_1$

$$f'(1, 1) = \begin{pmatrix} 1 & 0 \\ 1 & 1 \\ 1 & 1 \end{pmatrix}.$$

The tangent flat has equation

$$\begin{pmatrix} y_1 - 1 \\ y_2 - 2 \\ y_3 - 1 \end{pmatrix} = \begin{pmatrix} 1 & 0 \\ 1 & 1 \\ 1 & 1 \end{pmatrix} \begin{pmatrix} x_1 - 1 \\ x_2 - 1 \end{pmatrix}$$

i.e.

$$\left. \begin{aligned} y_1 - 1 &= x_1 - 1 \\ y_2 - 2 &= (x_1 - 1) + (x_2 - 1) \\ y_3 - 1 &= (x_1 - 1) + (x_2 - 1). \end{aligned} \right\}$$

(iv) $f_1^1(x_1, x_2, x_3) = x_2^2$ $f_2^1(x_1, x_2, x_3) = 2x_1 x_2$ $f_3^1(x_1, x_2, x_3) = 0$

$\qquad f_1^2(x_1, x_2, x_3) = 0$ $f_2^2(x_1, x_2, x_3) = 2x_2 x_3^3$ $f_3^2(x_1, x_2, x_3) = 3x_2^2 x_3^2$

$$f'(1, 2, 3) = \begin{pmatrix} 4 & 4 & 0 \\ 0 & 108 & 108 \end{pmatrix}.$$

The tangent flat has equation

$$\begin{pmatrix} y_1 - 4 \\ y_2 - 108 \end{pmatrix} = \begin{pmatrix} 4 & 4 & 0 \\ 0 & 108 & 108 \end{pmatrix} \begin{pmatrix} x_1 - 1 \\ x_2 - 2 \\ x_3 - 3 \end{pmatrix}$$

i.e.

$$\left. \begin{aligned} y_1 - 4 &= 4(x_1 - 1) + 4(x_2 - 2) \\ y_2 - 108 &= 108(x_2 - 2) + 108(x_3 - 3). \end{aligned} \right\}$$

(v)

$$f'(x) = \begin{pmatrix} 0 \\ 1 \\ 2x \end{pmatrix} \qquad f'(0) = \begin{pmatrix} 0 \\ 1 \\ 0 \end{pmatrix}.$$

The tangent flat has equation

$$\begin{pmatrix} y_1 - 1 \\ y_2 - 0 \\ y_3 - 0 \end{pmatrix} = \begin{pmatrix} 0 \\ 1 \\ 0 \end{pmatrix} x$$

i.e.

$$\left. \begin{matrix} y_1 - 1 = 0 \\ y_2 = x \\ y_3 = 0. \end{matrix} \right\}$$

3. A unit vector **u** which points in the direction of $(1, -1)$ is $(1/\sqrt{2}, -1/\sqrt{2})$. The slope at $(1, 2)$ in the direction $(1, -1)$ is therefore

$$D_{\mathbf{u}}f(1, 2) = f(1, 2)u = (6, 14)\begin{pmatrix} 1/\sqrt{2} \\ -1/\sqrt{2} \end{pmatrix} = -8/\sqrt{2}.$$

The slope is greatest in the direction in which the derivative (or gradient) points, i.e. in the direction of $(6, 14)$. The value of the slope in this direction is the length of the derivative (or gradient), i.e. $\|f'(1, 2)\| = \{36 + 196\}^{1/2} = \sqrt{232}$. The required normal is again the derivative (or gradient) at $(1, 2)$, i.e. $(6, 14)$. The tangent line therefore has the equation

$$6(x_1 - 1) + 14(x_2 - 2) = 0.$$

4 (I) This formula makes sense when L is an $m \times n$ matrix, c is an $m \times 1$ column vector and x is an $n \times 1$ column vector. The formula is then the evident assertion that

$$\frac{\partial}{\partial x_k}\left(\sum l_{ij}x_j + c_j \right) = l_{ik}.$$

(III) This formula makes sense when L is an $m \times p$ matrix, M is an $m \times q$ matrix, y is a $p \times 1$ column vector and z is a $q \times 1$ column vector. From formula (I) and (V) we know that

$$\frac{d}{dx}(Ly) = L\frac{dy}{dx}$$

and it is therefore enough to show that

$$\frac{d}{dx}(u + v) = \frac{du}{dx} + \frac{dv}{dx}$$

where u and v are $m \times 1$ column vectors. But this is simply the assertion that

$$\frac{d}{\partial x_j}(u_i + v_i) = \frac{\partial u_i}{\partial x_j} + \frac{\partial v_i}{\partial x_j} .$$

(IV)' This formula makes sense when y and z are both $m \times 1$ column vectors. It is then the assertion that

$$\frac{\partial}{\partial x_i}\left(\sum_{j=1}^{m} y_j z_j\right) = \sum_{j=1}^{m}\left(z_j \frac{\partial y_j}{\partial x_i} + y_j \frac{\partial z_j}{\partial x_i}\right)$$

which again is immediate.

(Note that (I), (III) also make sense when both L and M are scalars. The proofs are then even easier. Formula (IV)' makes sense when y is a scalar and z is an $m \times 1$ column vector. Again the proof is easy.)

5 We have that

$$\frac{\partial y}{\partial x_1} = \phi_1(x_1, x_2) = f'(x_1 x_2)x_2$$

$$\frac{\partial y}{\partial x_2} = \phi_2(x_1, x_2) = f'(x_1 x_2)x_1$$

and hence the result. The partial derivatives may also be evaluated by writing $u = x_1 x_2$ and observing that

$$\left(\frac{\partial y}{\partial x_1}, \frac{\partial y}{\partial x_2}\right) = \frac{dy}{dx} = \frac{dy}{du} \cdot \frac{du}{dx} = f'(u)\left(\frac{\partial u}{\partial x_1}, \frac{\partial u}{\partial x_2}\right) .$$

6 From the chain rule

$$\frac{\partial y}{\partial x_1} = \frac{\partial y}{\partial r} \cdot \frac{\partial r}{\partial x_1} + \frac{\partial y}{\partial \theta} \cdot \frac{\partial \theta}{\partial x_1}$$

$$\frac{\partial y}{\partial x_2} = \frac{\partial y}{\partial r} \cdot \frac{\partial r}{\partial x_2} + \frac{\partial y}{\partial \theta} \cdot \frac{\partial \theta}{\partial x_2},$$

The partial derivatives of r and θ with respect to x_1 and x_2 can be found by using formula (VI). Writing

$$u = \begin{pmatrix} r \\ \theta \end{pmatrix}$$

we have that

$$
\begin{pmatrix} \dfrac{\partial r}{\partial x_1} & \dfrac{\partial r}{\partial x_2} \\[3mm] \dfrac{\partial \theta}{\partial x_1} & \dfrac{\partial \theta}{\partial x_2} \end{pmatrix} = \frac{du}{dx} = \left(\frac{dx}{du}\right)^{-1} = \begin{pmatrix} \dfrac{\partial x_1}{\partial r} & \dfrac{\partial x_1}{\partial \theta} \\[3mm] \dfrac{\partial x_2}{\partial r} & \dfrac{\partial x_2}{\partial \theta} \end{pmatrix}^{-1}
$$

i.e. $\qquad \begin{pmatrix} \dfrac{\partial r}{\partial x_2} & \dfrac{\partial r}{\partial x_2} \\[3mm] \dfrac{\partial \theta}{\partial x_1} & \dfrac{\partial \theta}{\partial x_2} \end{pmatrix} = \begin{pmatrix} \cos\theta & \sin\theta \\[2mm] -r\sin\theta & r\cos\theta \end{pmatrix}^{-1} = \dfrac{1}{r}\begin{pmatrix} r\cos\theta & -\sin\theta \\[2mm] r\sin\theta & \cos\theta \end{pmatrix}.$

Alternatively, observe that

$$
\left.\begin{aligned} r &= \{x_1^2 + x_2^2\}^{1/2} \\ \theta &= \arctan(x_2/x_1). \end{aligned}\right\}
$$

• *Exercise 19.47*

1 The required system of linear equations is obtained from the equation $f'(x) = 0$. From formula (IV)′ of §19.21,

$$
\frac{d}{dx}(Ax + a)^{\mathrm{T}}(Bx + b) = (Bx + b)^{\mathrm{T}}A + (Ax + a)^{\mathrm{T}}B.
$$

Transposing the right hand side and setting the result equal to the (column vector) 0, we obtain that

$$
A^{\mathrm{T}}(Bx + b) + B^{\mathrm{T}}(Ax + a) = 0
$$

$$
(A^{\mathrm{T}}B + B^{\mathrm{T}}A)x + (A^{\mathrm{T}}b + B^{\mathrm{T}}a) = 0.
$$

2 We have that

$$
f_1(x_1, x_2) = e^{-x_2} \qquad f_2(x_1, x_2) = -x_1 e^{-x_2}
$$

and so

$$
f_{11}(x_1, x_2) = 0 \qquad f_{12}(x_1, x_2) = -e^{-x_2}
$$

$$
f_{21}(x_1, x_2) = -e^{-x_2} \qquad f_{22}(x_1, x_2) = x_1 e^{-x_2}.
$$

Since the second order partial derivatives are continuous everywhere, the function is twice differentiable everywhere. The second derivative is given by

$$
f''(x_1, x_2) = \begin{pmatrix} 0 & -e^{-x_2} \\ -e^{-x_2} & x_1 e^{-x_2} \end{pmatrix}.
$$

3 We first calculate the first order partial derivatives. The point $(0, 0)$ must be treated separately but, since $f(x_1, 0) = 0$ and $f(0, x_2) = 0$ for all x_1 and x_2, it is trivial that $f_1(0, 0) = 0$ and $f_2(0, 0) = 0$. If $(x_1, x_2) \neq (0, 0)$, some calculations are necessary. We have that

$$f_1(x_1, x_2) = \frac{3x_1^2 x_2(x_1^2 + x_2^2) - x_1^3 x_2^2 2x_1}{(x_1^2 + x_2^2)^2} = \frac{x_1^4 x_2 + 3x_1^2 x_2^3}{(x_1^2 + x_2^2)^2}$$

$$f_2(x_1, x_2) = \frac{x_1^3(x_1^2 + x_2^2) - x_1^3 x_2 2x_2}{(x_1^2 + x_2^2)^2} = \frac{x_1^5 - x_1^3 x_2^2}{(x_1^2 + x_2^2)^2}.$$

These partial derivatives are continuous everywhere except possibly at $(0, 0)$. To prove continuity at $(0, 0)$, we need to show that $f_1(x) \to 0$ as $x \to 0$ and $f_2(x) \to 0$ as $x \to 0$. This is quite simple because $|x_1| \leqslant \|x\| = \{x_1^2 + x_2^2\}^{1/2}$ and $|x_2| \leqslant \|x\| = \{x_1^2 + x_2^2\}^{1/2}$. Hence, for $x \neq 0$,

$$|f_1(x)| \leqslant \frac{\|x\|^5 + 3\|x\|^5}{\|x\|^4} = 4\|x\| \to 0 \text{ as } x \to 0$$

$$|f_2(x)| \leqslant \frac{\|x\|^5 + \|x\|^5}{\|x\|^4} = 2\|x\| \to 0 \text{ as } x \to 0.$$

It follows that f has a continuous first derivative f' at every point.

We now turn to the second order partial derivatives at $(0, 0)$. These are easily computed because $f_1(x_1, 0) = 0$, $f_1(0, x_2) = 0$, $f_2(x_1, 0) = x_1$ and $f_2(0, x_2) = 0$ for all x_1 and x_2. Thus $f_{11}(0, 0) = 0$, $f_{21}(0, 0) = 0$, $f_{12}(0, 0) = 1$ and $f_{22}(0, 0) = 0$. Note that $f_{21}(0, 0) \neq f_{12}(0, 0)$.

This last result does not contradict proposition 19.32 because f_{12} is not continuous at $(0, 0)$. It is evident that the formula for $f_{12}(x_1, x_2)$ when $(x_1, x_2) \neq (0, 0)$ will contain a factor of x_1 and hence $f_{12}(0, x_2) = 0$ for all $x_2 \neq 0$. Hence $f_{12}(x) \to 0$ as $x \to 0$ along the x_2-axis.

4 From example 19.10, the stationary points are $(0, 0)$, $(1, 0)$, $(0, 1)$, $(0, -1)$, $(2/5, 1/\sqrt{5})$, $(2/5, -1/\sqrt{5})$. The second derivative is given by

$$f''(x_1, x_2) = \begin{pmatrix} 2x_2 & 2x_1 + 3x_2^2 - 1 \\ 2x_1 + 3x_2^2 - 1 & 6x_1 x_2 \end{pmatrix}.$$

(a) The point $(0, 0)$ is a saddle point. We have that

$$f''(0, 0) = \begin{pmatrix} 0 & -1 \\ -1 & 0 \end{pmatrix}$$

and hence $\det f''(0, 0) = -1 < 0$.

(b) The point $(1, 0)$ is a saddle point. We have that

$$f''(1, 0) = \begin{pmatrix} 0 & 1 \\ 1 & 0 \end{pmatrix}$$

and hence $\det f''(1, 0) = -1 < 0$.

(c) The points $(0, 1)$ and $(0, -1)$ are also saddle points. We have that

$$f''(0, 1) = \begin{pmatrix} 2 & 2 \\ 2 & 0 \end{pmatrix}$$

and hence $\det f''(0, 1) = -4 < 0$. Similarly $\det f''(0, -1) = -4$.

(d) The point $(2/5, 1/\sqrt{5})$ is a local minimum. We have that

$$f''(2/5, 1/\sqrt{5}) = \begin{pmatrix} 2/\sqrt{5} & 2/5 \\ 2/5 & 12/5\sqrt{5} \end{pmatrix}$$

which is positive definite because $2/\sqrt{5} > 0$ and $\det f''(2/5, 1/\sqrt{5}) = (24 - 4)/25 = 4/5 > 0$.

(e) The point $(2/5, -1/\sqrt{5})$ is a local maximum. We have that

$$f''(2/5, -1/\sqrt{5}) = \begin{pmatrix} -2/\sqrt{5} & 2/5 \\ 2/5 & 12/5\sqrt{5} \end{pmatrix}$$

which is negative definite because $-2/\sqrt{5} < 0$ and $\det f''(2/5, -1/\sqrt{5}) = (24 - 4)/25 = 4/5 > 0$.

5 We have that

$$f(1) - f(0) = \begin{pmatrix} 1 \\ 1 \end{pmatrix}$$

and hence, if $f(1) - f(0) = f'(\xi)(1 - 0)$, then

$$\begin{pmatrix} 2\xi \\ 3\xi^2 \end{pmatrix} = f'(\xi) = \begin{pmatrix} 1 \\ 1 \end{pmatrix}.$$

But the equations $2\xi = 1$ and $3\xi^2 = 1$ are incompatible.

6 Let λ_n be the maximum eigenvalue of M. Then

$$\lambda_n = \max_{h \neq 0} \frac{h^T M h}{\|h\|^2}.$$

We have that $h^T M h \leqslant \lambda_n \|h\|^2$ from inequality (1) of §19.37. Equality is attained with $h = u^n$. To obtain the result required, we then appeal to the inequality obtained at the beginning of §19.40.

FURTHER PROBLEMS

Solutions to the following problems are not provided but many of the problems are similar to exercises set in the text and it therefore may be worth looking for such an exercise and examining its solution if a problem does not seem to be tractable. If written answers to the problems are required as part of a taught course, remember that it is not enough that the solution is correct. It is also essential that the manner of presentation be clear, unambiguous and detailed.

Problem set 1 (This is based on chapters 1 and 2)

1 Find all real values of x such that

(i) $\dfrac{x+1}{x^2+3} < \dfrac{2}{x}$ (ii) $\left| \dfrac{1}{x+1} - 1 \right| < 2.$

2 Suppose that, for any $\epsilon > 0$, $|x| < \epsilon$. Prove that $x = 0$.

3 Prove that, for any $y \in (4, 5)$, there exists an $x \in (4, 5)$ such that $x < y$. Deduce that $(4, 5)$ has no minimum.

4 For each of the following sets, find the sup, inf, max and min whenever these exist.

(i) $[0, 3)$ (ii) $[-1, \infty)$ (iii) $\{0, -1, 3, -6\}$ (iv $[1, 2]$

(v) $\{x: x^2 - 2x - 1 < 0\}$ (vi) $\{x^2 - 2x - 1: x \in \mathbb{R}\}$

(vii) $\{x: x = x + 1\}$.

5. Suppose that a and b are real numbers with $a > 0$. If S is non-empty and bounded above, prove that

$$\sup_{x \in S} (ax + b) = a(\sup_{x \in S} x) + b.$$

6 Find the distance between the number 2 and the set S in the following cases. (See exercise 2.13(4).)

(i) $S = \{-1, 0, 3\}$ (ii) $S = (0, 3)$ (iii) $S = [2, 4]$ (iv) $S = (2, 4)$.

Problem set 2 (This is based on chapters 3 and 4)

1 Suppose that $x > -1$ and $x \neq 0$. Prove by induction that, for any natural number $n \geq 2$, $(1 + x)^n > 1 + nx$. Deduce that

$$\left(1 + \frac{1}{n}\right)^n > 2$$

for all natural numbers $n \geq 2$.

2 Find the following limits

(i) $\displaystyle\lim_{n \to \infty} \left\{ \frac{4n^5 + 5n^3 + 6n}{2n^5 + 1} \right\}$

(ii) $\displaystyle\lim_{n \to \infty} \left\{ \frac{3^n + (-2)^n}{3^n - 2^n} \right\}$

(iii) $\displaystyle\lim_{n \to \infty} \left\{ \frac{n}{n^2} + \frac{(n+1)}{n^2} + \ldots + \frac{2n}{n^2} \right\}$.

3 Suppose that $0 < k < 1$ and $\langle x_n \rangle$ satisfies

$$|x_{n+1}| < k |x_n| \quad (n = 1, 2, 3, \ldots).$$

Prove that $x_n \to 0$ as $n \to \infty$. Explain why the same conclusion holds if it is only known that $|x_{n+1}| < k|x_n|$ when $n > N$ for some natural number N.

4 Suppose that $y_n \to l$ as $n \to \infty$. If $l < k$, prove that there exists an N such that $y_n < k$ for any $n > N$. Suppose that $0 < l < 1$ and

$$\left| \frac{a_{n+1}}{a_n} \right| \to l, \text{ as } n \to \infty.$$

Use question 3 to show that $a_n \to 0$ as $n \to \infty$. Hence show that, for any real number α and any x satisfying $|x| < 1$,

$$\lim_{n \to \infty} \frac{\alpha(\alpha - 1) \ldots (\alpha - n + 1)}{n!} x^n = 0.$$

5 Let $x_n < 0$ $(n = 1, 2, \ldots)$. Prove that $x_n \to 0$ as $n \to \infty$ if and only if $1/x_n \to -\infty$ as $n \to \infty$. Give an example of a sequence $\langle y_n \rangle$ such that $y_n \to 0$ as $n \to \infty$ and $\langle 1/y_n \rangle$ oscillates.

6 Define the sequence $\langle a_n \rangle$ by

$$a_n = \frac{1.3.5.\ldots.(2n-1)}{2.4.6.\ldots.(2n)} \quad (n = 1, 2, \ldots).$$

Show that the sequence $\langle b_n \rangle$ given by $b_n = a_n(2n+1)^{1/2}$
$(n = 1, 2, \ldots)$ decreases. [Hint: the inequality of the arithmetic and
geometric means (example 3.10) may be helpful.] Explain why it
follows that $\langle b_n \rangle$ converges. Deduce that $\langle a_n \rangle$ converges and find its
limit.

Problem set 3 (This is based on chapters 5 and 6 excluding 5.12–5.15
inclusive)

1 With the help of the identity $16 \times 18 = (17-1)(17+1) = 17^2 - 1$,
prove that

$$\tfrac{4}{3} < \sqrt{2} < \tfrac{17}{12}.$$

Take $a = 2$ and $x_1 = 17/12$ in example 5.6, and hence show that the
error in approximating to $\sqrt{2}$ by x_2 is at most 3.2^{-10}.

2 A sequence $\langle x_n \rangle$ is defined by $x_1 = 1$ and

$$x_{n+1} = \frac{2(2x_n + 1)}{x_n + 3} \quad (n = 1, 2, \ldots).$$

By considering $2 - x_{n+1}$, show that $x_n < 2$ $(n = 1, 2, \ldots)$. Deduce that
$\langle x_n \rangle$ increases. Explain why the sequence converges and find its limit.

3 A sequence $\langle x_n \rangle$ satisfies

$$x_{n+1} = (3x_n - 2)^{1/2} \quad (n = 1, 2, \ldots).$$

Given that $1 < x_1 \leqslant 2$, show that $\langle x_n \rangle$ converges and find the limit.
What happens when (a) $x_1 < 1$, (b) $x_1 = 1$ and (c) $x_1 > 2$?

4 A sequence $\langle x_n \rangle$ satisfies

$$x_{n+1} = \tfrac{1}{5}(3x_n + 2 - 2x_n^3) \quad (n = 1, 2, \ldots).$$

Given that $0 \leqslant x_1 \leqslant 1$, show that $0 \leqslant x_n \leqslant 1$ $(n = 1, 2, \ldots)$. Deduce
that $-3 \leqslant 3 - 2(x_n^2 + x_n x_{n-1} + x_{n-1}^2) \leqslant 3$ $(n = 2, 3, \ldots)$. By
considering $x_{n+1} - x_n$, prove that $\langle x_n \rangle$ is a Cauchy sequence. [Hint:
recall the formula for $a^n - b^n$ of exercise 3.11(2).] Deduce that the
equation $x^3 + x - 1 = 0$ has a solution x satisfying $0 \leqslant x \leqslant 1$.

5 Prove that

$$\sum_{n=1}^{\infty} \frac{1}{(n+1)(n+3)(n+5)} = \frac{23}{480}.$$

6 Discuss the convergence of the following series.

(i) $\sum_{n=1}^{\infty} \frac{3}{n^3+2}$ (ii) $\sum_{n=1}^{\infty} \frac{4}{3n+2}$ (ii) $\sum_{n=1}^{\infty} \frac{(n!)^3}{(3n)!}(26)^n$

(iv) $\sum_{n=1}^{\infty} \frac{(n!)^3}{(3n)!}(28)^n$ (v) $\sum_{n=1}^{\infty} (-1)^{n-1}n^{-1/4}$ (vi) $\sum \left(1+\frac{1}{n}\right)(-1)^n$.

Problem set 4 (This is based on chapters 7 and 8)

1 Draw a diagram illustrating the equation $x^2 + y^4 = 1$. Explain why:
   (i) the equation does *not* define a function $f: \mathbb{R} \to \mathbb{R}$
   (ii) the equation does *not* define a function $f: [-1, 1] \to [-1, 1]$
   (iii) the equation *does* define a function $f: [-1, 1] \to [0, 1]$.

2 Suppose that $f: (0, \infty) \to (0, \infty)$ is defined by $f(x) = 1/x$ $(x > 0)$ and that $g: (0, \infty) \to (0, \infty)$ is defined by $g(x) = x^2 - 2x + 2$ $(x > 0)$.
   (i) Which of these functions has an inverse? Find a formula for the inverse where it exists.
   (ii) Compute $f \circ g$ and $g \circ f$.

3 Evaluate the following limits.
   (i) $\lim_{x \to 3} \frac{x^3 + 5x + 7}{x^4 + 6x^2 + 8}$ (ii) $\lim_{x \to 0+} x^{1/2}$ (iii) $\lim_{x \to 0} \frac{(1+x)^{1/2} - (1-x)^{1/2}}{x}$.

[Hint: in the case of (iii), recall the formula $a^2 - b^2 = (a - b)(a + b)$.]

4 Let $f: \mathbb{R} \to \mathbb{R}$ be defined by

$$f(x) = \begin{cases} (x-1)^2 & (x < 1) \\ 1 & (x = 1) \\ 3x + 2 & (x > 1). \end{cases}$$

Use the definitions to show that $f(x) \to 0$ as $x \to 1-$ and $f(x) \to 5$ as $x \to 1+$. Is this function
   (i) continuous
   (ii) continuous on the right
   (iii) continuous on the left
   at the point 1?

5 Let $f\colon \mathbb{R} \to \mathbb{R}$ be defined by

$$f(x) = \begin{cases} \dfrac{1}{x^2} & (x < 0) \\ x^2 & (x \geqslant 0). \end{cases}$$

Show that $f(x) \to +\infty$ as $x \to 0-$ and $f(x) \to 0$ as $x \to 0+$ using only the appropriate definitions. Answer questions (i), (ii) and (iii) of the previous problem in respect of the point 0.

6 Suppose that $f\colon \mathbb{R} \to \mathbb{R}$ and $f(x) \to l$ as $x \to \xi$. Prove that there exists an $h > 0$ such that f is bounded on $(\xi - h, \xi + h)$.

Problem set 5 (This is based on chapters 9, 10 and 11)

1 Each of the following expressions defines a function on $(0, \infty)$. Decide in each case whether or not the function is continuous on (i) $(0, \infty)$, (ii) $[1, 2]$, (iii) $(1, 2)$.

(a) $f(x) = \dfrac{x-1}{x+1}$ (b) $f(x) = |2x - 3|$ (c) $f(x) = \begin{cases} 0 & (x < 1) \\ 1 & (x \geqslant 1) \end{cases}$

(d) $f(x) = \begin{cases} 0 & (x \leqslant 1) \\ 1 & (x > 1) \end{cases}$ (e) $f(x) = x + \dfrac{1}{x}$

(f) $f(x) = \begin{cases} (1-x)(x-2) & (0 < x < 1) \\ (x-1)(x-2) & (1 \leqslant x \leqslant 2) \\ (x-1)(2-x) & (x > 2). \end{cases}$

2 Show that the equation $x^{16} + x^7 - 1 = 0$ has a solution $\xi \in (0, 1)$.

3 A function $f\colon [a, b] \to \mathbb{R}$ has the property that $f(x) \geqslant 0$ for $a \leqslant x \leqslant b$ and $f(a) = 0$, $f(b) = 0$. If, for each $x \in [a, b]$ there exists exactly one distinct $y \in [a, b]$ such that $f(x) = f(y)$, prove that f cannot be continuous on $[a, b]$. [Hint: if f is continuous on $[a, b]$ there are points on $[a, b]$ at which it achieves its maximum. These can be used to obtain a contradiction via the intermediate value theorem.] Is the restriction on f given in the opening sentence necessary for the conclusion?

4 Show that the function $g\colon \mathbb{R} \to \mathbb{R}$ defined by

$$g(x) = \begin{cases} x^3 & (x < 0) \\ x^{1/2} & (x \geqslant 0) \end{cases}$$

is continuous at 0 but that g is *not* differentiable at 0. Show that the function h: $\mathbb{R} \to \mathbb{R}$ defined by $h(x) = xg(x)$ is differentiable at every point including 0. Finally show that h: $\mathbb{R} \to \mathbb{R}$ is continuous at 0 but *not* differentiable at 0.

5 Suppose that $a < c < d < b$ and f: $(a, b) \to \mathbb{R}$ is differentiable on (a, b). If $f'(c) > 0$ and $f'(d) < 0$, prove that there exists a $\xi \in (c, d)$ such that $f'(\xi) = 0$. [Hint: look at the proof of Rolle's theorem.] Deduce that the image of (a, b) under f' is an interval [Hint: look at the proof of the mean value theorem. Consider $F(x) = hx - f(x)$ for a suitable h.]

6. Let f: $\mathbb{R} \to \mathbb{R}$ be differentiable at every point and satisfy $f'(x) > 0$ for all values of x. Prove that the equation $f(x) = 0$ can have at most one solution. If $f''(x) > 0$ for all values of x show that $f(x) = 0$ can have at most two solutions. State and prove the generalisation involving derivatives of order n. [Hint: use Rolle's theorem.]

Problem set 6 (This is based on chapters 12 and 13)

1 Let f: $\mathbb{R} \to \mathbb{R}$ be defined by $f(x) = 1 + x^3 + x^5$. Prove that f is strictly increasing on $\mathbb{R}$ and explain why $f(\mathbb{R}) = \mathbb{R}$. Deduce that f has an inverse function f^{-1}: $\mathbb{R} \to \mathbb{R}$. Calculate the value of $Df^{-1}(y)$ when $y = -1$.

2 Let f be differentiable and convex on $\mathbb{R}$. Prove that one of the following three alternatives holds.
(i) f increases on $\mathbb{R}$ (ii) f decreases on $\mathbb{R}$
(iii) For some $\xi \in \mathbb{R}$, f decreases on $(-\infty, \xi]$ and increases on $[\xi, \infty)$
[Hint: theorem 12.18.]

3 Evaluate

$$\int_0^1 x^3 \, dx$$

by the method used in example 13.17.

4 Suppose that $f^{(n+1)}$ is continuous on $[a, b]$. Prove that

$$\frac{1}{n!} \int_a^b x^n f^{(n+1)}(x) \, dx = \left[\sum_{k=0}^n (-1)^{n-k} \frac{x^k}{n!} f^{(k)}(x) \right]_a^b.$$

5 Suppose that $f: [0, \infty) \to \mathbb{R}$ is continuous and positive on $[0, \infty)$ and that

$$\int_0^x f(t)\, dt \leqslant x^{-1}\{f(x)\}^3 \quad (x > 0).$$

Prove that

$$f(x) \geqslant \frac{x}{\sqrt{2}} \quad (x \geqslant 0).$$

[Hint: See exercise 13.26(3).]

6 Show that

$$\lim_{n \to \infty} \sum_{k=0}^{\infty} \frac{n}{(n+k)^2} = 1$$

in the following way. Begin by proving that

$$\sum_{k=0}^{\infty} \frac{1}{n}\left(1 + \frac{k+1}{n}\right)^{-2} \geqslant \int_1^{\to \infty} \frac{dx}{x^2} \leqslant \sum_{k=0}^{\infty} \frac{1}{n}\left(1 + \frac{k}{n}\right)^{-2}.$$

Observe that $(n+k)^2 > n^{1/2}k^{3/2}$ $(n > 0, k > 0)$ and hence obtain the inequality

$$0 < \sum_{k=0}^{\infty} \frac{n}{(n+k)^2} - 1 < \frac{2}{n^{1/2}}\left\{1 + \sum_{k=1}^{\infty} k^{-3/2}\right\}.$$

Problem set 7 (This is based on chapters 14–17 inclusive, excluding §17.4 *et seq*.)

1 Suppose that $\alpha > 0$. Find the value of ξ such that the function $f: (0, \infty) \to \mathbb{R}$ defined by

$$f(x) = \log x - \alpha x^2 \quad (x > 0)$$

is strictly increasing on $(0, \xi]$ and strictly decreasing on $[\xi, \infty)$.
   Prove that the equation $\log x = \alpha x^2$ has solutions if and only if $2\alpha e \leqslant 1$. Show that the equation has a unique solution if and only if $2\alpha e = 1$ and find the value of this solution.
   Suppose that $x_1 > \sqrt{e}$ and $x_{n+1} = \{2e \log x_n\}^{1/2}$ $(n = 1, 2, \ldots)$. Prove that $x_n \to \sqrt{e}$ as $n \to \infty$.

2 Find the interval of convergence of the following power series.

(i) $\displaystyle\sum_{n=1}^{\infty} \frac{(n!)^2}{(2n)!}(x-1)^n$ (ii) $\displaystyle\sum_{n=1}^{\infty} \left(\frac{n}{n+1}\right)^{n^2}(x+1)^n$

(iii) $\sum_{n=1}^{\infty} e^{\sqrt{n}} x^n$ (iv) $\sum_{n=1}^{\infty} e^{-\sqrt{n}} x^n$

(Do not neglect to consider what happens at the endpoints of the interval of convergence.)

3 Prove that the function $f: \mathbb{R} \to \mathbb{R}$ defined by $f(x) = x - \sin x$ increases on $\mathbb{R}$ and that the function $g: (-\frac{1}{2}\pi, \frac{1}{2}\pi) \to \mathbb{R}$ defined by $g(x) = \tan x - x$ increases on $(-\frac{1}{2}\pi, \frac{1}{2}\pi)$. Prove that the function $h: (0, \pi) \to \mathbb{R}$ defined by $h(x) = x/\sin x$ increases on $(0, \pi)$.

4 Find the following limits:

(i) $\lim_{x \to 0} (\cos x)^{\operatorname{cosec}^2 x}$ (ii) $\lim_{x \to 0} \dfrac{x^2 \sin(1/x)}{e^x - 1}$.

5 Prove that

(i) $\displaystyle\int_1^{+\infty} \frac{\cos t}{t} \, dt$ (ii) $\displaystyle\int_1^{+\infty} \frac{\sin t}{t} \, dt$

exist; [Hint: exercise 16.3(6).] Show that

(iii) $\displaystyle\int_1^{+\infty} \left| \frac{\cos t}{t} \right| dt$ (iv) $\displaystyle\int_1^{+\infty} \left| \frac{\sin t}{t} \right| dt$

do not exist. [Hint: $|\cos t| \geqslant \cos^2 t$, $|\sin t| \geqslant \sin^2 t$.]

6 If

$$a_k = \frac{(2k)!}{4^k (k!)^2}$$

prove that $(2k + 1)a_k = 2(k + 1)a_{k+1}$ $(k = 0, 1, 2, \ldots)$. Deduce that $\langle a_k \rangle$ decreases and find the radius of convergence of the power series

$$f(x) = \sum_{k=0}^{\infty} a_k x^k.$$

Using Stirling's formula, show that $k^{1/2} a_k \to l$ as $k \to \infty$ for some $l > 0$. Explain why these results imply that the interval of convergence of the power series is $[-1, 1)$. By differentiating $(1 - x)^{1/2} f(x)$ on $(-1, 1)$, prove that $f(x) = (1 - x)^{-1/2}$ for $-1 < x < 1$. By integrating $f(x^2)$, prove that

$$\arcsin y = \sum_{k=0}^{\infty} \frac{1}{2k + 1} a_k x^{2k+1} \quad (-1 < y < 1).$$

Explain why this formula also holds when $y = 1$ and hence obtain a series expansion for $\frac{1}{2}\pi$.

Problem set 8 (This is based on chapters 18 and 19 excluding the algebraic material of 18.1–18.25 and 19.37–19.39. Also excluded is 19.48.)

1 Find the following limits where these exist.

(i) $\lim_{k \to \infty} \left(\dfrac{\cos k}{k}, \dfrac{\sin k}{k} \right)$ (ii) $\lim_{k \to \infty} (\cos k, \sin k)$

(iii) $\lim_{(x_1, x_2) \to (1, 1)} \dfrac{x_1^2}{x_1^2 + x_2^2}$ (iv) $\lim_{(x_1, x_2) \to (0, 0)} \dfrac{x_1^2}{x_1^2 + x_2^2}$

(v) $\lim_{(x_1, x_2) \to (1, 1)} \left(\dfrac{1}{x_1 - 1}, x_2 - 1 \right)$ (vi) $\lim_{(x_1, x_2) \to (0, 0)} \left(\dfrac{1}{x_1 - 1}, x_2 - 1 \right)$.

2 Let $f: \mathbb{R}^2 \to \mathbb{R}$ be defined by

$$f(x_1, x_2) = \begin{cases} \dfrac{x_1 x_2}{x_1^2 + x_2^2} & (x_1, x_2) \neq (0, 0) \\ 0 & (x_1, x_2) = (0, 0). \end{cases}$$

Prove that
$$\lim_{x_1 \to 0} \left(\lim_{x_2 \to 0} f(x_1, x_2) \right) = \lim_{x_2 \to 0} \left(\lim_{x_1 \to 0} f(x_1, x_2) \right) = 0$$

but that

$$\lim_{(x_1, x_2) \to (0, 0)} f(x_1, x_2)$$

does not exist. Explain why f has directional derivatives in all directions at $(0, 0)$ but is not differentiable at $(0, 0)$.

3 Explain why each of the following functions is differentiable at the given point and compute its derivative there.

(i) $f: \mathbb{R}^2 \to \mathbb{R}$ at $(1, 1)$, where $f(x_1, x_2) = x_1 + x_2^2$

(ii) $f: \mathbb{R} \to \mathbb{R}^3$ at 2, where

$$f(x) = \begin{pmatrix} x + 1 \\ x^2 + 1 \\ 1 - x \end{pmatrix}$$

(iii) $f: \mathbb{R}^2 \to \mathbb{R}^2$ at $(0, 1)$ where

$$f(x_1, x_2) = \begin{pmatrix} e^{x_2} \\ e^{-x_2} \end{pmatrix}.$$

For each function, write down the equation of the appropriate tangent flat.

4 What is the slope of the function f of question 3(i) at the point $(1, 1)$ in the direction $(-4, 3)$? What is the direction at $(1, 1)$ in which the slope is greatest? What is the value of the slope in this direction? Write down a vector which is normal to the contour $x_1 + x_2^2 = 2$ at $(1, 1)$. Also write down the equation of the tangent line at this point.

5 A function $f: \mathbb{R}^n \to \mathbb{R}$ is homogeneous of degree α. This means that, for all $\mathbf{x} \in \mathbb{R}^n$ and all $\in \mathbb{R}$.

$$f(\lambda \mathbf{x}) = \lambda^\alpha f(\mathbf{x}).$$

If f is differentiable at every point, prove that

$$\langle \mathbf{x}, \nabla f(\mathbf{x}) \rangle = \alpha f(\mathbf{x})$$

for all $\mathbf{x} \in \mathbb{R}^n$. [Hint: consider the derivative of the function $g: \mathbb{R} \to \mathbb{R}$ defined by $g(t) = f(t\mathbf{x})$.]

6 Classify the stationary points of the real-valued functions f defined by

(i) $f(x_1, x_2) = x_1^2 x_2 + x_2^3 x_1 - x_1 x_2^2$

(ii) $f(x_1, x_2) = e^{x_1 + x_2}(x_1^2 - 2x_1 x_2 + 3x_2^2)$

(iii) $f(x_1, x_2) = x_2^3 + 3x_1^2 x_2 - 3x_1^2 - 3x_2^2 + 2$

(iv) $f(x_1, x_2) = 8x_1^2 x_2 - x_1^3 x_2 - 5x_2^3 x_1$

(v) $f(x_1, x_2, x_3) = x_1^2 x_2 + x_2^2 x_3 + x_3^2 x_1$

(vi) $f(x_1, x_2, x_3, x_4, x_5, x_6) = (x_1^3 - 3x_1 x_2^2 + x_2^4) + (x_3^3 - 3x_3 x_4^2 + x_4^4)$
$$+ (x_5^3 - 3x_5 x_6^2 + x_6^4).$$

SUGGESTED FURTHER READING

1. Further analysis

J. C. Burkill and H. Burkill, *A Second Course in Mathematical Analysis* (Cambridge University Press, 1970).
This book is notable in the clarity of its exposition but is perhaps a little old-fashioned in its treatment of some topics.

W. Rudin, *Principles of Mathematical Analysis* (McGraw-Hill, 1964).
This is a well-tried and popular text. The treatment is rather condensed.

T. M. Apostol, *Mathematical Analysis* (Addison-Wesley, 1957).
This is a useful text, especially in its treatment of vector methods.

R. V. Churchill, *Complex Variables and Applications* (McGraw-Hill, 1960).
All the books above include a discussion of complex analysis (i.e. analysis involving the entity $i = \sqrt{-1}$), but this book is a particularly easy introduction to the topic.

H. L. Royden, *Real Analysis* (Macmillan, 1968).
This is an excellent book. The choice of material and manner of presentation has been carefully considered. Its treatment of the Lebesque integral is particularly good. It is, however, a book which makes considerable demands on the reader.

G. F. Simmons, *Introduction to Topology and Modern Analysis* (McGraw-Hill, 1963).
The title of the book describes its content. It is very well-written indeed.

2. Logic, set theory and foundations

K. G. Binmore, *Foundations of Analysis* (Cambridge University Press, 1980).
Book 1. *Logic, Sets and Numbers.* This was written as a companion volume to the current book. Assuming very little previous knowledge, it fills in the logical, set-theoretic and algebraic background which was taken for granted here.
Book 2. *Topological Ideas.* This book is intended to link the more advanced texts on analysis with elementary texts like the current one. Its basic theme is the development of the ideas of limits and continuity in spaces other than $\mathbb{R}$ with an emphasis on the relation these ideas have to those already mastered rather than to abstract notions yet to be encountered.

E. A. Maxwell, *Fallacies in Mathematics* (Cambridge University Press, 1963).
This is for entertainment purposes. But it also contains some very instructive examples.

N. Ya Vilenkin, *Stories about Sets* (Academic Press, 1968).
This is another entertaining book in which the paradoxes of the infinite are explored at length. It is highly recommended.

P. R. Halmos, *Naive Set Theory* (Van Nostrand, 1960).
This is the standard reference for elementary set theory.

K. Kuratowski, *Introduction to Set Theory and Topology* (Pergamon Press, 1972).
This is a very good book which covers some difficult material.

A. Margaris, *First Order Mathematical Logic* (Blaisdell, 1967).
Formal mathematical logic is a fascinating and expanding field. This book is a good introduction.

E. Moise, *Elementary Geometry from an Advanced Standpoint* (Addison-Wesley, 1963).
The real number system was developed largely in response to geometric needs. This fascinating and clearly written book describes this process and many other topics.

E. Landau, *Foundations of Analysis* (Chelsea, 1951).
This remains an excellent and simple account of the algebraic foundations of the real number system.

E. Sondheimer and A. Rogerson, *Numbers and Infinity* (Cambridge University Press, 1981).
This is a lucid account of the development of number concepts from their earliest beginnings up to and including the idea of transfinite numbers.

J. Dieudonné, *Foundations of Analysis* (Academic Press, 1963).
In §19.48 we began to say something about calculus in an abstract setting. This very influential book is the standard reference for these ideas.

NOTATION

INDEX